Lecture Notes in Computer Science   8952

Commenced Publication in 1973
Founding and Former Series Editors:
Gerhard Goos, Juris Hartmanis, and Jan van Leeuwen

More information about this series at http://www.springer.com/series/7407

Evripidis Bampis · Ola Svensson (Eds.)

Approximation and Online Algorithms

12th International Workshop, WAOA 2014
Wrocław, Poland, September 11–12, 2014
Revised Selected Papers

Springer

Editors
Evripidis Bampis
Université Pierre et Marie Curie, LIP6
Paris
France

Ola Svensson
EPFL-IC
Lausanne
Switzerland

ISSN 0302-9743   ISSN 1611-3349 (electronic)
Lecture Notes in Computer Science
ISBN 978-3-319-18262-9   ISBN 978-3-319-18263-6 (eBook)
DOI 10.1007/978-3-319-18263-6

Library of Congress Control Number: 2015937010

LNCS Sublibrary: SL1 – Theoretical Computer Science and General Issues

Springer Cham Heidelberg New York Dordrecht London

Printed on acid-free paper

Springer International Publishing AG Switzerland is part of Springer Science+Business Media
(www.springer.com)

Preface

The 12th Workshop on Approximation and Online Algorithms (WAOA 2014) focused on the design and analysis of algorithms for online and computationally hard problems. Both kinds of problems have a large number of applications from a variety of fields. WAOA 2014 took place in Wrocław, Poland, during September 11–12, 2014. The workshop was part of the ALGO 2014 event that also hosted ESA, ALGOSENSORS, ATMOS, IPEC, MASSIVE, and WABI. The previous WAOA workshops were held in Budapest (2003), Rome (2004), Palma de Mallorca (2005), Zurich (2006), Eilat (2007), Karlsruhe (2008), Copenhagen (2009), Liverpool (2010), Saarbrücken (2011), Ljubljana (2012), and Sophia Antipolis (2013). The proceedings of these previous WAOA workshops have appeared as LNCS volumes.

Topics of interest for WAOA 2014 were: coloring and partitioning, competitive analysis, network design, packing and covering, paradigms for design and analysis of approximation and online algorithms, randomization techniques, real-world applications, and scheduling problems. In response to the call for papers, we received 49 submissions. Each submission was reviewed by at least three referees. The submissions were mainly judged on originality, technical quality, and relevance to the topics of the conference. Based on the reviews, the Program Committee selected 22 papers.

We are grateful to Aleksander Madry for his invited talk and to Monaldo Mastrolilli for his tutorial. We would also like to thank Andrei Voronkov for providing the EasyChair conference system, which was used to manage the electronic submissions, the review process, and the electronic PC meeting. It made our task much easier. We would also like to thank all the authors who submitted papers to WAOA 2014 as well as the local organizers of ALGO 2014.

August 2014   Evripidis Bampis
                                                                     Ola Svensson

Organization

Program Committee

Evripidis Bampis (Co-chair)	Sorbonne Universités, Université Pierre et Marie Curie, France
Nikhil Bansal	Eindhoven University of Technology, The Netherlands
Ioannis Caragiannis	University of Patras, Greece
Marek Chrobak	University of California, Riverside, USA
Alina R. Ene	Princeton University, USA
Moran Feldman	EPFL, Switzerland
Laurent Gourvès	CNRS, Université Paris-Dauphine, France
Nguyen Kim Thang	Université d'Evry-Val-d'Essonne, France
Alejandro López-Ortiz	University of Waterloo, Canada
Giorgio Lucarelli	Sorbonne Universités, Université Pierre et Marie Curie, France
Monaldo Mastrolilli	IDSIA, Lugano, Switzerland
Julian Mestre	The University of Sydney, Australia
Seffi Naor	Technion, Israel
Neil Olver	CWI and VU University Amsterdam, The Netherlands
Kirk Pruhs	University of Pittsburgh, USA
Thomas Rothvoss	University of Washington, USA
Laura Sanità	University of Waterloo, Canada
Ola Svensson (Co-chair)	EPFL, Switzerland
José Verschae	Universidad de Chile, Chile
Andreas Wiese	Max-Planck-Institut für Informatik, Germany

Invited Speaker

Aleksander Madry   EPFL, Switzerland

Tutorial

Monaldo Mastrolilli   IDSIA, Switzerland

Subreviewers

Antonios Antoniadis
Ashwinkumar Badanidiyuru
Niv Buchbinder
Martin Derka
Yann Disser
Christoph Dürr
Aristotelis Giannakos
Michael Goldwasser
Arpita Ghosh
Sandy Heydrich
Wiebke Höhn
Chien-Chung Huang
Christian Icking
Sungjin Im
Shahin Kamali
Panagiotis Kanellopoulos
Hans Kellerer
Isaac Keslassy
Kim-Manuel Klein
Alexander Kononov
Maria Kyropoulou
Adam Kurpisz

Luigi Laura
Julien Lesca
Dimitrios Letsios
Shi Li
Daniela Maftuleac
Luke Mathieson
Ioannis Milis
Marco Molinaro
Jérôme Monnot Kiyohito Nagano
Viswanath Nagarajan
Krzysztof Onak
Dror Rawitz
Alejandro Salinger
Roy Schwartz
Uwe Schwiegelshohn
Matteo Seminaroti
Florian Sikora
Lydia Tlilane
Erik Jan van Leeuwen
David Wajc
Justin Ward
Neal Young

Contents

Improved Approximations for the Max k-Colored Clustering Problem

Alexander Ageev[1] and Alexander Kononov[1,2](✉)

[1] Sobolev Institute of Mathematics, Novosibirsk, Russia
{ageev,alvenko}@math.nsc.ru
[2] Novosibirsk State University, Novosibirsk, Russia

Abstract. In the Max k-Colored Clustering Problem we are given an undirected graph $G = (V, E)$. Each edge e of G has a nonnegative weight $w(e)$ and a color $c(e) \in \mathcal{C} = \{1, 2, \ldots, k\}$. It is required to assign a color from $\mathcal{C}$ to each vertex of G so as to maximize the total weight of edges whose both endpoints have the same color as the color of the edge. Angel et al. [1] show that the problem is strongly NP-hard and present a randomized constant-factor approximation algorithm for solving it. We improve this result in two directions. First, we give a more careful analysis of the algorithm in [1], which significantly improves on its approximation bound (0.25 instead of $1/e^2 \approx 0.135$). Second, we present a different algorithm with a better worst case performance guarantee of $7/23 \approx 0.304$. Both algorithms are based on using similar randomized rounding schemes for a natural LP relaxation of the problem. They can be derandomized in a standard way by computing conditional expectations for some estimate function.

Keywords: Clustering · Edge-colored graph · Linear relaxation · Randomized rounding · Worst case behavior analysis

1 Introduction

We consider the *Max k-Colored Clustering Problem* introduced in [1].

In this problem we are given an undirected graph $G = (V, E)$. Each edge e of G has a nonnegative weight $w(e)$ and a color $c(e) \in \mathcal{C} = \{1, 2, \ldots, k\}$. It is required to assign a color from $\mathcal{C}$ to each vertex of G so as to maximize the total weight of edges whose both endpoints have the same color as the color of the edge.

Given a vertex coloring, we say that an edge of G is *matched* if both its endpoints have the same color as the color of the edge. By using this term the goal is to color the vertices of G in such a way that the total weight of the matched vertices is maximized.

Note that in the case where each edge has its own color the problem coincides with the edge packing problem which is equivalent to finding a maximum weight perfect matching.

© Springer International Publishing Switzerland 2015
E. Bampis and O. Svensson (Eds.): WAOA 2014, LNCS 8952, pp. 1–10, 2015.
DOI: 10.1007/978-3-319-18263-6_1

The model has similarities with the centralized version of the information-sharing model introduced by Kleinberg and Ligett [3,6]. In their model, the edges are not colored and two adjacent nodes share information only if they are colored with the same color. As they mention, one interesting extension of their model would be the incorporation of different categories of information. In the problem considered in our paper every edge-color corresponds to a different information category and two adjacent vertices share information if their color is the same as the color of the edge that connects them. *Max k-Colored Clustering Problem* is also related to the classical correlation clustering problem [2,5]. Another observation is that the *Max k-colored clustering problem* can be formulated as a combinatorial allocation problem [4]. We can consider each color as a player and each vertex as an item, where items have to be allocated to competing players by a central authority, with the goal of maximizing the total utility provided to the players. Every player (each color) has utility functions derived from the different subsets of vertices. Feige and Vondrak [4] consider subadditive, fractional subadditive and submodular functions. It is easy to see that in our case the function is supermodular and therefore their method cannot be applied.

While the centralized version of the information-sharing problem of Kleinberg and Ligget is easy to solve, *Max k-colored clustering problem* is strongly NP-hard (Angel et al. [1]). Angel et al. [1] also present a randomized constant-factor approximation for this problem with an expected approximation ratio bounded by $\frac{1}{e^2} \approx 0.135$. We improve this result in two directions. First, we give a more careful analysis of the algorithm in [1], which significantly improves on its approximation bound (0.25 instead of $1/e^2$). Second, we present a different algorithm based on the same ideas with a better approximation ratio of $7/23 \approx 0.304$.

As in [1], we formulate the problem as an integer linear program and develop randomized rounding schemes for the linear programming relaxation. Notice here that straitforward rounding schemes apparently do not lead to constant-factor approximations for this problem.

In Sect. 2 we give a description of the first algorithm (RR-2) and develop theoretical tools for analyzing the worst case behavior of both algorithms. The key ingredients of these tools are Lemmas 4 and 5. Note that Algorithm RR-2 slightly differs from algorithm RR in [1] but as can be easily seen their analysis and approximation bounds are just the same.

In Sect. 3 we describe the second algorithm and derive its approximation bound using the theoretical background developed in the previous section.

Finally, in Sect. 4 we say a few words about the derandomization of the developed algorithms and the integrality gap of the equivalent integer program.

2   A 0.25-Approximation Algorithm

For every vertex i of the graph and for every available color $c \in \{1, \ldots, k\}$ we introduce a variable x_{ic} which is equal to one if i is colored with color c and zero otherwise. Also, for every edge $e = [i, j]$ we introduce a variable z_{ij} which is equal to one if both endpoints of e are colored with the same color as e and zero otherwise. For any edge $e \in E$ denote by $c(e)$ its color. An edge is called c-colored if it has color c.

By using this notation *Max k-Colored Clustering Problem* can be formulated as the following integer program:

$$\max \sum_e w_e z_e \tag{1}$$

subject to

$$z_e \leq x_{ic(e)} \quad \forall e = [i,j] \in E \tag{2}$$

$$z_e \leq x_{jc(e)} \quad \forall e = [i,j] \in E \tag{3}$$

$$\sum_{c \in \mathcal{C}} x_{ic} = 1, \quad \forall i \in V \tag{4}$$

$$x_{ic}, z_e \geq 0 \quad \forall i \in V, c \in \mathcal{C}, e \in E \tag{5}$$

$$x_{ic}, z_e \in \{0,1\} \quad \forall i \in V, c \in \mathcal{C}, e \in E \tag{6}$$

Denote by LP the linear relaxation (1)–(5) of this program. Our algorithm starts with solving LP and then works in k iterations, by considering each color c, $1 \leq c \leq k$, independently from the others, and so the order in which the colors are considered does not matter. When an edge is chosen, this means that both its endpoints get the color of this edge. Since in general a vertex can be adjacent to differently colored edges, it may get more than one colors. In this case we choose randomly one of these colors. The algorithm is given below.

Algorithm 1. Algorithm RR-2
___
1: Solve the linear program LP, and let z_e^* be the values of variables z_e.
2: **for each** color c **do**
3:    Order, non decreasingly, the c-colored edges $e_1, \ldots, e_{l(c)}$ according to their z_e^* values.
4:    Let us assume that we have $z_{e_1}^* \leq z_{e_2}^* \leq \cdots \leq z_{e_{l(c)}}^*$.
5:    Let r be a random value in $[0,1]$.
6:    Choose the c-colored edges e with $z_e^* > r$
7: **for each** vertex $v \in V$ **do**
8:    **if** v gets l colors **then**
9:        assign randomly one of them to v, each one with probability $\frac{1}{l}$.
___

Notice that the above algorithm differs from Algorithm RR presented in [1]. In the second FOR-cycle Algorithm RR removes vertices with more than two colors and for each vertex with two colors randomly chooses one of them with probability $\frac{1}{2}$. However, we derive the same worst-case performance guarantee for both algorithms.

We now proceed to the worst-case analysis of Algorithm RR-2. Lemmas 1–3 are essentially Lemmas 1–3 in [1].

Lemma 1. *For any edge e, the probability that e is chosen in line 6 is z_e^*.*

Notice that for a vertex v, it may be the case that none of its adjacent edges are chosen. In that case, v gets no color. But in general, several of its adjacent edges can be chosen, and the vertex v can get more than one colors. We denote by X_{vc} (resp. $\bar{X}_{vc}$) the event that v gets (resp. does not get) color c after a random choice on Step 6. We denote by Y_{vc} (resp. $\bar{Y}_{vc}$) the event that v is colored by color c in line 9.

Lemma 2. *For every vertex v if there exists at least one c-colored edge incident to v, then $Pr(X_{vc}) = z_{e'}^*$ where e' is a c-colored edge having the maximal value of z_e^* among all c-colored edges e incident to v.*

Lemma 3. *For every vertex $v \in V$ $\sum_c Pr(X_{vc}) \leq 1$.*

As stated above, a vertex v can get more than one colors during the execution of the algorithm. However, in general this number will be small. Let $\gamma(v) = \sum_{i \in C} Pr(X_{vi})$. The following lemma gives a lower bound for the probability that color c was assigned in lines 7–9 of the algorithm.

Lemma 4. *Assume that a vertex v gets color c at step 6 of Algorithm RR-2 and $\mu(v,c) = \max_{i \in C \setminus \{c\}} Pr(X_{vi})$. Then for different values of μ and γ we have the following results:*

(a) $Pr(Y_{vc}) \geq \frac{1}{2}$;
(b) if $\gamma(v) \leq \frac{1}{2}$ then $Pr(Y_{vc}) \geq \frac{3}{4}$;
(c) if $\mu(v,c) \leq \frac{1}{2}$ then $Pr(Y_{vc}) \geq \frac{7}{12}$.

Proof. For simplicity of notation we assume that a vertex v meets colors $1, \ldots, t$ and c in line 8 of Algorithm RR-2, i.e., $l = t + 1$. The probability that vertex v is colored by color c depends on how many colors were assigned to vertex v at step 6. The more additional colors were assigned, the less the probability that vertex v is colored by color c. By the formula of total probability we have

$$Pr(Y_{vc}) \geq \prod_{i=1}^{t}(1 - Pr(X_{vi})) + \frac{1}{2}\sum_{i=1}^{t} Pr(X_{vi}) \prod_{i' \neq i}(1 - Pr(X_{vi'}))$$

$$+ \frac{1}{3}\sum_{i=1}^{t}\sum_{j \neq i} Pr(X_{vi})Pr(X_{vj}) \prod_{i' \neq i, i' \neq j}(1 - Pr(X_{vi'})). \qquad (7)$$

The first term is the probability that vertex v has no additional color at step 6. The second term is the probability that vertex v is colored by color c under the condition that vertex v gets one additional color at step 6. The third term is the probability that vertex v is colored by color c under the condition that vertex v gets two additional colors at step 6. We dropped all the remaining terms of the formula because they are equal to zero in the worst case.

Denote the right part of (7) by f_{vc}.

We now show that the minimum of f_{vc} as a function of the vector $X = (X_{vi}, i = 1, \ldots, t)$ is attained at some vector X^* each entry of whom except at most two is equal to 0. To simplify computations we set $\chi_i = Pr(X_{vi})$, $i = 1, \ldots, t$. Without loss of generality we may assume that $\chi_1 \geq \chi_i$ for all $i = 2, \ldots, t$. We have

$$f_{vc} = \prod_{i=1}^{t}(1 - \chi_i) + \frac{1}{2}\sum_{i=1}^{t}\chi_i\prod_{i' \neq i}(1 - \chi_{i'}) + \frac{1}{3}\sum_{i=1}^{t}\sum_{j \neq i}\chi_i\chi_j\prod_{i' \neq i, i' \neq j}(1 - \chi_{i'})$$

$$= (1 - \chi_1)(1 - \chi_2)\prod_{j \geq 3}(1 - \chi_j) + \frac{1}{2}(\chi_1(1 - \chi_2) + \chi_2(1 - \chi_1))\prod_{j \geq 3}(1 - \chi_j)$$

$$+ \frac{1}{2}(1 - \chi_1)(1 - \chi_2)\sum_{i \geq 3}\chi_i\prod_{j \geq 3, j \neq i}(1 - \chi_j) + \frac{1}{3}\chi_1\chi_2\prod_{j \geq 3}(1 - \chi_j)$$

$$+ \frac{1}{3}(\chi_1(1 - \chi_2) + \chi_2(1 - \chi_1))\sum_{i \geq 3}\chi_i\prod_{j \geq 3, j \neq i}(1 - \chi_j)$$

$$+ \frac{1}{3}(1 - \chi_1)(1 - \chi_2)\sum_{i=3}^{t}\sum_{j \geq 3, j \neq i}\chi_i\chi_j\prod_{i' \geq 3, i' \neq i, i' \neq j}(1 - \chi_{i'}).$$

By setting $A = \prod_{j \geq 3}(1 - \chi_j)$, $B = \sum_{i \geq 3}\chi_i\prod_{j \geq 3, j \neq i}(1 - \chi_j)$, and $C = \sum_{i=3}^{t}\sum_{j \geq 3, j \neq i}\chi_i\chi_j\prod_{i' \geq 3, i' \neq i, i' \neq j}(1 - \chi_{i'})$ we get

$$f_{vc} = (1 - \chi_1)(1 - \chi_2)A + \frac{1}{2}(\chi_1(1 - \chi_2) + \chi_2(1 - \chi_1))A + \frac{1}{2}(1 - \chi_1)(1 - \chi_2)B$$

$$+ \frac{1}{3}\chi_1\chi_2 A + \frac{1}{3}(\chi_1(1 - \chi_2) + \chi_2(1 - \chi_1))B + \frac{1}{3}(1 - \chi_1)(1 - \chi_2)C$$

$$= (1 - \frac{1}{2}\chi_1 - \frac{1}{2}\chi_2)A + (\frac{1}{2} - \frac{1}{6}\chi_1 - \frac{1}{6}\chi_2)B + (\frac{1}{3} - \frac{1}{3}\chi_1 - \frac{1}{3}\chi_2)C$$

$$+ \frac{1}{6}\chi_1\chi_2(2A - B + 2C).$$

We now consider f_{vc} as a function of two variables χ_1 and χ_2. Let $\chi_1 + \chi_2$ be equal to some constant κ. Assume that $\chi_1, \chi_2 > 0$ and we increase χ_1 by $\delta > 0$ and decrease χ_2 by the same number δ. Then the values of the first three terms in the last expression for the function f_{vc} do not change. If $2A - B + 2C > 0$, then the last term decreases and therefore the function f_{vc} decreases as well.

Now we consider two cases.

Case 1: $\chi_1 \geq 1/2$. Let us show that $2A - B + 2C \geq 0$. Taking into account $C \geq 0$ we obtain

$$2A - B + 2C \geq 2A - B \geq 2\prod_{j \geq 3}(1 - \chi_j) - \sum_{i \geq 3}\chi_i\prod_{j \geq 3, j \neq i}(1 - \chi_j)$$

$$= \prod_{j \geq 3}(1 - \chi_j)\left(2 - \sum_{i \geq 3}\frac{\chi_i}{1 - \chi_i}\right) \geq \prod_{j \geq 3}(1 - \chi_j)\left(2 - \frac{\sum_{i \geq 3}\chi_i}{1 - \chi_1}\right).$$

From (4), we have $\sum_{i\geq 3} \chi_i \leq 1 - \chi_1 - \chi_2$. It follows that $\frac{\sum_{i\geq 3}\chi_i}{1-\chi_1} \leq 1$ and

$$\prod_{j\geq 3}(1-\chi_j)\left(2 - \frac{\sum_{i\geq 3}\chi_i}{1-\chi_1}\right) > 0.$$

Since $A \geq 0$, $B \geq 0$, and $C \geq 0$ then the minimum of f_{vc} is attained at $\chi_1 = \kappa$ and $\chi_2 = 0$. By repeating this argument we get that the minimum of f_{vc} is attained when all $X_{vi}, i = 1,\ldots,t$ except one of them are equal to 0. In this case we have $A = 1$, $B = 0$, $C = 0$ and $Pr(Y_{vc}) \geq 1 - \frac{1}{2}\gamma(v)$. This inequality immediately implies (a) and (b).

Case 2: $\chi_1 \leq 1/2$. We note that $1 - \chi_i \geq 1/2$. It follows that

$$2A - B + 2C \geq \prod_{j\geq 3}(1-\chi_j)\left(2 - \sum_{i\geq 3}\frac{\chi_i}{1-\chi_i}\right) \geq \prod_{j\geq 3}(1-\chi_j)\left(2 - 2\sum_{i\geq 3}\chi_i\right) \geq 0.$$

Let $1/2 > \chi_1 \geq \chi_2 > 0$ and $\chi_1 + \chi_2 = \kappa$. Setting $\chi_1 = \min\{1/2,\kappa\}$ and $\chi_2 = \kappa - \min\{1/2,\kappa\}$ we decrease the value of f_{vc}.

By repeating this argument for any two nonnegative values χ_i and χ_j such that $1/2 > \chi_i \geq \chi_j > 0$ we get that a lower bound for f_{vc} is the value of this function at $\chi_1 = \chi_2 = 1/2$ and $\chi_i = 0$, $i = 3,\ldots,t$. In this case we have $A = 1$, $B = 0$, $C = 0$ and

$$Pr(Y_{vc}) \geq 1 - \frac{1}{2}\cdot\frac{1}{2} - \frac{1}{2}\cdot\frac{1}{2} + 2\cdot\frac{1}{6}\cdot\frac{1}{2}\cdot\frac{1}{2} = \frac{7}{12}. \qquad \square$$

The following lemma is similar to Proposition 1 in [1]. We prove it by the same method but our proof is a bit shorter and treats a more general case when the number of colors is arbitrary.

Lemma 5. *Suppose that edge $e = [u, v]$ has a color c and it is chosen at step 6 of Algorithm RR-2. Denote by p_e the probability that both extremities of e also get the color c. Then, $p_e \geq Pr(Y_{uc})Pr(Y_{vc})$.*

Proof. To prove the lemma we consider a sequence of algorithms denoted by $\Sigma_0,\ldots,\Sigma_k$ where Σ_0 is algorithm RR-2. The difference among these algorithms comes from the way in which the vertices get a color. Let us fix a color x. We consider two different procedures for assigning colors to the vertices. Procedure 1 assigns the colors in the same way as our algorithm does. Let us recall how our algorithm works for just two vertices. Without loss of generality we assume that there exists an edge e' incident to u colored by c and an edge e'' incident to v colored by x. Moreover, we suppose e' (resp. e'') is the edge with the maximal value of $z^*_{e'}$ (resp. $z^*_{e''}$) among all x-colored edges incident to u (resp. v). Let us assume that $z^*_{e'} \leq z^*_{e''}$. Let p be the probability that u gets color x in the algorithm (we know that it is $z^*_{e'}$ from Lemma 2), and let q be the probability that v gets color x assuming that u does not get color x. Using the procedure 1, we color both vertices u and v (with color x) with probability p, and we color only vertex v with probability $(1 - p)q$. Procedure 2 colors the vertices with color x independently. More precisely, we color vertex u with probability p, and we color vertex v with probability $(1 - p)q + p := y$. In the algorithm Σ_0, for each color x, $1 \leq x \leq k$ we use Procedure 1 to assign colors to vertices. In the algorithm Σ_i, $1 \leq i \leq k$, for colors x such that $1 \leq x \leq i$ (resp. $i+1 \leq x \leq k$) we use Procedure 2 (resp. Procedure 2) for assigning those colors to vertices. Thus, in algorithm Σ_k, all colors are assigned to vertices using Procedure 2.

Let us consider two consecutive algorithms Σ_i and Σ_{i+1}. The algorithms Σ_i and Σ_{i+1} differ only in the way they assign color $i + 1$ to vertices. If there is no $(i+1)$-colored edge incident to either u or v, then $p_e(\Sigma_i) = p_e(\Sigma_{i+1})$, i.e., these two algorithms have the same behavior. Recall that we denote by X_{vc} (resp. $\overline{X}_{vc}$) the event that v gets (resp. does not get) color c. We have the following probabilities:

	When $\Sigma = \Sigma_i$	When $\Sigma = \Sigma_{i+1}$
$Pr_\Sigma(X_{u,i+1} \wedge \overline{X}_{v,i+1})$	0	$p(1 - y)$
$Pr_\Sigma(\overline{X}_{u,i+1} \wedge X_{v,i+1})$	$(1 - p)q$	$(1 - p)y$
$Pr_\Sigma(\overline{X}_{u,i+1} \wedge \overline{X}_{v,i+1})$	$(1 - p)(1 - q)$	$(1 - p)(1 - y)$
$Pr_\Sigma(X_{u,i+1} \wedge X_{v,i+1})$	p	py

Let $C' = C \setminus \{c, i + 1\}$. Denote by A_i (resp. B_i) the event which corresponds to the situation where vertex u (resp. v) gets i colors from the set C'. It is clear that the probabilities of the event $A_i \wedge B_j$ are the same for both algorithms Σ_i and Σ_{i+1}, i.e., $Pr_{\Sigma_i}(A_i \wedge B_j) = Pr_{\Sigma_{i+1}}(A_i \wedge B_j)$.

For $\Sigma \in \{\Sigma_0, \ldots, \Sigma_k\}$, we have $p_e(\Sigma) = \sum_{ij} Pr_\Sigma(A_i \wedge B_j)\phi(\Sigma)$, where

$$\phi(\Sigma) = \frac{1}{ij} Pr_\Sigma(\overline{X}_{u,i+1} \wedge \overline{X}_{v,i+1}) + \frac{1}{(i+1)j} Pr_\Sigma(X_{u,i+1} \wedge \overline{X}_{v,i+1})$$

$$+ \frac{1}{i(j+1)} Pr_\Sigma(\overline{X}_{u,i+1} \wedge X_{v,i+1}) + \frac{1}{(i+1)(j+1)} Pr_\Sigma(\overline{X}_{u,i+1} \wedge \overline{X}_{v,i+1}).$$

We claim that $\phi(\Sigma_i) \geq \phi(\Sigma_{i+1})$. Taking into account notation in the table we have

$$\phi(\Sigma_i) = \frac{(1 - p)(1 - q)}{ij} + \frac{(1 - p)q}{i(j + 1)} + \frac{p}{(i + 1)(j + 1)}$$

$$= \frac{ij + i(1 - y) + j(1 - p) + (1 - y)}{(i + 1)(j + 1)ij}$$

and

$$\phi(\Sigma_{i+1}) = \frac{(1 - p)(1 - y)}{ij} + \frac{p(1 - y)}{(i + 1)j} + \frac{(1 - p)y}{i(j + 1)} + \frac{py}{(i + 1)(j + 1)}$$

$$= \frac{ij + i(1 - y) + j(1 - p) + (1 - p)(1 - y)}{(i + 1)(j + 1)ij}.$$

Since $1 - p \leq 1$ we obtain $\phi(\Sigma_i) \geq \phi(\Sigma_{i+1})$ and a term-by-term comparison for $p_e(\Sigma)$ implies the result of the lemma. $\square$

Lemmas 1, 4 and 5 imply the following result.

Theorem 1. *The expected approximation ratio of Algorithm RR-2 is bounded by $\frac{1}{4}$.*

Proof. Let OPT denote the sum of the weights of the matched edges in an optimal solution. Since the linear program LP is a linear relaxation, we have $\sum_{e \in E} w_e z_e^* \geq OPT$. Consider an edge $e \in E$ chosen at step 6. This occurs with the probability z_e^* according to Lemma 1. Suppose that $e = [u, v]$ has a color c. By Lemma 5 the probability p_e that both endpoints of e get the color c greater than or equal to $Pr(X_{uc})Pr(X_{vc})$. On the hand, Lemma 4(a) implies that each edge is matched with probability $\frac{z_e^*}{4}$. Thus the expected cost of the solution returned by the algorithm is at least

$$\sum_{e \in E} \frac{w_e z_e^*}{4} \geq \frac{OPT}{4}.$$

                                                                                                   $\square$

3   A 7/23-Approximation Algorithm

Proceeding from an optimal solution of LP we separate edges of E into big edges and small edges. An edge e is called *big* if $z_e^* > \frac{1}{2}$ or it is adjacent to edge e' with $z_{e'}^* > \frac{1}{2}$ and edges e and e' have the same color; otherwise an edge e is *small*. Lemma 3 implies that all adjacent big edges have the same color. We say that a vertex v is *heavy* if it is incident to at least one big edge; otherwise we say that vertex v is *light*. We say that a color c is a dominating color for a heavy vertex v if vertex v is incident to a c-colored big edge. Now we partition the set of edges into four sets. Denote by Z_1 the set of big edges, by Z_2 the set of small edges whose both endpoints are light, by Z_3 the set of small edges whose both endpoints are heavy, and by Z_4 the set of small edges with one endpoint heavy and the other light. Let $W(Z_i) = \sum_{e \in Z_i} w_e z_e^*$ for $i = 1, \dots, 4$. Then

$$W(Z_1) + W(Z_2) + W(Z_3) + W(Z_4) \geq OPT.$$

Next we present two algorithms. The first algorithm put all big edges in the solution and then add some small edges. The second algorithm ignore big edges and find a solution for small edges. It is clear that both algorithms may give arbitrarily bad solutions. However, we show that the best of these solution has a performance bound better than that of algorithm RR-2.

Algorithm 2 matches all big edges and randomly choose edges from the set Z_2. Let $e = [u, v] \in Z_2$. It follows that the vertices u and v are light, and $z_{e'}^* \leq 1/2$ for any incident edge e'. Let an edge e chosen at step 9 of algorithm 2. This occurs with the probability z_e^* according to Lemma 1. Suppose that $e = [u, v]$ has a color c. By Lemma 5 the probability p_e that both endpoints of e get the color c greater

---

Algorithm 2. Algorithm "Big Edges"

1: Solve the linear program LP, and let z_e^* be the values of variables z_e.
2: **for each** heavy vertex v **do**
3:    Color vertex v by its dominating color.
4:    Remove vertex v and all the edges incident to v.
5: **for each** color c **do**
6:    Order, non decreasingly, the c-colored edges $e_1, \ldots, e_{l(c)}$ according to their z_e^* values.
7:    Let us assume that we have $z_{e_1}^* \leq z_{e_2}^* \leq \cdots \leq z_{e_{l(c)}}^*$.
8:    Let r be a random value in $[0, 1]$.
9:    Choose the c-colored edges e with $z_e^* > r$.
10: **for each** vertex $v \in V$ **do**
11:    **if** v gets l colors **then**
12:       assign randomly one of them to v, each one with probability $\frac{1}{l}$.
13: Output this assignment as ψ_1.

---

than or equal to $Pr(X_{uc})Pr(X_{vc})$. Since v and u are light we have $\mu(v, c) \leq 1/2$ and $\mu(u, c) \leq 1/2$. Lemma 4(c) implies that $Pr(Y_{vc}) \geq \frac{7}{12}$ and $Pr(Y_{uc}) \geq \frac{7}{12}$. Thus the expected cost W_1 obtained by the algorithm is at least

$$W_1 \geq W(Z_1) + \frac{7}{12} \cdot \frac{7}{12} \cdot W(Z_2) \geq W(Z_1) + \frac{49}{144} W(Z_2).$$

Algorithm 3. Algorithm "Small Edges"

1: Solve the linear program LP, and let z_e^* be the values of variables z_e.
2: Remove all big edges.
3: **for each** color c **do**
4:    Order, non decreasingly, the c-colored edges $e_1, \ldots, e_{l(c)}$ according to their z_e^* values.
5:    Let us assume that we have $z_{e_1}^* \leq z_{e_2}^* \leq \cdots \leq z_{e_{l(c)}}^*$.
6:    Let r be a random value in $[0, 1]$.
7:    Choose the c-colored edges e with $z_e^* > r$
8: **for each** vertex $v \in V$ **do**
9:    **if** v gets l colors **then**
10:       assign randomly one of them to v, each one with probability $\frac{1}{l}$.
11: Output this assignment as ψ_2.

Algorithm 3 removes all big edges and randomly choose edges from the set $Z_2 \bigcup Z_3 \bigcup Z_4$. After removing big edges for each heavy vertex v we have $\gamma(v) \leq 1/2$. Suppose that $e = [u, v]$ has a color c and v is heavy. Lemma 4(b) implies that $Pr(Y_{vc}) \geq \frac{3}{4}$. Let W_2 be the expected cost of the solution returned by this algorithm. According to Lemma 4(b),(c) and Lemma 5 we have

$$W_2 \geq \frac{7}{12} \cdot \frac{7}{12} \cdot W(Z_2) + \frac{3}{4} \cdot \frac{3}{4} \cdot W(Z_3) + \frac{7}{12} \cdot \frac{3}{4} \cdot W(Z_4)$$

$$\geq \frac{49}{144} W(Z_2) + \frac{9}{16} W(Z_3) + \frac{7}{16} W(Z_4).$$

Choosing the best of two solutions and taking into account the inequality

$$W(Z_1) + W(Z_2) + W(Z_3) + W(Z_4) \geq OPT$$

we obtain that

$$\max\{W_1, W_2\} \geq \frac{7}{23}OPT,$$

which is achieved at $W(Z_1) = \frac{7}{23}OPT$, $W(Z_2) = 0$, $W(Z_3) = 0$, and $W(Z_4) = \frac{16}{23}OPT$. Eventually we have the following result.

Theorem 2. *The best of the two solutions ψ_1 and ψ_2 has expected weight at least $\frac{7}{23}OPT$.* □

4   Final Remarks

In the paper, we present two randomized approximations algorithms for *Max k-Colored Clustering Problem*.

First, we notice that both algorithms can be derandomized in a standard way by computing conditional expectations for some estimate function (for the method of conditional probabilities, see [7]).

Second, note that Theorem 2 provides a lower bound for the integrality gap IG of the integer program (1)–(6): $IG \geq 7/23$. On the other hand, a trivial example of triangle with edge weights 1 and a unique color for each edge provides an upper bound $IG \leq 2/3$. Therefore there may well exist rounding schemes providing better approximations than those presented in this paper, which stimulates further work.

Acknowledgements. The authors would like to thank the anonymous referees whose many useful comments and suggestions helped to improve the presentation of the paper.

Research of the first author is partially supported by RFBR grant 12-01-00184. Research of the second author is partially supported by RFH grant 13-22-10002.

References

1. Angel, E., Bampis, E., Kononov, A., Paparas, D., Pountourakis, E., Zissimopoulos, V.: Clustering on k-edge-colored graphs. In: Chatterjee, K., Sgall, J. (eds.) MFCS 2013. LNCS, vol. 8087, pp. 50–61. Springer, Berlin (2013)
2. Bansal, N., Blum, A., Chawla, S.: Correlation clustering. Mach. Learn. **56**, 89–113 (2004)
3. Ducoffe, G., Mazauric, D., Chaintreau A.: Convergence of Coloring Games with Collusions. CoRR. abs/1212.3782 (2012)
4. Feige, U., Vondrak, J.: Approximation algorithm for allocation problems improving the factor of $1 - \frac{1}{e}$. FOCS **2006**, 667–676 (2006)
5. Jain, A.K., Dubes, R.C.: Algorithms for Clustering Data. Prentice-Hall, Inc., Upper Saddle River (1988)
6. Kleinberg, J.M., Ligett, K.: Information-Sharing and Privacy in Social Networks. CoRR. abs/1003.0469 (2010)
7. Motwani, R., Raghavan, P.: Randomized Algorithms. Cambridge University Press, Cambridge (1995)

A $o(n)$-Competitive Deterministic Algorithm for Online Matching on a Line

Antonios Antoniadis[2], Neal Barcelo[1], Michael Nugent[1](✉), Kirk Pruhs[1], and Michele Scquizzato[3]

[1] Department of Computer Science, University of Pittsburgh, Pittsburgh, USA
{ncb30,mpn1}@pitt.edu,
kirk@cs.pitt.edu
[2] Max-Planck Institut für Informatik, Saarbrücken, Germany
aantonia@mpi-inf.mpg.de
[3] Department of Computer Science, University of Houston, Houston, USA
michele@cs.uh.edu

Abstract. Online matching on a line involves matching an online stream of items of various sizes to stored items of various sizes, with the objective of minimizing the average discrepancy in size between matched items. The best previously known upper and lower bounds on the optimal deterministic competitive ratio are linear in the number of items, and constant, respectively. We show that online matching on a line is essentially equivalent to a particular search problem, that we call k-lost cows. We then obtain the first deterministic sub-linearly competitive algorithm for online matching on a line by giving such an algorithm for the k-lost cows problem.

1 Introduction

The classic Online Metric Matching problem (OMM) is set in a metric space (V, d), containing a set of servers $S = \{s_1, s_2, \ldots, s_n\} \subseteq V$. A set of requests $R = \{r_1, r_2, \ldots, r_n\} \subseteq V$ arrive one by one online. When a request r_i arrives it must irrevocably be matched to some previously unmatched server s_j. The cost of matching request r_i to s_j is $d(r_i, s_j)$, and the objective is to minimize the total (equivalently, average) cost of matching all requests. There is a deterministic $(2n - 1)$-competitive algorithm, and this competitive ratio is optimal for deterministic algorithms [7,11].

A. Antoniadis—Part of the work was done while the author was at the University of Pittsburgh, supported by a fellowship within the Postdoc-Programme of the German Academic Exchange Service (DAAD).
N. Barcelo—This material is based upon work supported by the National Science Foundation Graduate Research Fellowship under Grant No. DGE-1247842.
K. Pruhs—Supported in part by NSF grants CCF-1115575, CNS-1253218, CCF-1421508, and an IBM Faculty Award.
M. Scquizzato—Work done while at the University of Pittsburgh, supported in part by a fellowship of "Fondazione Ing. Aldo Gini", University of Padova.

© Springer International Publishing Switzerland 2015
E. Bampis and O. Svensson (Eds.): WAOA 2014, LNCS 8952, pp. 11–22, 2015.
DOI: 10.1007/978-3-319-18263-6_2

The Online Matching on a Line problem (OML) is a special case of OMM where V is the real line and $d(r_i, s_j) = |r_i - s_j|$. The original motivation for considering OML came from applications where there is an online stream of items of various sizes, and the goal is to match each item as it arrives to a stored item of approximately the same size; For example, matching skiers, as they arrive in a ski rental shop, to skis of approximately their height. It is acknowledged that OML is perhaps the most interesting instance of OMM (see, e.g., [12]). Despite some efforts, there has been no progress in obtaining a better deterministic upper bound for this special case, and thus the best known upper bound on the competitive ratio for deterministic algorithms is inherited from the upper bound for OMM, namely $2n - 1$.

In the classical cow-path problem, also known as the Lost Cow problem (LC), a short-sighted cow is standing at a fence (formally, the real line) that contains a single gate at some unknown distance. The cow needs to traverse the fence until she finds the gate (formally, the algorithm needs to specify a walk on the real line). The objective is to minimize the distance traveled until the gate is found. There is a 9-competitive algorithm for LC, and this is optimal for deterministic algorithms [1]. Kalyanasundaram and Pruhs [8] observed that LC is a special case of OML where there is an optimal matching with only one positive cost edge.

In 1996, [8] conjectured that the hardest instances for OML are LC instances, and thus that there should be a 9-competitive algorithm for OML. In 2003, [5] refuted this conjecture by giving a rather complicated adversarial strategy that gives a lower bound of 9.001 on the competitive ratio of any deterministic algorithm for OML. This is currently the best known lower bound on the deterministic competitive ratio for OML.

1.1   Our Results

Upon further reflection, the lower bound in [5] can be intuitively understood as giving a lower bound on the competitive ratio for a search problem involving two lost cows (instead of one), and showing that the optimal deterministic competitive ratio for OML is at least the optimal deterministic competitive ratio for this two lost cows problem. This motivates us to ask the question of whether there is some natural search problem that is equivalent to OML. As search problems seem easier to reason about than online matching, we hypothesize that perhaps one can make progress on online matching by attacking the equivalent search problem. We show that the following search problem is essentially equivalent to OML:

k-Lost Cows (k-LC): k short-sighted cows arrive at a fence (formally, the real line) at potentially different times. The fence contains k gates in unknown positions. At each point in time, the online algorithm can specify a particular cow that has already arrived, and a direction, and then that cow will move one unit in that direction.[1] When a cow finds a gate, she will cross the fence, and this

---

[1] Allowing the cows to instead move simultaneously would not affect our results.

gate cannot be used by other cows. Each cow must cross the fence through a gate. The objective is to minimize the total distance traveled by the k cows.

More precisely we show that:

- If there is a deterministic (resp., randomized) $f(k)$-competitive algorithm for k-LC then there is a deterministic (resp., randomized) $f(n)$-competitive algorithm for OML.
- If there is a deterministic (resp., randomized) $f(p)$-competitive algorithm for OML, where the parameter p is the minimum number of positive cost edges one can have in an optimal matching, then there is a deterministic (resp., randomized) $f(k)$-competitive algorithm for k-LC.

This shows that OML is essentially equivalent to a search problem involving many lost cows, instead of one lost cow (modulo the difference in the parameters n and p).

We give the first sublinearly-competitive, $O\left(n^{\log_2(3+\epsilon)-1}/\epsilon\right)$-competitive for any $\epsilon > 0$ to be more precise, deterministic online algorithm for OML, which we obtain by first giving a deterministic $O\left(k^{\log_2(3+\epsilon)-1}/\epsilon\right)$-competitive algorithm for k-LC. Our algorithm for k-LC is a reasonably natural greedy algorithm, but the resulting OML algorithm is not particularly intuitive. This provides mild support for the hypothesis that it is easier to reason about online matching via search rather than online matching directly. We also obtain a lower bound of $\Omega\left(n^{\log_2(3+\epsilon)-1}\right)$ for our algorithm, showing that this analysis is essentially tight.

1.2 Other Related Work

For OML, it had been conjectured [8] that the generalized Work Function Algorithm (WFA) of [13] is $O(1)$-competitive, but this was disproved in [12], where it was shown that the WFA has a competitive ratio of $\Omega(\log n)$.

Randomized algorithms for OML have also been investigated. In 2006, [15] gave the first randomized algorithm and analysis giving a competitive ratio of $o(n)$ for general metric spaces (and thus for the line). More precisely, [15] obtained an $O\left(\log^3 n\right)$-competitive randomized algorithm using randomized embeddings into trees [4]. Bansal et al. [2] refined the approach in [15] to obtain an $O\left(\log^2 n\right)$-competitive randomized algorithm for general metrics. Finally, [6] gave two different $O(\log n)$-competitive randomized algorithms for the line metric, one again using randomized embeddings, and one being the natural harmonic algorithm. Kao et al. [10] gave a randomized algorithm for LC with competitive ratio of approximately 4.5911, and proved a matching lower bound. Many variants of searching problems such as the LC problem have been extensively studied (e.g., [14]).

OMM also has been studied within the framework of resource augmentation, where the online algorithm is given additional servers. Kalyanasundaram and Pruhs [9] showed that a modified greedy algorithm is $O(1)$-competitive if the online algorithm gets twice as many servers. Chung et al. [3] showed that poly-log competitiveness is achievable if the online algorithm gets an additive number of additional servers.

Our algorithm for k-LC is similar to the natural offline greedy algorithm, which repeatedly matches the two closest points. More precisely, if all the cows arrived at the same time and ϵ was zero, then our algorithm for k-LC would give the same matching as the offline greedy algorithm. Reingold and Tarjan [16] showed that the approximation ratio of the offline greedy algorithm for *non-bipartite* matching is essentially the same as the competitive ratio as our algorithm for k-LC. The first step of the two analyses is the same, looking at the cycles formed by the algorithm's matching and the optimal matching, but they diverge from there. A corollary of our analysis is that the natural offline greedy algorithm is a $\Theta(n^{\log_2 3 - 1})$-approximation for *bipartite* matching.

2   Overview

In this section we present an informal overview of both our algorithms and analyses for k-LC and OML.

In Sect. 3 we consider the problem of k-Lost Cows Without Arrivals (k-LCWA), which is a restriction of k-LC in which each cow arrives at time $t = 0$. Our algorithm for OML is based on simulating an algorithm for k-LC, which in turn is based on simulating an algorithm for k-LCWA. Recall that the optimal deterministic algorithm for 1-LC switches directions at increasing powers of 2 (i.e., switch directions at points $-1, 2, -4$, etc.). For k-LCWA, we consider the algorithm A where each cow independently and in parallel uses this optimal single cow algorithm, but switches directions at powers of $1 + \epsilon$ instead of 2. One nice feature of A is that for any $\epsilon \leq 1$, the cost for A will be within a factor of $O(1/\epsilon)$ of the cost of the final matching M between cows' starting positions and the corresponding gates that the cows used in A. To analyze the cost of M we consider the union of M and the optimal matching OPT. It is easy to see that these edges can be decomposed into a set of disjoint cycles. We give directions to edges in M and OPT based on whether a cow's starting position is on the left or on the right of the matched gate. We then prove some structural properties regarding the directions of edges in M and OPT. We can then charge the cost of M's edges to the cost of OPT's edges based on the order in which A matches cows.

As an example, consider the base case of the first cow c that finds a gate in this cycle, and let ℓ denote the length of the edge corresponding to this matching. Since c's search is never biased more than $1 + \epsilon$ in either direction from its origin, we know that the closest gate to c is at least distance $\ell/(1 + \epsilon)$ away. Also since A has all cows walking in parallel and no other cow has found a gate, we know that no other cow has a gate closer than $\ell/(1 + \epsilon)$. Using this argument we can charge the cost of this edge to any edge in OPT. As we proceed inductively, the inequalities become more complicated since we now may have to charge to multiple edges in both M and OPT. To aid our analysis we define a weighted binary tree for each cycle, with the property that the sum of the leaf costs is OPT's cost, and the sum of the internal nodes is an upper bound on M's cost. We show that if each tree is perfect (complete and balanced) then M is $O(k^{\log_2(3+\epsilon)-1})$-competitive. The last step is to show that perfect trees are the worst case. Although this is somewhat intuitive, this is by far the most technical aspect of the analysis, and involves showing that given an arbitrary tree, we can make a sequence of transformations such that the resulting tree is perfect, and at each step we do not decrease the competitive ratio.

In Sect. 4 we then show how to extend the algorithm for k-LCWA to an algorithm for k-LC. Dealing with the online arrival of cows is a bit tricky since the charging argument used in the analysis of the algorithm A for k-LCWA is delicately based on the order in which cows find their gates. To cope with this, we simulate the state that A would be in had all the cows arrived at time 0, and use this state to change how the cows walk. More specifically, if a cow c is walking in the k-LC setting and finds a gate occupied by some cow c', the algorithm determines which cow would have found this gate first in the no arrivals case, and allows this cow to stay there, "kicking out" the other cow to continue walking. It is then relatively straightforward to see that the matching produced by the simulation is identical to the matching that A would have produced had all the cows arrived at time 0.

In Sect. 5 we show how to reduce k-LC to OML, and OML to k-LC. To convert an algorithm for k-LC into an algorithm for OML, one can release a cow for every request r in the OML instance, and wait until this cow hits an unoccupied server s. Then, by matching r to s, the matchings in the two settings are equivalent, and the number of cows k will be equal to the number of servers n. To convert an algorithm for OML into an algorithm for k-LC one can continually issue requests at a cow's current location until a request is matched with a server corresponding to an unoccupied gate.

Due to space constraints, many of the proofs are deferred to the full version of the paper.

3   Analysis of Parallel Cows Algorithm for k-Lost Cows Without Arrivals

We now define the $(1 + \epsilon)$-Parallel Cows algorithm for the k-LCWA problem. The algorithm is to have every cow move according to the $(1 + \epsilon)$-cow algorithm independently and in parallel (i.e., simulate moving all cows at the same time by choosing cows to move in a round-robin fashion, and have each cow make one step to their left, then $1 + \epsilon$ steps to the right of their starting location, etc.). In particular, this means that every cow ignores every other cow or any used gates that she finds. Throughout this section, we assume ϵ is some fixed parameter and so remove reference to it when possible to lighten notation (e.g., Parallel Cows algorithm is a shorthand for $(1 + \epsilon)$-Parallel Cows algorithm).

In this section we analyze the Parallel Cows algorithm, and prove the following theorem:

Theorem 1. *For* $\epsilon \leq 1$*, the* $(1+\epsilon)$*-Parallel Cows algorithm is* $O(k^{\log_2(3+\epsilon)-1}/\epsilon)$*-competitive for* k*-LCWA.*

In Subsect. 3.1 we prove that we can view the matchings of cows to gates found by the Parallel Cows algorithm and OPT as a set of disjoint weighted cycles. We show these cycles have special structure, which allows us to use weighted binary trees to analyze the total cost of the Parallel Cows algorithm in relation to OPT, where the leaves of a tree correspond to the edges in OPT and the internal nodes correspond to edges in the Parallel Cows matching. In Subsect. 3.2 we analyze this tree in the case when it is a perfect binary tree and all of OPT's edges have the same cost, which intuitively seems like worst case. In Subsect. 3.3 we prove that we do indeed obtain the worst case matching for the Parallel Cows algorithm when the cycle analysis yields a perfect binary tree. The competitive ratio for the Parallel Cows algorithm then follows from observing that the walking costs are only $O(1/\epsilon)$ times the matching costs.

3.1   Cycle Property

In this subsection we will show some useful properties about the combination of the matchings produced by OPT and by the Parallel Cows algorithm, denoted with A. In addition to building some intuition for the problem, these properties will show that our analysis is tight.

We first define some notation. Let $C = \{c_1, \ldots, c_k\} \subseteq \mathbb{Z}$ denote the set of cow starting locations. We use c_i to both refer to the cow as it is walking and as the starting location, specifying when it is not clear from context. Let $G = \{g_1, \ldots, g_k\} \subseteq \mathbb{Z}$ denote the set of gate locations. When referring to a specific algorithm, we use $g(c_i) : C \to G$ to denote the gate that cow c_i matches to. Consider the graph with vertices $V = C \cup G$. Let $E^{\mathrm{OPT}} = \{e_1^{\mathrm{OPT}}, e_2^{\mathrm{OPT}}, \ldots, e_k^{\mathrm{OPT}}\}$ be OPT's edges in the graph, where $e_i^{\mathrm{OPT}} = (c_i, g(c_i))$ and has weight $|c_i - g(c_i)|$. We can similarly define E^A for A. It is easy to see that the graph with vertices V and edges $E^{\mathrm{OPT}} \cup E^A$ is a disjoint union of cycles.

We now argue that there exists an optimal solution OPT where, loosely speaking, edges e^{OPT} do not *cross*, where by crossing we mean a pair of edges (e_i, e_j) such that exactly one of the two endpoints of one edge lies between the two endpoints of the other edge. Formally, we have the following lemma.

Lemma 1. *There exists an optimal matching* $e_1^{\mathrm{OPT}}, e_2^{\mathrm{OPT}}, \ldots, e_k^{\mathrm{OPT}}$ *where for each* $i, j \in \{1, 2, \ldots, k\}$, *with* $i \leq j$, *the following holds:*

$$\min\{c_j, g(c_j)\} \leq \min\{c_i, g(c_i)\} \Rightarrow$$
$$\text{either} \max\{c_j, g(c_j)\} \leq \min\{c_i, g(c_i)\} \text{or} \max\{c_j, g(c_j)\} \geq \max\{c_i, g(c_i)\}.$$

Henceforth, we may assume that OPT does not have crossings.

It is easy to show that, by the definition of the algorithm, the same non-crossing property holds for the Parallel Cows algorithm. Given an input instance, consider both the solutions produced by OPT and by A. By leveraging the above property, we can partition the arcs of the two solutions into single cycles, where by cycle we mean a set of arcs from both A and OPT such that the undirected graph induced by them is a cycle. Figure 1 depicts one such cycle.

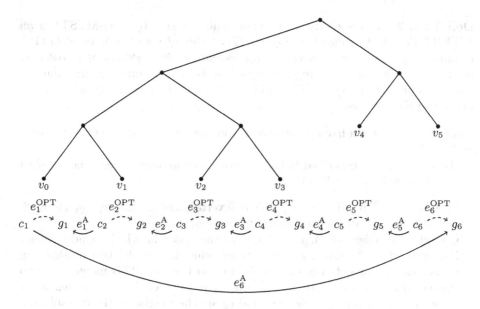

Fig. 1. A cycle and the corresponding MVST.

Finally, we will argue that any such cycle looks like the one in Fig. 1. We say that two edges e_i and e_j are in the *same direction* if either (i) $c_i \le g(c_i)$ and $c_j \le g(c_j)$ or (ii) $c_i \ge g(c_i)$ and $c_j \ge g(c_j)$, and the *opposite direction* if this is not the case.

Lemma 2. *Fix a cycle in the graph of A's and OPT's matchings. Let A's edges be $e_1^A, e_2^A, \ldots, e_\ell^A$ and OPT's edges be $e_1^{OPT}, \ldots, e_\ell^{OPT}$. Then $e_1^A, \ldots, e_{\ell-1}^A$ are all in one direction and $e_\ell^A, e_1^{OPT}, \ldots, e_\ell^{OPT}$ are all in the opposite direction.*

3.2   Perfect Tree Case

In this subsection, we show how to associate a cycle as described in the previous section with a weighted binary tree, where the sum of the weights of the leaves of the tree is the cost of the optimal matching in that cycle, and the sum of the weights of the internal nodes of the tree provides an upper bound on the cost of the matching found by the Parallel Cows algorithm. We additionally analyze the cost of this tree when the tree is perfect.

Definition 1. *A Full Weighted/Valued Binary Tree (FWVBT) T is a full binary tree, whose vertices are each associated with a weight and a value.*

- *The weight of T is defined as the sum of the values of all the vertices in T.*
- *The cost of T is defined as the sum of the values of all internal (non-leaf) vertices of T.*
- *The total leaf value of T is defined as the sum of the values of all the leaves of T.*

Definition 2. A $(1 + \epsilon)$-*Minimum Value Subtree Tree* $((1 + \epsilon)$-*MVST) is an FWVBT T with the following property: The value of a node i is equal to $(1 + \epsilon) \min(v_1, v_2)$, where v_1 and v_2 are the weights of the subtrees of i rooted at the left and right child of i respectively (there is no constraint on the values of leaves). A* perfect $(1+\epsilon)$-*MVST is a* $(1+\epsilon)$-*MVST where each leaf has the same depth and the same value.*

Since we assume ϵ is a fixed parameter, throughout we abbreviate $(1+\epsilon)$-MVST by MVST.

Given a cycle as described in the previous section, one can associate it with a MVST, as follows (see Fig. 1):

- Each edge e_i^{OPT} of OPT corresponds to a vertex/leaf of value $|c_i - g(c_i)|$. Each such leaf forms a distinct (connected) component.
- Consider the edges of A, except e_ℓ^A, in the order in which A adds them. For each edge e_i^A added, a vertex is introduced, and the two neighboring connected components become its children in the tree. This merges the two connected components into a single one. The value of this new vertex is set to be $(1 + \epsilon) \min(v_1, v_2)$, where v_1 and v_2 are the weights of the two subtrees of the vertex.
- It can be easily verified that for each edge e_i^A the total number of connected components gets decreased by one, and that the tree is indeed binary and full.

We have the following lemma:

Lemma 3. *With respect to this cycle and the corresponding MVST, the cost of the optimal matching OPT is the total leaf value of the tree, while the cost of A's matching is upper bounded by the cost of the tree.*

Let us now assume that the cycle produced by A has a length that is a power of two, and that the above construction produces a perfect binary tree, where each leaf has a value of 1. We prove the following lemma:

Lemma 4. *A perfect MVST where each leaf has a value of* 1 *has cost* $\Theta(k^{\log_2(3+\epsilon)})$.

Combined with Lemma 2, the above lemma also implies that the cost of the matching returned by the algorithm for this particular class of cycles (the ones corresponding to a perfect binary tree with leaf values of 1) is a $\Theta(n^{\log_2(3+\epsilon)-1})$-approximation with respect to the optimal matching. As we will see in the next subsection, this particular class of cycles is actually the worst case.

3.3 Perfect Trees are the Worst Case

The result of this section is the following lemma:

Lemma 5. *The cost of any MVST T is at most the cost of the smallest (in terms of number of nodes) perfect MVST with size at least that of T and total leaf value the same as T.*

To prove this, one can show that any MVST that has maximum cost for a fixed tree with fixed total leaf value must have certain structure. One can then show that a series of transformations can be performed on the tree such that the final tree is a perfect MVST and that each transformation results in a new tree with same or greater cost. The first transformation removes any nodes in the tree with value 0, and the second transformation creates a balanced tree by replacing leaves of maximum depth in the tree with leaves at a minimum possible depth in the tree. The final transformation creates a perfect tree by adding leaves until the tree is perfect.

3.4   Proof of Parallel Cows Competitiveness

We can now prove Theorem 1.

Proof (Theorem 1). Fix some instance of k-LCWA where the value of the optimal solution is OPT, and let A be the cost of the Parallel Cows algorithm on this instance and M_A be the cost of the matching found by the Parallel Cows algorithm on this instance. Note that the cost of the optimal matching and the cost of the optimal solution are the same. By Lemma 3, M_A is the sum of the costs of the MVSTs built on the cycles induced by the algorithm's and optimal's matchings, while OPT is the sum of the leaves of those same MVSTs. Fix one cycle, and let T be the MVST for that cycle. By Lemma 5, the cost of T can be upper bounded by the cost of the minimum perfect MVST larger than T, while the cost of the optimal solution on that cycle remains fixed. Thus we have by Lemma 4 that the cost of the algorithm's matching on this cycle is at most $O\left(k_c^{\log_2(3+\epsilon)-1}\right)$ times that of the optimal's, where k_c is the number of cows and gates in the cycle. Thus M_A is at most $O\left(k^{\log_2(3+\epsilon)-1}OPT\right)$.

Since each cow only stops when it finds an unused gate, the walking cost of each cow in the Parallel Cows algorithm is the same as if that cow and the gate it finds were the only ones present. Thus when $\epsilon \leq 1$ its walking cost is $O(1/\epsilon)$ times its matching cost and we obtain that $A = O\left(k^{\log_2(3+\epsilon)-1}OPT/\epsilon\right)$.   □

4   Extending the Algorithm for k-Lost Cows Without Arrivals to k-Lost Cows

In this section we show how to extend the solution for the k-LCWA problem to the k-LC problem. The basic idea is to maintain the state of our Parallel Cows algorithm assuming all cows arrived at time 0. When a cow finds a gate (occupied or unoccupied) we stop the cow and figure out who would match there in the Parallel Cows algorithm. Based on this we update our state and continue searching. We formalize this below.

While there is a cow c that is released and not matched to a gate, have c walk as a $(1 + \epsilon)$-biased cow. If during this walk c finds a gate g there are two cases to consider. If g does not yet have a cow matched to it, then match c to g. Otherwise, if there is some cow c' currently matched to h, we calculate which cow would have reached location h first if both c and c' were released at time 0. If c' would reach this location first, then c continues her walk. If c would find h first, then c "kicks" c' out who continues his walk. Since we cannot actually remove c' from this gate once it is matched, we simply have c walk as c', and record that c is really matched to g.

The first observation is that the offline optimal matching does not change if there are release times associated with the cows. So to show that this algorithm has the same competitive ratio of the Parallel Cows algorithm, we only need to show that the matching cost of this simulation is equal to the matching cost if all cows were released at time 0. The result then follows from Theorem 1.

Lemma 6. *Let $I = (C, G)$ be an instance to the generalized cow problem with release time a_i for cow c_i. The walking cost of the above simulation algorithm on I is equal to the walking cost of the parallel cow algorithm on I with zero release times.*

Proof. Let $M \subset C \times G$ be the matching found by the Parallel Cows algorithm on I with zero release times. For each $c \in C$, recall that $g(c) \in G$ is the gate that c matches to in M, that is $(c, g(c)) \in M$. We first show that if c ever reaches the point $g(c)$ in the simulation, then it will remain there for the rest of the simulation, i.e., if c reaches its gate it will match there. To see this, assume by contradiction that some cow c reaches $g(c)$ but later leaves this location. Further, assume c is the first cow to do this, and let c' be the cow that causes c to leave $g(c)$. By definition of the simulation, this means that c' would reach location $g(c)$ first in the parallel cows algorithm. However, since c' does not match to $g(c)$, this means that c' has already reached and left location $g(c')$ contradicting that c was the first cow to leave its gate.

With this we can now prove the desired lemma, that the matching is indeed the same. Let $M_s \subset C \times G$ denote the matching found by the simulation. Let $c_1, c_2, \ldots, c_k \in C$ such that c_i is the i-th cow to find its gate in the parallel cows algorithm. We show by induction that, for all i, $(c_i, g(c_i)) \in M_s$. For the base case, note that since c_1 is the first cow to find a gate in the parallel algorithm, c_1 will find $g(c_1)$ before any other gate in G. However by the above argument, once c_1 finds $g(c_1)$ it will match there in M_s. Now for the inductive hypothesis, assume the first $j - 1$ cows match the same in M and M_s. Note that since c_j is the j-th cow to find its gate in the parallel algorithm, it sees at most $j - 1$ gates before reaching $g(c_j)$ in the parallel algorithm, and these all belong to $g(c_1), g(c_2), \ldots, g(c_{j-1})$. However, by the inductive hypothesis, $c_1, c_2, \ldots, c_{j-1}$ will match there in M_s, so c_j can not match to any of those. This means that c_j must reach $g(c_j)$ and by the above argument $(c_j, g(c_j)) \in M_s$. $\qquad\square$

5   Comparing k-Lost Cows and Online Matching on a Line

In this section we explore the relationship between the k-LC and OML problems. In particular we show that positive results in the k-LC setting carry over to positive results in the OML setting (assuming the competitive ratio is defined in terms of the number of servers, n). We also show that lower bounds in the k-LC setting carry over to lower bounds in the OML setting, however here the competitive ratio for OML is defined in terms of the minimum number of positive requests in an optimal matching.

Theorem 2. *Let p be the minimum number of positive requests in an optimal solution to OML. The following two implications hold.*

1. *If there is an $f(p)$-competitive algorithm for OML then there is an $f(k)$-competitive algorithm for k-LC.*
2. *If there is an $f(k)$-competitive algorithm for k-LC then there is an $f(n)$-competitive algorithm for OML.*

Proof. For the first implication, assume that we have an $f(p)$-competitive algorithm A for OML and let $I = (C, G)$ be an instance to the k-LC problem. To obtain an $f(k)$-competitive algorithm, we create an instance I' in the OML problem with a server at every integer. When a cow arrives, at some location c we have two requests for location c in I'. It can easily be shown that A can be converted to an algorithm where every request is matched with the left most or right most server with no increase in cost. Assuming this, we now have c walk to the server that A matches the second request with, which will either be one unit to the left or right. While c has not yet found a gate, we continue to have a request in I' at c's current location, always having c walk to the server that is used to match the latest request. Once c finds a gate, we stop requesting to c's location. We continue to do this until all cows have been matched.

First note that by construction the total walking cost of the cows is within a constant factor of the matching cost of A. Further, we claim that there is an optimal solution to I' with k positive requests. To see this assume by contradiction that the minimum (in terms of positive requests) optimal matching to I' has more than k positive requests. Since there are k gates, at least one positive request, say r_i, must be matched to a non-gate location, say $s(r_i)$. However every used server that is not a gate receives a request. So there is another request at location $s(r_i)$, say r_j, that is matched to some positive cost sever $s(r_j)$. Note that the optimal solution that matches r_i to $s(r_j)$ and r_j as cost 0 does not increase the cost of the matching and has one less positive request. This contradicts our choice of OPT. This shows that there is an algorithm A that is $f(k)$-competitive on I' and therefore the corresponding cow algorithm is $f(k)$-competitive on I.

To prove the second implication assume that we have an $f(k)$-competitive algorithm for the k-LC problem and as input to the OML problem we are given an input I consisting of a set of servers S. Whenever a request r_i arrives, we release a cow c at location r_i and it begins to walk. Whenever c reaches a location corresponding to a server $s \in S$ that no previous request has been matched to s we match r_i to s and place a gate location at s so c stops walking. It is clear that the cost of the matching is at most the total walking cost of the cows and further the number of cows is equal to the number of servers.   □

Theorem 3. *There is an $O\bigl(n^{\log_2(3+\epsilon)-1}/\epsilon\bigr)$-competitive algorithm for OML.*

Proof. By Theorem 1 and Lemma 6, we have an $O\bigl(k^{\log_2(3+\epsilon)-1}/\epsilon\bigr)$-competitive algorithm for k-LC. The result follows from Theorem 2. $\square$

We note that, in a manner similar to that presented in Sect. 4, it is possible to extend the Parallel Cows algorithm for k-LCWA to OML directly and obtain an $O\bigl(p^{\log_2(3+\epsilon)-1}/\epsilon\bigr)$-competitive algorithm for OML.

References

1. Baeza-Yates, R.A., Culberson, J.C., Rawlins, G.J.E.: Searching in the plane. Inf. Comput. **106**(2), 234–252 (1993)
2. Bansal, N., Buchbinder, N., Gupta, A., Naor, J.: A randomized $O(\log^2 k)$-competitive algorithm for metric bipartite matching. Algorithmica **68**(2), 390–403 (2014)
3. Chung, C., Pruhs, K., Uthaisombut, P.: The online transportation problem: On the exponential boost of one extra server. In: Laber, E.S., Bornstein, C., Nogueira, L.T., Faria, L. (eds.) LATIN 2008. LNCS, vol. 4957, pp. 228–239. Springer, Heidelberg (2008)
4. Fakcharoenphol, J., Rao, S., Talwar, K.: A tight bound on approximating arbitrary metrics by tree metrics. J. Comput. Syst. Sci. **69**(3), 485–497 (2004)
5. Fuchs, B., Hochstättler, W., Kern, W.: Online matching on a line. Theor. Comput. Sci. **332**(1–3), 251–264 (2005)
6. Gupta, A., Lewi, K.: The online metric matching problem for doubling metrics. In: Czumaj, A., Mehlhorn, K., Pitts, A., Wattenhofer, R. (eds.) ICALP 2012, Part I. LNCS, vol. 7391, pp. 424–435. Springer, Heidelberg (2012)
7. Kalyanasundaram, B., Pruhs, K.: Online weighted matching. J. Algorithms **14**(3), 478–488 (1993)
8. Kalyanasundaram, B., Pruhs, K.: Online network optimization problems. Online Algorithms 1996. LNCS, vol. 1442, pp. 268–280. Springer, Heidelberg (1998)
9. Kalyanasundaram, B., Pruhs, K.: The online transportation problem. SIAM J. Discret. Math. **13**(3), 370–383 (2000)
10. Kao, M.-Y., Reif, J.H., Tate, S.R.: Searching in an unknown environment: an optimal randomized algorithm for the cow-path problem. Inf. Comput. **131**(1), 63–79 (1996)
11. Khuller, S., Mitchell, S.G., Vazirani, V.V.: On-line algorithms for weighted bipartite matching and stable marriages. Theor. Comput. Sci. **127**(2), 255–267 (1994)
12. Koutsoupias, Elias, Nanavati, Akash: The online matching problem on a line. In: Solis-Oba, Roberto, Jansen, Klaus (eds.) WAOA 2003. LNCS, vol. 2909, pp. 179–191. Springer, Heidelberg (2004)
13. Koutsoupias, E., Papadimitriou, C.H.: On the k-server conjecture. J. ACM **42**(5), 971–983 (1995)
14. López-Ortiz., A.: On-line target searching in bounded and unbounded domains. Ph.D thesis, University of Waterloo (1996)
15. Meyerson, A., Nanavati, A., Poplawski, L.J.: Randomized online algorithms for minimum metric bipartite matching. In: Proceedings of the 17th Annual ACM-SIAM Symposium on Discrete Algorithms (SODA), pp. 954–959 (2006)
16. Reingold, E.M., Tarjan, R.E.: On a greedy heuristic for complete matching. SIAM J. Comput. **10**(4), 676–681 (1981)

Better Algorithms for Online Bin Stretching

Martin Böhm[1]([✉]), Jiří Sgall[1], Rob van Stee[2], and Pavel Veselý[1]

[1] Computer Science Institute of Charles University, Prague, Czech Republic
{bohm,sgall,vesely}@iuuk.mff.cuni.cz
[2] Department of Computer Science, University of Leicester, Leicester, UK
rob.vanstee@leicester.ac.uk

Abstract. ONLINE BIN STRETCHING is a semi-online variant of bin packing in which the algorithm has to use the same number of bins as the optimal packing, but is allowed to slightly overpack the bins. The goal is to minimize the amount of overpacking, i.e., the maximum size packed into any bin.

We give an algorithm for ONLINE BIN STRETCHING with a stretching factor of 1.5 for any number of bins. We also show a specialized algorithm for three bins with a stretching factor of $11/8 = 1.375$.

1 Introduction

The most famous algorithmic problem dealing with online assignment is arguably ONLINE BIN PACKING. In this problem, known since the 1970s, items of size between 0 and 1 arrive in a sequence and the goal is to pack these items into the least number of unit-sized bins, packing each item as soon as it arrives.

ONLINE BIN STRETCHING, which has been introduced by Azar and Regev in 1998 [2], deals with a similar online scenario. Again, items of size between 0 and 1 arrive in a sequence, and the algorithm needs to pack them as soon as each item arrives, but it has two advantages: (i) The packing algorithm knows m, the number of bins that an optimal offline algorithm would use, and must also use only at most m bins, and (ii) the packing algorithm can use bins of capacity R for some $R \geq 1$. The goal is to minimize the stretching factor R.

While formulated as a bin packing variant, ONLINE BIN STRETCHING can also be thought of as a semi-online scheduling problem, in which we schedule jobs in an online manner on exactly m machines, before any execution starts. We have a guarantee that the optimum offline algorithm could schedule all jobs with makespan 1. Our task is to present an online algorithm with makespan of the schedule being at most R.

History. ONLINE BIN STRETCHING has been proposed by Azar and Regev [2]. The original lower bound of 4/3 for three bins has appeared even before that, in [10], for two bins together with a matching algorithm. Azar and Regev extended the same lower bound to any number of bins and gave an online algorithm with a stretching factor 1.625.

M. Böhm, J. Sgall and P. Veselý—Supported by the project 14-10003S of GA ČR and by the GAUK project 548214.

E. Bampis and O. Svensson (Eds.): WAOA 2014, LNCS 8952, pp. 23–34, 2015.
DOI: 10.1007/978-3-319-18263-6_3

The problem has been revisited recently, with both lower bound improvements and new efficient algorithms. On the algorithmic side, Kellerer and Kotov [9] have achieved a stretching factor $11/7 \approx 1.57$ and Gabay et al. [7] have achieved $26/17 \approx 1.53$. In the case with only three bins, the previously best algorithm was due to Azar and Regev [2], with a stretching factor of 1.4.

On the lower bound side, the lower bound $4/3$ of [2] was surpassed only for the case of three bins by Gabay et al. [6], who show a lower bound of $19/14$, using an extensive computer search.

Our Contributions. In Sect. 2, we present a new algorithm for ONLINE BIN STRETCHING with a stretching factor of 1.5. We build on the techniques of [7,9] who designed two-phase algorithms where the first phase tries to fill some bins close to $R - 1$ and achieve a fixed ratio between these bins and empty bins, while the second phase uses the bins in blocks of fixed size and analyzes each block separately. This technique, with some case analysis, seemed to be able to lead to improved results approaching 1.5. To actually reach 1.5, we needed to significantly improve the analysis using amortization techniques (represented by a weight function in our presentation) to amortize among blocks and bins of different types.

In Sect. 3, we focus on the case of three bins. For this case, there is a recent lower bound of $19/14 \approx 1.357$ [6]. We present an algorithm for three bins of capacity $11/8 = 1.375$. This is the first improvement of the stretching factor 1.4 of Azar and Regev [2] for three bins and significantly decreases the remaining gap.

Related Work. The NP-hard problem BIN PACKING was originally proposed by Ullman [11] and Johnson [8] in the 1970s. Since then it has seen major interest and progress, see the survey of Coffman et al. [4] for many results on classical Bin Packing and its variants. While our problem can be seen as a variant of BIN PACKING, note that the algorithms cannot open more bins than the optimum and thus general results for BIN PACKING do not translate to our setting.

As noted, ONLINE BIN STRETCHING can be formulated as the online scheduling on m identical machines with known optimal makespan. Such algorithms were studied and are important in designing constant-competitive algorithms without the additional knowledge, e.g., for scheduling in the more general model of uniformly related machines [1,3,5].

Definitions and Notation. Our main problem, ONLINE BIN STRETCHING, can be described as follows:

Input: an integer m and a sequence of items $I = i_1, i_2, \ldots$ given online one by one. Each item has a *size* $s(i) \in [0, 1]$ and must be packed immediately and irrevocably.

Parameter: The *stretching factor* R, a limit of the capacity of all bins.

Output: Partitioning (packing) of I into bins $B_1, \ldots, B_m$ so that $\sum_{i \in B_j} s(i) \le R$ for all $j = 1, \ldots, m$.

Guarantee: there exists a packing of all items in I into m bins of capacity 1.

Goal: Design an online algorithm with the stretching factor R as small as possible which packs all input sequences satisfying the guarantee.

For a bin B, we define the *size of the bin* $s(B) = \sum_{i \in B} s(i)$. Unlike $s(i)$, $s(B)$ can change during the course of the algorithm, as we pack more and more items into the bin. To easily differentiate between items, bins and lists of bins, we use lowercase letters for items (i, b, x), uppercase letters for bins (A, B, X), and calligraphic letters for lists of bins $(\mathcal{A}, \mathcal{C}, \mathcal{L})$.

In both sections of our paper, we rescale the item sizes and bin capacities for simplicity. Therefore, in our setting, each item has an associated size $s(i) \in [0, k]$, where $k \in \mathbb{N}$ is also the capacity of the bins which the optimal offline algorithm uses. The online algorithm for ONLINE BIN STRETCHING uses bins of capacity $t \in \mathbb{N}$, $t \geq k$. The resulting stretching factor is thus t/k.

We omit some proofs due to space restrictions. Full version is available at http://arxiv.org/abs/1404.5569.

2 Upper Bound for an Arbitrary Number of Bins

We rescale the bin sizes so that the optimal bins have size 12 and the bins of the algorithm have size 18.

We follow the general two-phase scheme of recent results [7,9] which we sketch now. In the first phase of the algorithm we try to fill the bins so that their size is at most 6, as this leaves space for an arbitrary item in each bin. Of course, if items larger than 6 arrive, we need to pack them differently, namely in bins of size at least 12, whenever possible. We stop the first phase when the number of non-empty bins of size at most 6 is three times the number of empty bins. In the second phase, we work in blocks of three non-empty bins and one empty. The goal is to show that we are able to fill the bins so that the average size is at least 12, which guarantees we are able to pack the total size of $12m$ which is the upper bound on the size of all items.

The limitation of the previous results using this scheme was that the volume achieved in a typical block of four bins is slightly less than four times the size of the optimal bin, which then leads to bounds strictly above $3/2$. This is also the case in our algorithm: A typical block may contain in three bins items from the first phase of size just above 4 plus one item of size 7 from the second phase, while the last bin contains two items of size 7 from the second phase—a total of 47 instead of desired $4 \cdot 12$. However, we notice that such a block contains five items of size 7 which the optimum cannot fit into four bins. To take an advantage of this, we cannot analyze each block separately. Instead, we need to show that a bin with no item of size more than 6 typically has size at least 13 and amortize among the blocks of different types. Technically this is done using a weight function w that takes into account both the total size of items and the number of items larger than 6. This is the main new technical idea of our proof.

There are other complications. We need to guarantee that a typical bin of size at most 6 has size at least 4 after the first phase. However, this is impossible to guarantee if the items packed there have size between 3 and 4. Larger items are fine, as one per bin is sufficient, and the smaller ones are fine as well as we can always fit at least two of them and this guarantees that we have only two bins filled below 4. This motivates our classification of items: Only the regular items of size in $(0,3] \cup (4,6]$ are packed in the bins filled up to size 6. The medium items of size in $(3,4]$ are packed in their own bins (four or five per bin). Similarly, large items of size in $(6,9]$ are packed in pairs in their own bins. Finally, the huge items of size larger than 9 are handled similarly as in the previous papers: If possible, they are packed with the regular items, otherwise each in their own bin.

The introduction of medium size items in turn implies that we need to revisit the analysis of the first phase and also of the case when the first phase ends with no empty bin. These parts of the proof are similar to the previous works, but due to the new item type we need to carefully revisit it; it is now convenient to introduce another weight function v that counts the items according to their type. The analysis of the second phase when empty bins are present is more complicated, as we need to take care of various degenerate cases, and it is also here where the novel amortization is used.

Lower Bound. We note that this two-phase approach cannot give a better stretching factor than 1.5. Consider the following instance. Send two items of size 6 which are in the first phase packed separately into two bins. Then send $m-1$ items of size 12 and one of them must be put into a bin with an item of size 6, i.e., one bin receives items of size 18, while all the items can be packed into m bins of size 12. This instance and its modifications with more items of size 6 or slightly smaller items at the beginning thus show that decreasing the upper bound below 1.5 would need a significantly different approach, as we would be forced to pack these items in pairs. This also shows that the analysis of our algorithm is tight.

Now we are ready to proceed with the formal statement of the algorithm and proof.

Theorem 1. *There exists an algorithm for* ONLINE BIN STRETCHING *with a stretching factor of* 1.5 *for an arbitrary number of bins.*

We take an instance with an optimal packing into m bins of size 12 and, assuming that our algorithm fails, we derive a contradiction. One way to get a contradiction is to show that the size of all items is larger than $12m$. We also use two other bounds in the spirit of weight functions: weight $w(i)$ and value $v(i)$. The weight $w(i)$ is a slightly modified size to account for items of size larger than 6. The value $v(i)$ only counts the number of items with relatively large sizes. For our calculations, it is convenient to normalize the functions so that they are at most 0 for optimal bins (see Lemma 1).

We classify the items and define their value $v(i)$ as follows.

Type	Huge	Large	Medium	Regular
s(i)	(9,12]	(6,9]	(3,4]	(0,3]∪(4,6]
v(i)	3	2	1	0

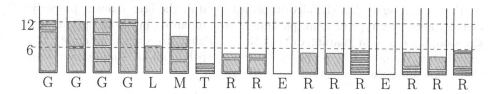

Fig. 1. A typical state of the algorithm after the first phase. The bin labels correspond to the bin types of the first phase. The non-complete bins (other than G) are ordered as in the list $\mathcal{L}$ at the beginning of the second phase with regular bins.

Definition 1. *For a set of items A, we define the value $v(A) = (\sum_{i \in A} v(i)) - 3$ and we define weight $w(A)$ as follows. Let $k(A)$ be the number of large and huge items in A. Then $w(A) = s(A) + k(A) - 13$. For a set of bins $\mathcal{A}$ we define $v(\mathcal{A}) = \sum_{A \in \mathcal{A}} v(A)$, $w(\mathcal{A}) = \sum_{A \in \mathcal{A}} w(A)$ and $k(\mathcal{A}) = \sum_{A \in \mathcal{A}} k(A)$.*

Lemma 1. *For any packing of a valid instance into m bins $\mathcal{A}$ of any size, we have $w(\mathcal{A}) \leq 0$ and $v(\mathcal{A}) \leq 0$.* □

First Phase. During the first phase, our algorithm maintains the invariant that only bins of the following types exist. See Fig. 1 for an illustration of the types of the bins.

E Empty bins: bins that have no item.
G Complete bins: all bins A that have $w(A) \geq 0$ and $s(A) \geq 12$;
H Huge-item bins: all bins A that contain a huge item (plus possibly some other items) and have $s(A) < 12$;
L One large-item bin: a bin containing only a single large item;
M One medium-item bin: a bin A with $s(A) \leq 13$ and only medium items;
T One tiny bin: a bin with $s(A) \leq 3$;
R Regular bins: all other bins with $s(A) \in (3, 6]$;

First-phase algorithm:

Let e be the number of empty bins and r the number of regular bins. If $r \geq 3e$, stop the first phase.

Assign the current item i according to its item type, using the first possible option in the particular column. The first letter in a cell indicates the required type of the bin before the assignment and the second column denotes the type of the bin after the assignment. If there are two types listed, the new bin type depends on the new size and weight of the bin.

Note: As an additional rule when packing regular items, the item is packed in a regular or tiny bin only if the total size packed into this bin does not exceed 6 afterwards.

Item type	huge	large	medium	regular
Option 1	$R \rightarrow G$	$L \rightarrow G$	$M \rightarrow G/M$	$H \rightarrow G/H$
Option 2	$T \rightarrow H$	$E \rightarrow L$	$E \rightarrow M$	$R \rightarrow R$
Option 3	$E \rightarrow H$			$T \rightarrow T/R$
Option 4				$E \rightarrow T/R$

First we observe that the algorithm described in the box above is properly defined. For every type of item, packing it into an empty bin is an option, and the stopping criterion guarantees that the algorithm stops when no empty bin is available. We now state properties of the algorithm; all are simple invariants that follow from the description of the algorithm.

Lemma 2. *At any time during the first phase the following holds:*

(i) *All bins used by the algorithm are of type* **E, G, H, L, M, T,** *or* **R.**
(ii) *All complete bins B have $s(B) \geq 12$, $v(B) \geq 0$, and $w(B) \geq 0$.*
(iii) *If there is a huge-item bin, there is no regular and no tiny bin.*
(iv) *There is at most one large-item bin and at most one medium-item bin.*
(v) *There is at most one tiny bin T. If T exists, then for any regular bin, $s(T) + s(R) > 6$. There is at most one regular bin R with $s(R) \leq 4$.*
(vi) *At the end of the first phase $3e \leq r \leq 3e + 3$.*   □

If the algorithm packs all items in the first phase, it stops. Otherwise according to Lemma 2 (iii) we split the algorithm in two very different branches. If there is no regular bin, follow the second phase with huge-item bins below. If there is at least one regular bin, follow the second phase with regular bins.

Let $\mathcal{G}$ be the set of all complete bins; we do not use these bins in the second phase. In addition to $\mathcal{G}$ and either huge-item bins, or the regular and empty bins, there may exist at most three *special bins* denoted and ordered as follows: the large-item bin L, the medium item bin M, and the tiny bin T.

Second Phase with Huge-Item Bins. Let the list of bins $\mathcal{L}$ contain first all the huge-item bins, followed by the special bins L, M, in this order, if they exist. There are no other non-empty bins by Lemma 2 and no empty bins because we have $3e \leq r = 0$. We use First Fit on $\mathcal{L}$, without allowing new bins to be opened. Suppose that we have an instance that has a packing into bins of capacity 12 and on which our algorithm fails. We may assume that the algorithm fails on the last item f. By considering the total volume, there always exists a bin with size at most 12. Thus $s(f) > 6$ and $v(f) \geq 2$.

If during the second phase an item n with $s(n) \leq 6$ is packed into the last bin in $\mathcal{L}$, we know that all other bins have size more than 12, thus all the remaining items fit into the last bin. Otherwise we consider $v(\mathcal{L})$. Any bin $B \in \mathcal{G}$ has $v(B) \geq 0$ by Lemma 2 (ii) and each huge-item bin gets nonnegative value too. Also $v(L) \geq -1$ if L exists. This shows that M must exist, since otherwise $v(\mathcal{L}) + v(f) \geq -1 + 2 \geq 1$, a contradiction.

M is the last bin of $\mathcal{L}$ and thus in this last case M contains only medium items from the first phase and possibly large and/or huge items from the second phase. We claim that $v(M) + v(f) \geq 2$ using the fact that f does not fit into M and M contains no item a with $v(a) = 0$: If f is huge we have $s(M) > 6$, thus M must contain either two medium items or one medium item and one large or huge item and $v(M) \geq -1$. If f is large, we have $s(M) > 9$; thus M contains either three medium items or one medium and one large or huge item and $v(M) \geq 0$. Thus we always have $v(\mathcal{L}) \geq -1 + v(M) + v(f) \geq 1$, a contradiction.

Second Phase with Regular Bins. Let $\mathcal{E}$ resp. $\mathcal{R}$ be the set of empty resp. regular bins at the beginning of the second phase, and let $e = |\mathcal{E}|$. Let $\lambda \in \{0,1,2,3\}$ be such that $|\mathcal{R}| = 3e + \lambda$; Lemma 4 (vi) implies that it exists.

We order the bins that are not complete into a list $\mathcal{L}$ as follows. We group the bins in $\mathcal{E} \cup \mathcal{R}$ into blocks of typically one empty and three regular bins as follows. Denote the empty bins $E_1, E_2, \ldots, E_e$. The regular bins are denoted by $R_{i,j}$, $i = 1, \ldots, e+1$, $j = 1,2,3$. The ith block $\mathcal{B}_i$ consists of bins $R_{i,1}, R_{i,2}, R_{i,3}, E_i$ in this order. There are two exceptions: The last block $\mathcal{B}_{e+1}$ has no empty bin, only exactly 3 regular bins. The first block contains only λ regular bins instead of 3 and an empty bin. As the first regular bin we choose the one with size less than 4, if there is such a bin. By Lemma 2 (v) there exists at most one such bin and all the remaining $R_{i,j}$ have size at least 4. Denote the first regular bin by $\bar{R}$ if $\mathcal{R} \neq \emptyset$; note that $\bar{R}$ is either the first bin in $\mathcal{B}_1$ or the first bin in $\mathcal{B}_2$ if $\lambda = 0$.

The list of bins $\mathcal{L}$ we use in the second phase contains first the special bins and then all the blocks $\mathcal{B}_1, \ldots, \mathcal{B}_{e+1}$. Thus the list $\mathcal{L}$ is (some or all of the first six bins may not exist):

$$L, M, T, R_{1,1}, R_{1,2}, R_{1,3}, E_1, R_{2,1}, R_{2,2}, R_{2,3}, E_2, \ldots, E_e, R_{e+1,1}, R_{e+1,2}, R_{e+1,3}.$$

Whenever we refer to the ordering of the bins, we mean the ordering in $\mathcal{L}$. See Fig. 1 for an illustration.

We use First Fit on the reversed list $\mathcal{L}$ for huge items (that is, we pack each huge item to the last bin in $\mathcal{L}$ where it fits) and we use First Fit on $\mathcal{L}$ for all other items.

Suppose that we have an instance that has a packing into bins of capacity 12 and on which our algorithm fails. We may assume that the algorithm fails on the last item. Let us denote this item by f. Call the items that arrived in the second phase *new* (including f), the items from the first phase are *old*. See Fig. 2 for an illustration of a typical final situation. Our overall strategy is to obtain a contradiction by showing that

$$w(\mathcal{L}) + w(f) > 0.$$

In some cases, we instead argue that $v(\mathcal{L}) + v(f) > 0$ or $s(\mathcal{L}) + s(f) > 12|\mathcal{L}|$. Any of these is sufficient for a contradiction, as all bins in $\mathcal{G}$ have both volume and weight nonnegative and size larger than 12. Note also that $s(f) > 6$ since by considering the total volume, there always exists a bin with size at most 12.

Let $\mathcal{H}$ denote all the bins from $\mathcal{L}$ with a huge item, and let $h = |\mathcal{H}| \mod 4$. First we show that the average size of bins in $\mathcal{H}$ is large and exclude some degenerate cases.

Lemma 3. *Let ρ be the total size of old items in $\bar{R}$ if $\bar{R} \in \mathcal{H}$, otherwise set $\rho = 4$.*

(i) *The bins $\mathcal{H}$ are a final segment of the list and $\mathcal{H} \subsetneq \mathcal{E} \cup \mathcal{R}$.*
(ii) *We have $s(\mathcal{H}) \geq 12|\mathcal{H}| + h + \rho - 4$.*
(iii) *If $\mathcal{H}$ does not include $\bar{R}$, then $s(\mathcal{H}) \geq 12|\mathcal{H}| + h \geq 12|\mathcal{H}|$.*
(iv) *If $\mathcal{H}$ includes $\bar{R}$, then $s(\mathcal{H}) \geq 12|\mathcal{H}| + h - 1 \geq 12|\mathcal{H}| - 1$.*   □

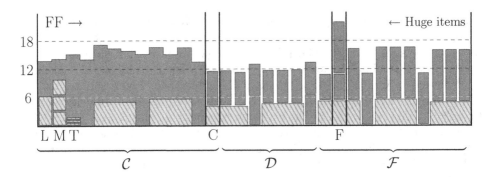

Fig. 2. A typical state of the algorithm after the second phase with regular bins. The gray (hatched) areas denote the old items (i.e., packed in the first phase), the red (solid) regions and rectangles denote the new items (i.e., packed in the second phase). The bins that are complete at the end of the first phase are not shown. The item f on which the algorithm fails is shown as packed into the final bin F and exceeding the capacity of the bin (Color figure online).

Let F, the *final bin* be the last bin in $\mathcal{L}$ before $\mathcal{H}$, or the last bin if $\mathcal{H} = \emptyset$; by Lemma 3 we have $F \in \mathcal{E} \cup \mathcal{R}$. Now modify the packing so that f is put into F, f is also considered a new item. Thus $s(F) > 18$ and f as well as all the new items packed in F or a bin before F satisfy the property that they do not fit into any previous bin. Let C, the *critical bin*, be the first bin in $\mathcal{L}$ of size at most 12; such a bin exists, as otherwise the total size is more than $12m$.

We start by some easy observations. Only items of size at most 9 are packed in bins before F, in F itself only the item f can be larger. All the new items in the bins after C are large; f can be also huge. Each bin, possibly with the exception of L and M, contains a new item, as it enters the phase with size at most 6, and the algorithm failed. Each bin in $\mathcal{E}$ before F contains two new items. The bin F always has two new items, one that did fit into it and f. More observations are given in the next two lemmata.

Lemma 4. *(i) Let B be any bin before F. Then $s(B) > 9$.*

(ii) Let B, B', B'' be any three bins in this order before or equal to F and let B'' contain two new items. Then $s(B) + s(B') + s(B'') > 36 + o$, where o is the size of old items in B''.

(iii) Let B be arbitrary and $B' \in \mathcal{R}$ after both B and C.
If $B' \neq \bar{R}$ then $s(B) + s(B') > 22$, in particular $s(B) > 11$ or $s(B') > 11$.
If $B' = \bar{R}$ then $s(B) + s(B') > 21$. □

Lemma 5. *The critical bin C is before F, there are at least two bins between C and F and C is not in the same block as F.* □

Now we partition $\mathcal{L}$ into several parts, see Fig. 2 for an illustration of these parts. Let $\mathcal{F} = \mathcal{B}_i \cup \mathcal{H}$, where $F \in \mathcal{B}_i$. Let $\mathcal{D}$ be the set of all bins after C and before $\mathcal{F}$. Let $\mathcal{C}$ be the set of all bins before and including C. Lemma 5 shows that the parts are non-overlapping. We analyze the weight of the parts separately, essentially block by block. The proof is relatively straightforward if C is not special (and thus also $\mathcal{F} \not\subseteq \mathcal{B}_1$), which is the most important case driving our choices for w. A typical block has nonnegative weight, we gain more weight in the block of F which exactly compensates the loss of weight in C, which occurs mainly in C itself.

Lemma 6. *If F is not in the first block then $w(\mathcal{F}) > 5$, else $w(\mathcal{F}) > 4$.* □

Lemma 7. *If $C \in \mathcal{R}$ then $w(C) \geq -6$. If $C \in \mathcal{E}$ then $w(C) \geq -5$. If C is a special bin then $w(C) \geq -4$.* □

Lemma 8. (i) *For every block $\mathcal{B}_i \subseteq \mathcal{D}$ we have $w(\mathcal{B}_i) \geq 0$.*
(ii) *If there is no special bin in $\mathcal{D}$, then $w(\mathcal{D}) \geq 0$. If also $C \in \mathcal{R}$ then $w(\mathcal{D}) \geq 1$.*

We are now ready to derive the final contradiction. If $\mathcal{D}$ does not contain a special bin, we add the appropriate bounds from Lemmata 7, 8 and 6. If $C \in \mathcal{R}$ then F is not in the first block and $w(\mathcal{L}) = w(C) + w(\mathcal{D}) + w(\mathcal{F}) > -6 + 1 + 5 = 0$. If $C \in \mathcal{E}$ then F is not in the first block and $w(\mathcal{L}) = w(C) + w(\mathcal{D}) + w(\mathcal{F}) > -5 + 0 + 5 = 0$. If C is the last special bin then $w(\mathcal{L}) = w(C) + w(\mathcal{D}) + w(\mathcal{F}) > -4 + 0 + 4 = 0$. In all subcases $w(\mathcal{L}) > 0$, a contradiction.

If $\mathcal{D}$ does contain a special bin we need to analyze several cases depending on the number of special bins and regular bins in $\mathcal{B}_1$.

In all of the cases we can derive a contradiction, which implies that our algorithm cannot fail. This concludes the proof of Theorem 1. □

3 Bin Stretching for Three Bins

We scale the input sizes by 16. The stretched bins in our setting therefore have capacity 22 and the optimal offline algorithm can pack all items into three bins of capacity 16 each. The three bins of our setting are named A, B, and C. We prove the following theorem.

Theorem 2. *There exists an algorithm that solves* ONLINE BIN STRETCHING *for three bins with stretching factor $1 + 3/8 = 1.375$.*

A natural idea is to try to pack first all items in a single bin, as long as possible. In general, this is the strategy that we follow. However, somewhat surprisingly, it turns out that from the very beginning we need to put items in two bins even if the items as well as their total size are relatively small.

It is clear that we have to be very cautious about exceeding a load of 6. For instance, if we put 7 items of size 1 in bin A, and 7 such items in B, then if two items of size 16 arrive, the algorithm will have a load of at least 23 in some bin. Similarly, we cannot assign too much to a single bin: putting 20 items of size 0.5 all in bin A gives a load of 22.5 somewhere if three items of size 12.5 arrive next.

On the other hand, it is useful to keep one bin empty for some time; many problematic instances end with three large items such that one of them has to be placed in a bin that already has high load. Keeping one bin free ensures that such items must have size more than 11 (on average), which limits the adversary's options, since all items must still fit into bins of size 16.

Deciding when exactly to start using the third bin and when to cross the threshold of 6 for the first time was the biggest challenge in designing this algorithm: both of these events should preferably be postponed as long as possible, but obviously they come into conflict at some point.

Good Situations. Before stating the algorithm itself, we list several *good situations* (GS). These are configurations of the three bins which allow us to complete the packing regardless of the following input. Obviously the identities of the bins are not important here; for instance, in the first good situation, all that we need is that *any* two bins together have items of size at least 26. We have used names only for clarity of presentation and of the proofs.

Good Situation 1. *Given a partial packing such that $s(A) + s(B) \geq 26$ and $s(C)$ is arbitrary, there exists an online algorithm that packs all remaining items into three bins of capacity 22.*

Proof. Since the optimum can pack into three bins of size 16, the total size of items in the instance is at most $3 \cdot 16 = 48$. If two bins have size $s(A) + s(B) \geq 26$, all the remaining items (including the ones already placed on C) have size at most 22. Thus we can pack them all into bin C. □

Good Situation 2. *Given a partial packing such that $s(A) \in [4, 6]$ and $s(B)$ and $s(C)$ are arbitrary, there exists an online algorithm that packs all remaining items into three bins of capacity 22.*

Proof. Let A be the bin with size between 4 and 6 and B be one of the other bins (choose arbitrarily). Put all the items greedily into B. When an item does not fit, put it into A, where it fits, as originally $s(A)$ is at most 6. Now the size of all items in B plus the last item is at least 22. In addition, A has items of size at least 4 before the last item by the assumption. Together we have $s(A) + s(B) \geq 26$, allowing us to apply GS1. □

Good Situation 3. *Given a partial packing such that $s(A) \in [15, 22]$ and either (i) $s(C) \leq 6$ and $s(B)$ is arbitrary or (ii) $s(B) + s(C) \geq 22$, there exists an online algorithm that packs all remaining items into three bins of capacity 22.*

Proof. If $s(B) + s(C) \geq 22$, then $\max(s(B), s(C)) \geq 11$, so we are in GS1 on bins A and B or on bins A and C. Else, if $s(C) \leq 6$, we pack arriving items into B. If $s(B) \geq 11$, we apply GS1 on bins A and B. Thus we can assume $s(B) < 11$ and we cannot continue packing into B any further. This implies that an item i arrives such that $s(i) > 11$. As $s(C) \leq 6$, we pack i into it and apply GS1 on bins A and C. □

Good Situation 4. *Given a partial packing such that $s(A) + s(B) \geq 15 + \frac{1}{2}s(C)$, $s(B) < 4$, and $s(C) < 4$, there exists an online algorithm that packs all remaining items into three bins of capacity 22.* □

Good Situation 5. *Given a partial packing such that an item a with $s(a) > 6$ is packed into bin A, $s(B) \in [3, 6]$, and C is empty, there exists an algorithm that packs all remaining items into three bins of capacity 22.* □

Good Situation 6. *If $s(C) \leq 6 \leq s(B)$ and $s(A) \geq s(B) + 4 - s(C)$, there exists an algorithm that packs all remaining items into 3 bins of capacity 22.* □

Good Situation 7. *Suppose $s(C) \leq s(B) < 6 < s(A)$. If $s(A) \leq 9 + \frac{1}{2}(s(C) + s(B))$ and for some item x we have $s(A) + x > 22$, there exists an online algorithm that packs all remaining items into three bins of capacity 22.* □

The Algorithm. We now proceed to describe the bin packing algorithm itself. Its analysis is omitted due to space restrictions. The algorithm will often use a special variant of FIRST FIT, described as follows:

Definition 2. *Let $\mathcal{L} = (X|_k, Y|_l, \dots)$ be a list of bins $X, Y, \dots$ where each bin X has an associated integral capacity k satisfying $s(X) \leq k$. GSFF$\mathcal{L}$ (Good Situation First Fit) is an online algorithm for bin stretching that works as follows:*

Algorithm GSFF($\mathcal{L}$): *For each item i:*
If it is possible to pack i into any bin (including bins not in $\mathcal{L}$, and using capacities of 22 for all bins) such that a good situation is reached, do so and continue with the algorithm of the relevant good situation.
Otherwise, traverse the list $\mathcal{L}$ in order and pack i into the first bin X such that $X|_k \in \mathcal{L}$ and $s(X) + s(i) \leq k$.

For example, GSFF$A|_4, B|_{22}$ checks whether either $(A \cup \{j\}, B, C)$, $(A, B \cup \{j\}, C)$ or $(A, B, C \cup \{j\})$ is a partial packing of any good situation. If this is not the case, the algorithm packs j into bin A provided that $s(A) + s(j) \leq 4$. If $s(A) + s(j) > 4$, the algorithm packs j into bin B with capacity 22. If j cannot be placed into B, GSFF$A|_4, B|_{22}$ reports failure and another online algorithm must be applied.

In the first phase, we pack items into two bins so that either an item of size 6 arrives relatively early in the input sequence, or we can reach a good situation.

Algorithm FIRST PHASE:

(1) GSFF($A|_4, B|_4$). Rename the bins so that $s(A) \geq s(B)$.
(2) Let the item on which Step (1) failed be j. If $s(j) > 6$ place j into A, and go to the SECOND PHASE.
(3) Place j into B and rename the bins so that $s(A) > s(B) \geq s(C) = 0$.
(4) GSFF($B|_4, A|_q, C|_4$) where $q := 9 + \frac{1}{2}(s(B) + s(C))$. As soon as C receives its first item, rename the bins B and C so that $s(B) \geq s(C)$ and continue. Note that also the value of q may change between packing of different items.

We start the algorithm SECOND PHASE only after Step (2) of FIRST PHASE fails when an item j of size $s(j) > 6$ arrives. We have not entered GS5 before placing j on A, and so we have $s(A \setminus \{j\}) \leq 3$.

Algorithm SECOND PHASE:

(1) GSFF$(A|_q, B|_4)$, where $q := 6 + s(j)$.

(2) If the next item x fits into A, apply GSFF$(A|_{22}, B|_{22}, C|_{22})$.

(3) Else: Place x into B. Let j' be the smallest item of $\{j, x\}$.

(4)   Reorder the bins A and B so that $j' \in A$.

(5)   GSFF$(A|_q, B|_{22})$, where $q := 6 + s(j')$.

(6)   Place next item y into C. Let j'' be the smallest item of $\{j', y\}$.

(7)   Reorder the bins A and C so that $j'' \in A$.

(8)   GSFF$(A|_q, B|_{22}, C|_{22})$, where $q := 6 + s(j'')$.

Conclusions. With our algorithm for $m = 3$, the remaining gap is small. For arbitrary m, we have seen at the beginning of Sect. 2 that a significantly new approach would be needed for an algorithm with a better stretching factor than 1.5. Thus, after the previous incremental results, our algorithm is the final step of this line of study. It is quite surprising that there are no lower bounds for $m > 3$ larger than the easy bound of $4/3$.

Acknowledgment. The authors thank Emese Bittner for useful discussions during her visit to Charles University.

References

1. Aspnes, J., Azar, Y., Fiat, A., Plotkin, S., Waarts, O.: On-line load balancing with applications to machine scheduling and virtual circuit routing. J. ACM **44**, 486–504 (1997)

2. Azar, Y., Regev, O.: On-line bin-stretching. In: Rolim, J.D.P., Serna, M., Luby, M. (eds.) RANDOM 1998. LNCS, vol. 1518, pp. 71–81. Springer, Heidelberg (1998)

3. Berman, P., Charikar, M., Karpinski, M.: On-line load balancing for related machines. J. Algorithms **35**, 108–121 (2000)

4. Coffman Jr., E., Csirik, J., Galambos, G., Martello, S., Vigo, D.: Bin packing approximation algorithms: survey and classification. In: Pardalos, P.M., Du, D.-Z., Graham, R.L. (eds.) Handbook of Combinatorial Optimization, pp. 455–531. Springer, New York (2013)

5. Ebenlendr, T., Jawor, W., Sgall, J.: Preemptive online scheduling: optimal algorithms for all speeds. Algorithmica **53**, 504–522 (2009)

6. Gabay, M., Brauner, N., Kotov, V.: Computing lower bounds for semi-online optimization problems: application to the bin stretching problem. HAL preprint hal-00921663 (2013)

7. Gabay, M., Kotov, V., Brauner, N.: Semi-online bin stretching with bunch techniques. HAL preprint hal-00869858 (2013)

8. Johnson, D.: Near-optimal Bin Packing Algorithms. Massachusetts Institute of Technology, project MAC. Massachusetts Institute of Technology (1973)

9. Kellerer, H., Kotov, V.: An efficient algorithm for bin stretching. Oper. Res. Lett. **41**(4), 343–346 (2013)

10. Kellerer, H., Kotov, V., Speranza, M.G., Tuza, Z.: Semi on-line algorithms for the partition problem. Oper. Res. Lett. **21**, 235–242 (1997)

11. Ullman, J.: The performance of a memory allocation algorithm. Technical report **100** (1971)

Online Colored Bin Packing

Martin Böhm, Jiří Sgall$^{(\boxtimes)}$, and Pavel Veselý$^{(\boxtimes)}$

Computer Science Institute of Charles University, Prague, Czech Republic
{bohm,sgall,vesely}@iuuk.mff.cuni.cz

Abstract. In the Colored Bin Packing problem a sequence of items of sizes up to 1 arrives to be packed into bins of unit capacity. Each item has one of $c \geq 2$ colors and an additional constraint is that we cannot pack two items of the same color next to each other in the same bin. The objective is to minimize the number of bins.

In the important special case when all items have size zero, we characterize the optimal value to be equal to color discrepancy. As our main result, we give an (asymptotically) 1.5-competitive algorithm which is optimal. In fact, the algorithm always uses at most $\lceil 1.5 \cdot OPT \rceil$ bins and we show a matching lower bound of $\lceil 1.5 \cdot OPT \rceil$ for any value of $OPT \geq 2$. For items of arbitrary size we give a lower bound of 2.5 and an absolutely 3.5-competitive algorithm. We also show that classical algorithms First Fit, Best Fit and Worst Fit are not constant competitive.

In the case of two colors—the Black and White Bin Packing problem—we prove that all Any Fit algorithms have the absolute competitive ratio 3. When the items have sizes of at most $1/d$ for a real $d \geq 2$ we show that the Worst Fit algorithm is absolutely $(1 + d/(d-1))$-competitive.

1 Introduction

In the *Online Black and White Bin Packing* problem proposed by Balogh et al. [1,2] as a generalization of classical bin packing, we are given a list of items of size in $[0,1]$, each item being either black, or white. The items are coming one by one and need to be packed into bins of unit capacity. The items in a bin are ordered by their arrival time. The additional constraint to capacity is that the colors inside the bins are alternating, i.e., no two items of the same color can be next to each other in the same bin. The goal is to minimize the number of bins used.

Online Colored Bin Packing is a natural generalization of Black and White Bin Packing in which items can have more than two colors. As before, the only additional condition to unit capacity is that we cannot pack two items of the same color next to each other in one bin.

Observe that optimal offline packings with and without reordering the items differ in this model. The packings even differ by a non-constant factor: Let the input sequence have n black items and then n white items, all of size zero. The offline optimal number of bins with reordering is 1, but an offline packing without

This work was supported by the project 14-10003S of GA ČR and by the GAUK project 548214.

E. Bampis and O. Svensson (Eds.): WAOA 2014, LNCS 8952, pp. 35–46, 2015.
DOI: 10.1007/978-3-319-18263-6_4

reordering (or an online packing) needs n bins, since the first n black items must be packed into different bins. Hence we need to use the offline optimum without reordering in the analysis of online colored bin packing algorithms.

There are several well-known and often used algorithms for classical Bin Packing. We investigate the *Any Fit* family of algorithms (AF). These algorithms pack an incoming item into some already open bin whenever it is possible with respect to the size and color constraints. The choice of the open bin (if more are available) depends on the algorithm. AF algorithms thus open a new bin with an incoming item only when there is no other possibility. Among AF algorithms, *First Fit* (FF) packs an incoming item into the first bin where it fits (in the order by creation time), *Best Fit* (BF) chooses the bin with the highest level where the item fits and *Worst Fit* (WF) packs the item into the bin with the lowest level where it fits.

Next Fit (NF) is more restrictive than Any Fit algorithms, since it keeps only a single open bin and puts an incoming item into it whenever the item fits, otherwise the bin is closed and a new one is opened.

Previous Results. Balogh et al. [1,2] introduced the Black and White Bin Packing problem. As the main result, they give an algorithm *Pseudo* with the absolute competitive ratio exactly 3 in the general case and $1 + d/(d-1)$ in the parametric case, where the items have sizes of at most $1/d$ for a real $d \geq 2$. They also proved that there is no deterministic or randomized online algorithm whose asymptotic competitiveness is below $1 + \frac{1}{2\ln 2} \approx 1.721$.

Concerning specific algorithms, they proved that Any Fit algorithms are at most 5-competitive and even optimal for zero-size items. They show input instances on which FF and BF create asymptotically $3 \cdot OPT$ bins. For WF there are sequences of items witnessing that it is at least 3-competitive and $(1 + d/(d-1))$-competitive in the parametric case for an integer $d \geq 2$. Furthermore, NF is not constant competitive.

The idea of the algorithm Pseudo, on which we build as well, is that we first pack the items regardless of their size, i.e., treating their size as zero. This can be done optimally for two colors, and the optimum equals the maximal discrepancy in the sequence of colors (to be defined below). Then these bins are partitioned by NF into bins of level at most 1.

In the offline setting, Balogh et al. [2] gave a 2.5-approximation algorithm with $\mathcal{O}(n \log n)$ time complexity and an asymptotic polynomial time approximation scheme, both when reordering is allowed.

Very recently and independent of us Dósa and Epstein [7] studied Colored Bin Packing. They improved the lower bound for online Black and White Bin Packing to 2 for deterministic algorithms, which holds for more colors as well. For 3 colors and more they proved an asymptotic lower bound of 1.5 for zero-size items. They designed a 4-competitive algorithm based on Pseudo and a balancing algorithm for zero-size items. They also showed that BF, FF and WF are not competitive at all (with non-zero sizes).

Our Results. We completely solve the case of Colored Bin Packing for zero-size items. As we have seen, this case is important for constructing general algorithms. The offline optimum (without reordering) is actually not only lower bounded by the color discrepancy, but equal to it for zero-size items (see Sect. 2). For online algorithms, we give an (asymptotically) 1.5-competitive algorithm which is optimal (see Sect. 3.2). In fact, the algorithm always uses at most $\lceil 1.5 \cdot OPT \rceil$ bins and we show a matching lower bound of $\lceil 1.5 \cdot OPT \rceil$ for any value of $OPT \geq 2$ (see Sect. 3.1). This is significantly stronger than the asymptotic lower bound of 1.5 of Dósa and Epstein [7], in particular it shows that the absolute ratio of our algorithm is 5/3, and this is optimal.

For items of arbitrary size and three colors, we show a lower bound of 2.5, which breaks the natural barrier of 2 (see Sect. 4.1). We use the optimal algorithm for zero-size items and the algorithm Pseudo to design an (absolutely) 3.5-competitive algorithm which is also (asymptotically) $(1.5+d/(d-1))$-competitive in the parametric case, where the items have sizes of at most $1/d$ for a real $d \geq 2$ (see Sect. 4.2). (Note that for $d < 2$ we have $d/(d-1) > 2$ and the bound for arbitrary items is better.)

We show that algorithms BF, FF and WF are not constant competitive, in contrast to their 3-competitiveness for two colors. Their competitiveness cannot be bounded by any function of the number of colors even for only three colors and very small items. Instances showing that are omitted due to space limitations.

For Black and White Bin Packing (Sect. 5), we improve the upper bound on the absolute competitive ratio of Any Fit algorithms in the general case to 3 which is tight for BF, FF and WF. For WF in the parametric case, we prove that it is absolutely $(1 + d/(d-1))$-competitive for a real $d \geq 2$ which is tight for an integral d. Therefore, WF has the same competitive ratio as the Pseudo algorithm. The proofs of both results are also omitted.

In Sects. 2, 3 and 4 we provide as much intuition as we can, although we have to omit some proofs of our results.

Related Work. In the classical Bin Packing problem, we are given items with sizes in $(0, 1]$ and the goal is to assign them into the minimum number of unit capacity bins. The problem was proposed by Ullman [13] and by Johnson [10] and it is known to be NP-hard. See the survey of Coffman et al. [5] for the many results on classical Bin Packing and its many variants.

For the online problem, there is no online algorithm which is better than $248/161 \approx 1.540$-competitive [3]. The currently best algorithm is Harmonic++ by Seiden [12], approximately 1.589-competitive. Regarding AF algorithms, NF is 2-competitive and both FF and BF have the absolute competitive ratio exactly 1.7 [8,9]. This is similar to Black and White Bin Packing in which FF and BF have the absolute competitive ratio of 3 and the hard instances proving tightness of the bound are the same for both algorithms.

In the context of Colored Bin Packing, we are interested in variants that further restrict the allowed packings. Of particular interest is Bounded Space Bin Packing where an algorithm can have only $K \geq 1$ open bins in which it is allowed to put incoming items. When a bin is closed an algorithm cannot pack any further item in the bin or open it again. Such algorithms are called K-*bounded-space*. The champion among these algorithms is K-Bounded Best Fit, i.e., Best Fit with at most K open bins, which is (asymptotically) 1.7-competitive for all $K \geq 2$ [6]. Lee and Lee [11] presented Harmonic(K) which is K-bounded-space with the asymptotic ratio of 1.691 for K large enough. Lee and Lee also proved that there is no bounded space algorithm with a better asymptotic ratio.

The Bounded Space Bin Packing is an especially interesting variant in our context due to the fact that it matters whether we allow the optimum to reorder the input instance or not. If we allow reordering for Bounded Space Bin Packing, we get the same optimum as classical Bin Packing. In fact, all the bounds on online algorithms in the previous paragraph hold if the optimum with reordering is considered, which is a stronger statement than comparing to the optimum without reordering. This is a very different situation than for Colored Bin Packing, where no online algorithms can be competitive against the optimum with reordering, as we have noted above.

The bounded space offline optimum without reordering was studied by Chrobak et al. [4]. It turns out that the computational complexity is very different: There exists an offline $(1.5 + \varepsilon)$-approximation algorithm for 2-bounded-space Bin Packing with polynomial running time for every constant $\varepsilon > 0$, but exponential in ε. No polynomial time 2-bounded-space algorithm can have its approximation ratio better than $5/4$ (unless $P = NP$). In the online setting it is open whether there exists a better algorithm than 1.7-competitive K-Bounded Best Fit when compared to the optimum without reordering; the current lower bound is $3/2$.

Motivation. Suppose that a television or a radio station maintains several channels and wants to assign a set of programs to them. The programs have types like "documentary", "thriller", "sport", on TV, or music genres on radio. To have a fancy schedule of programs, the station does not want to broadcast two programs of the same type one after the other. Colored Bin Packing can be used to create such a schedule. Items here correspond to programs, colors to genres and bins to channels. Moreover, the programs can appear online and have to be scheduled immediately, e.g., when listeners send requests for music to a radio station via the Internet.

Another application of Colored Bin Packing comes from software which renders user-generated content (for example from the Internet) and assigns it to columns which are to be displayed. The content is in boxes of different colors and we do not want two boxes of the same color to be adjacent in a column, otherwise they would not be distinguishable for the user.

Moreover, Colored Bin Packing with all items of size zero corresponds to a situation in which we are not interested in loads of bins (lengths of the schedule, sizes of columns, etc.), but we just want some kind of diversity or colorfulness.

2   Preliminaries and Offline Optimum

Definitions and Notation. There are three settings of Colored Bin Packing: In the *offline setting* we are given the items in advance and we can pack them in an arbitrary order. In the *restricted offline setting* we also know sizes and colors of all items in advance, but they are given as a sequence and they need to be packed in that order. In the *online setting* the items are coming one by one and we do not know what comes next or even the total number of items. Moreover, an online algorithm has to pack each incoming item immediately and it is not allowed to change its decisions later.

We focus mostly on the online setting. To measure the effectiveness of online algorithms for a particular instance L, we use the restricted offline optimum denoted by $OPT(L)$ or OPT when the instance L is obvious from the context. Let $ALG(L)$ denote the number of bins used by the algorithm ALG. The algorithm is *absolutely r-competitive* if for any instance $ALG(L) \leq r \cdot OPT(L)$ and *asymptotically r-competitive* if for any instance $ALG(L) \leq r \cdot OPT(L) + o(OPT(L))$; typically the additive term is just a constant. We say that an algorithm has the (absolute or asymptotic) competitive ratio r if it is (absolutely or asymptotically) r-competitive and it is not r'-competitive for $r' < r$.

For Colored Bin Packing, let C be the set of all colors. For $c \in C$, the items of color c are called c-items and bins with the top (last) item of color c are called c-bins. By a non-c-item we mean an item of color $c' \neq c$ and similarly a non-c-bin is a bin of color $c' \neq c$. The *level of a bin* means the cumulative size of all items in the bin.

Lower Bounds on the Restricted Offline Optimum. We use two lower bounds on the number of bins in any packing. The first bound LB_1 is the sum of sizes of all items.

The second bound LB_2 is the maximal color discrepancy inside the input sequence. In Black and White Bin Packing, the color discrepancy introduced by Balogh et al. [1] is simply the difference of the number of black and white items in a segment of input sequence, maximized over all segments. It is easy to see that it is a lower bound on the number of bins.

In the generalization of color discrepancy for more than two colors we count the difference between c-items and non-c-items for all colors c and segments. It is easy to see that this is a lower bound as well. Formally, let $s_{c,i} = 1$ if the i-th item from the input sequence has color c, and $s_{c,i} = -1$ otherwise. We define

$$LB_2 = \max_{c \in C} \max_{i,j} \sum_{\ell=i}^{j} s_{c,\ell}.$$

For Black and White Bin Packing, equivalently $LB_2 = \max_{i,j} |\sum_{\ell=i}^{j} s_\ell|$, where $s_i = 1$ if the i-th item is white, and $s_i = -1$ otherwise; the absolute value replaces the maximization over colors.

In Black and White Bin Packing, when all the items are of size zero, all Any Fit algorithms create a packing into the optimal number of bins [1]. For more than two colors this is not true and in fact no deterministic online algorithm can have a competitive ratio below 1.5. However, in the restricted offline setting a packing into LB_2 bins is still always possible, even though this fact is not obvious. This shows that the color discrepancy fully characterizes the combinatorial aspect of the color restriction in Colored Bin Packing.

Theorem 1. *Let all items have size equal to zero. Then a packing into LB_2 bins is possible in the restricted offline setting, i.e., items can be packed into LB_2 bins without reordering.*

3 Algorithms for Zero-Size Items

3.1 Lower Bound on Competitiveness of Any Online Algorithm

Theorem 2. *For zero-size items of at least three colors, there is no deterministic online algorithm with an asymptotic competitive ratio less than 1.5. Precisely, for each $n > 1$ we can force any deterministic online algorithm to use at least $\lceil 1.5n \rceil$ bins using three colors, while the optimal number of bins is n.*

Proof. We show that if an algorithm uses less than $\lceil 1.5n \rceil$ bins, the adversary can send some items and force the algorithm to increase the number of black bins or to use at least $\lceil 1.5n \rceil$ bins, while the maximal discrepancy stays n. Applying Theorem 1 we know that $OPT = n$, but the algorithm is forced to open $\lceil 1.5n \rceil$ bins using finitely many items as the number of black bins is increasing. Moreover, the adversary uses only three colors throughout the whole proof, denoted by black, white and red and abbreviated by b, w and r in formulas.

We introduce the current discrepancy of a color c which basically tells us how many c-items have come recently and thus how many c-items may arrive without increasing the overall discrepancy. Formally, we define the current discrepancy after packing the k-th item as $CD_{c,k} = \max_{i \le k+1} \sum_{\ell=i}^{k} s_{c,\ell}$, i.e., the discrepancy on an interval which ends with the last packed item (the k-th). Note that $CD_{c,k}$ is at least zero as we can set $i = k + 1$. We omit the k index in $CD_{c,k}$ when it is obvious from the context.

Initially the adversary sends n black items, then he continues by phases and ends the process whenever the algorithm uses $\lceil 1.5n \rceil$ bins at the end of a phase. When a phase starts, there are less than $\lceil 1.5n \rceil$ black bins and possibly some other white or red bins. We also guarantee $CD_w = 0$, $CD_r = 0$, and $CD_b \le n$. Let N_b be the number of black bins when a phase starts. In each phase the adversary forces the algorithm to use $\lceil 1.5n \rceil$ bins or to have more than N_b black bins, while $CD_w = 0$, $CD_r = 0$, and $CD_b \le n$ at the end of each phase in which N_b increases.

We now present how a phase works. Let new items be items from the current phase and old items be items from previous phases. The adversary begins the phase by sending n new items of colors alternating between white and red, starting by white, so he sends $\lceil n/2 \rceil$ white items and $\lfloor n/2 \rfloor$ red items. After these new items, the current discrepancy is one either for red if n is even, or for white if n is odd, and it is zero for the other colors.

If some new item is not put on an old black item, the adversary sends n black items. Since the new items are packed into less than n black bins (more precisely, black at the beginning of the phase), the number of black bins increases. Moreover, $CD_w = 0$, $CD_r = 0$, and $CD_b = n$, hence the adversary finishes the phase and continues with the next phase if there are less than $\lceil 1.5n \rceil$ bins.

Otherwise all new red and white items are put on old black items. If n is even, $CD_w = 0$ and the adversary sends additional n white items. After that there are at least $1.5n$ white bins, so the adversary reaches his goal.

If n is odd, $CD_w = 1$ and the adversary can send only $n - 1$ white items forcing $\lceil 1.5n \rceil - 1$ white bins. This suffices to prove the result in the asymptotic sense, but for the precise lower bound of $\lceil 1.5n \rceil$ for an odd n we need a somewhat more complicated construction.

Therefore if all new red and white items are put on old black items and n is odd, the adversary sends a black item e. If e does not go on a new white item, he sends n white items forcing $\lceil n/2 \rceil + n$ white bins and it is done. Otherwise the black item e is put on a new white item. White and red have $\lfloor n/2 \rfloor$ new items on the top of bins, $CD_w = 0$, and $CD_r = 0$. The adversary sends another black item f. Since red and white are equivalent colors (considering only new items), w.l.o.g. f goes into a red bin or into newly opened bin.

Next he sends a white item g and a red item h. After packing g there are $\lceil n/2 \rceil$ bins with a new white item on the top and at least one bin with a new black item on the top. Moreover, after packing the red item h we have $CD_b = 0$ and $CD_w = 0$. So if h is not put on a new white item (i.e., it is put into a black bin, a new bin or on an old white item), the adversary sends n white items and the algorithm must use $\lceil 1.5n \rceil$ bins. Otherwise h is packed on a new white item and the adversary sends n black items. The number of black bins increases, because the adversary sent $n + 2$ new black items and at most $n + 1$ new non-black items were put into a black bin (at most n items at the beginning of the phase plus the item g). Since $CD_w = 0$, $CD_r = 0$, and $CD_b = n$, the adversary continues with the next phase.   □

The lower bound has additional properties that we use later in our lower bound for items of arbitrary size. Most importantly, we have at least $\lceil 1.5 \cdot OPT \rceil$ of c-bins at the end (and possibly some additional bins of other colors).

Lemma 1. *After packing the instance from Theorem 2 by an online algorithm there is a color c for which we have $\lceil 1.5 \cdot OPT \rceil$ of c-bins and $CD_c = OPT$, while $CD_{c'} = 0$ for all other colors $c' \neq c$. Moreover, in each restricted offline optimal packing of the instance all the bins have a c-item on the top.*

Proof. Let $n = OPT$ as in the previous proof. The adversary stops sending items when he finishes the last phase. In the last phase either the number of black bins increases to $\lceil 1.5n \rceil$, or the adversary forces $\lceil 1.5n \rceil$ white or red bins by sending n white or red items. In the former case the requirements of the lemma are satisfied, because the proof guarantees $CD_w = 0$ and $CD_r = 0$ at the end of each phase in which the number of black bins increases. Moreover $CD_b = n$, since n black items are sent just before the end of such phase. In the latter case, the last n white items cause $CD_b = 0$, $CD_r = 0$, and $CD_w = n$; the case of n red items is symmetric.

Since an optimal packing uses n bins and the last n items are of the same color (in each case of the construction), they must go into different bins. Hence each bin of a restricted offline optimal packing has a c-item on the top.   □

3.2   Optimal Algorithm for Zero-Size Items

The overall problem of FF, BF and WF is that they pack items regardless of the colors of bins. We address the problem by balancing the colors of top items in bins – we mostly put an incoming c-item into a bin of the most frequent other color. When we have more choices of bins where to put an item we use First Fit. We call this algorithm *Balancing Any Fit* (BAF).

We define BAF for items of size zero and show that it opens at most $\lceil 1.5LB_2 \rceil$ bins which is optimal in the worst case by Theorem 2. Then we combine BAF with the algorithm Pseudo by Balogh et al. [1] for items of arbitrary size and prove that the resulting algorithm is absolutely 3.5-competitive.

After packing the k-th item from the sequence, let D_k be the maximal discrepancy so far, i.e., the discrepancy on an interval before the $(k+1)$-st item, and let $N_{c,k}$ be the number of c-bins after packing the k-th item. As in the proof of Theorem 2, we define the current discrepancy as $CD_{c,k} = \max_{i \le k+1} \sum_{\ell=i}^{k} s_{c,\ell}$, i.e., the discrepancy on an interval which ends with the last packed item (the k-th). Note that $CD_{c,k} \le D_k$ and that $CD_{c,k}$ is at least zero as we can set $i = k + 1$. The current discrepancy basically tells us how many c-items have come recently and thus how many c-items may arrive without increasing the overall discrepancy.

Let $\alpha_{c,k} = N_{c,k} - \lceil D_k/2 \rceil$ be the difference between the number of c-bins and the half of the maximal discrepancy so far. Observe that $\lceil D_k/2 \rceil$ is the number of bins which BAF may use in addition to OPT bins. We omit the index k in D_k, $N_{c,k}$, $CD_{c,k}$ and $\alpha_{c,k}$ when it is obvious from the context.

While processing the items, if D is the maximal discrepancy so far, the adversary can send $D - CD_c$ of c-items without increasing the maximal discrepancy, while forcing the algorithm to use $N_c + D - CD_c$ bins. Hence, to end with at most $\lceil 1.5D \rceil$ bins we try to keep $N_c - CD_c \le \lceil D/2 \rceil$ for all colors c. For simplicity, we use an equivalent inequality of $\alpha_c = N_c - \lceil D/2 \rceil \le CD_c$. If we can keep the inequality valid and it occurs that there is a color c with $N_c > \lceil 1.5D \rceil$, we get $CD_c \ge N_c - \lceil D/2 \rceil > \lceil 1.5D \rceil - \lceil D/2 \rceil = D$ which contradicts $CD_c \le D$. Let the **main invariant** for a color c be

$$\alpha_c = N_c - \left\lceil \frac{D}{2} \right\rceil \le CD_c. \tag{1}$$

As $CD_c \ge 0$, keeping the invariant is easy for all colors with at most $\lceil D/2 \rceil$ bins. Also when there is only one color c with $N_c > \lceil D/2 \rceil$, we just put all non-c-items into c-bins. Therefore, if a non-c-item comes, the number of c-bins N_c decreases and the current discrepancy CD_c decreases by at most one. (CD_c stays the same when it is zero.) Since both increase with an incoming c-item, we are keeping our main invariant (1) for the color c.

Moreover, there are at most two colors with strictly more than $\lceil D/2 \rceil$ bins, given that we have at most $\lceil 1.5D \rceil$ open bins. Thus we only have to deal with two colors having $N_c > \lceil D/2 \rceil$. We state the algorithm Balancing Any Fit for items of size zero.

Balancing Any Fit (BAF):

1. For an incoming c-item, if there are no bins or c-bins only, open a new bin and put the item into it.
2. Otherwise, if there is at most one color with the number of bins strictly more than $\lceil D/2 \rceil$, put an incoming c-item into a bin of color $c' = \arg\max_{c'' \neq c} N_{c''}$. If more colors have the same maximal number of bins, choose color c' arbitrarily among them, e.g., by First Fit. Among c'-bins, choose again arbitrarily.
3. Suppose that there are two colors b and w such that $N_{\mathrm{b}} > \lceil D/2 \rceil$ and $N_{\mathrm{w}} > \lceil D/2 \rceil$. If $c = \mathrm{w}$, put the item into a bin of color b. If $c = \mathrm{b}$, put the item into a bin of color w. Otherwise $c \notin \{\mathrm{b}, \mathrm{w}\}$; if $N_{\mathrm{b}} - \lceil D/2 \rceil < CD_{\mathrm{b}}$, put the item into a bin of color w, otherwise into a bin of color b.

As we discussed, keeping the main invariant (1) is easy in the first and the second case of the algorithm. Therefore we can conclude the following lemma.

Lemma 2. *Suppose that the main invariant holds for all colors before packing the t-th item and that there is at most one color c with $N_{c,t-1} > \lceil D_{t-1}/2 \rceil$ before the t-th item, i.e., the t-th item is packed using the first or the second case of the algorithm. Then the main invariant holds for all colors also after packing the t-th item.*

Most of the proof of 1.5-competitiveness of BAF thus deals with two colors having more than $\lceil D/2 \rceil$ bins. W.l.o.g. let these two colors be black and white in the following and let us abbreviate them by b and w.

In the third case of the algorithm we have to choose either black or white bin for items of other colors than black and white, but the current discrepancy decreases for both black and white, while the number of bins stays the same for the color which we do not choose. So if $\alpha_{\mathrm{b}} = CD_{\mathrm{b}}$ and $\alpha_{\mathrm{w}} = CD_{\mathrm{w}}$, the adversary can force the algorithm to open more than $\lceil 1.5D \rceil$ bins.

Therefore we need to prove that in the third case, i.e., when $N_{\mathrm{b}} > \lceil D/2 \rceil$ and $N_{\mathrm{w}} > \lceil D/2 \rceil$, at least one of inequalities $\alpha_{\mathrm{b}} \leq CD_{\mathrm{b}}$ and $\alpha_{\mathrm{w}} \leq CD_{\mathrm{w}}$ is strict. This motivates the following **secondary invariant:**

$$2\alpha_{\mathrm{b}} + 2\alpha_{\mathrm{w}} \leq CD_{\mathrm{b}} + CD_{\mathrm{w}} + 1. \tag{2}$$

If the secondary invariant holds, it is not hard to see that in the third case of the algorithm the choice of the bin maintains the main invariant. The tricky part of the proof is to prove the *base case* of the inductive proof of the secondary invariant. We prove that it holds already at the moment when b and w become the two strictly most frequent colors on top of the bins, i.e., $N_{\mathrm{b}} > N_c$ and $N_{\mathrm{w}} > N_c$ for all other colors c. After that, maintaining both invariants is relatively easy.

Theorem 3. *Balancing Any Fit is 1.5-competitive for items of size zero and an arbitrary number of colors. Precisely, it uses at most $\lceil 1.5 \cdot OPT \rceil$ bins.*

4   Algorithms for Items of Arbitrary Size

4.1   Lower Bound on Competitiveness of Any Online Algorithm

We use the construction by Dósa and Epstein [7] proving the lower bound 2 for two colors to get a lower bound 2.5 using three colors. We combine it with the hard instance that shows the lower bound 1.5 for zero-size items.

Theorem 4. *For items of at least three colors, there is no deterministic online algorithm with an asymptotic competitive ratio less than 2.5.*

Proof. Throughout the whole proof the adversary uses only three colors denoted by black, white and red and abbreviated by b, w and r in formulas. Let $n > 1$ be a large integer. The adversary starts with the hard instance for zero-size items from the proof of Theorem 2 with the optimum equal to n. By Lemma 1 there are at least $\lceil 1.5n \rceil$ bins of the same color, w.l.o.g. white. Let W be the set of bins that are white after the first part of the instance ($|W| \geq \lceil 1.5n \rceil$). We also know that $CD_w \leq n$, $CD_b = 0$, and $CD_r = 0$.

The second part of the instance is a slightly simplified construction by Dósa and Epstein [7]. Their idea goes as follows: The adversary sends the instance in phases, each starting with white and black small items. If the black item is put into an already opened bin, we send a huge white item that can be only put on said small black item. Therefore the algorithm has to put the huge white item in a new bin, but an optimal offline algorithm puts the small black item into a new bin and the huge white item on it. Otherwise, if the small black item is put into a new bin, the phase is finished: The online algorithm opened a bin in the phase, while an optimal offline algorithm does not need to. This way an online algorithm is forced to behave oppositely to an optimal offline algorithm.

We formalize this idea by the following adversarial algorithm. Let $\varepsilon = 1/(6n)$ and $\delta_i = 1/(5^i \cdot 6n)$. The adversary uses the items of the following types:

- regular white items of size ε,
- regular black items of size δ_i for some $i \geq 1$,
- special black items of size $3\delta_i$ for some $i \geq 1$,
- huge white items of size $1 - 2\delta_i$ for some $i \geq 1$.

Let i be the index of the current phase and let j be the number of huge white items in the instance so far. The adversarial algorithm is as follows:

1. Let $i = 0$ and $j = 0$.
2. If $j = n$ or if $i = \lceil 2.5n \rceil$, then stop.
3. Let $i = i + 1$. Send $\left(\begin{smallmatrix} \text{white} \\ \varepsilon \end{smallmatrix}, \begin{smallmatrix} \text{black} \\ \delta_i \end{smallmatrix} \right)$, i.e., a group consisting of a regular white item and a regular black item.
4. If the regular black item goes to a new bin or to a bin with level zero, go to the step 2 (continue with the next phase).
5. Let $j = j + 1$. Send $\left(\begin{smallmatrix} \text{black} \\ 3\delta_i \end{smallmatrix}, \begin{smallmatrix} \text{white} \\ 1-2\delta_i \end{smallmatrix}, \begin{smallmatrix} \text{black} \\ \delta_i \end{smallmatrix} \right)$. Then go to the step 2. $\square$

4.2 3.5-Competitive Algorithm

We now show that there is a constant competitive online algorithm even for items of sizes between 0 and 1. We combine algorithms Pseudo from [1] and our algorithm BAF that is 1.5-competitive for zero-size items. The algorithm Pseudo uses *pseudo bins* which are bins of unbounded capacity.

Pseudo-BAF:

1. First pack an incoming item into a pseudo bin using the algorithm BAF (treat the item as a zero-size item).
2. In each pseudo bin, items are packed into unit capacity bins using Next Fit.

Theorem 5. *The algorithm Pseudo-BAF for Colored Bin Packing is absolutely 3.5-competitive. In the parametric case when items have size at most $1/d$, for a real $d \geq 2$, it uses at most $\lceil (1.5 + d/(d-1))OPT \rceil$ bins. Moreover, the analysis is asymptotically tight.*

Proof. In the general case for items between 0 and 1 we know that two consecutive bins in one pseudo bin have total size strictly more than one, since no two consecutive items of the same color are in a pseudo bin. In each pseudo bin we match each bin with an odd index with the following bin with an even index, therefore we match all bins except at most one in each pseudo bin. Moreover, the total size of a pair of matched bins is more than one. Therefore the number of matched bins is strictly less than $2 \cdot LB_1 \leq 2 \cdot OPT$, i.e., at most $2 \cdot OPT - 1$. The number of not matched bins is at most the number of pseudo bins created by the algorithm BAF which uses at most $\lceil 1.5 \cdot LB_2 \rceil \leq \lceil 1.5 \cdot OPT \rceil \leq 1.5 \cdot OPT + 0.5$ bins. Summing both bounds, the algorithm Pseudo-BAF creates at most $3.5 \cdot OPT$ bins.

For the parametric case, inside each pseudo bin all real bins except the last one have level strictly more than $(d-1)/d$, so their number is strictly less than $d/(d-1) \cdot OPT$. The number of pseudo bins is still bounded by $\lceil 1.5 \cdot OPT \rceil$, thus the algorithm Pseudo opens at most $\lceil (1.5 + d/(d-1))OPT \rceil$ bins. □

5 Black and White Bin Packing

For sequences with only two colors, we improve the upper bound on the absolute competitive ratio of Any Fit algorithms from 5 to 3. Then we show that Worst Fit performs even better for items with size of at most $1/d$ (for $d \geq 2$) as it is $(1 + d/(d-1))$-competitive in this case. Both bounds are tight by the results of Balogh et al. [1] (more precisely, the bound for WF is tight only for an integer $d \geq 2$).

Theorem 6. *Any algorithm in the Any Fit family is absolutely 3-competitive for Black and White Bin Packing.*

Theorem 7. *Suppose that all items in the input sequence have sizes of at most $1/d$, for a real $d \geq 2$. Then Worst Fit is absolutely $(1 + d/(d-1))$-competitive for Black and White Bin Packing.*

Conclusions and Open Problems

The Colored Bin Packing for zero-size items is completely solved.

For items of arbitrary size, our online algorithm still leaves a gap between our lower bound 2.5 and our upper bound of 3.5. The upper bounds are only 0.5 higher than for two colors (Black and White Bin Packing) where a gap between 2 and 3 remains for general items.

Classical algorithms FF, BF and WF, although they maintain a constant approximation for two colors, start to behave badly when we introduce the third color. For two colors, we now know their exact behavior. Surprisingly, even the simple Worst Fit algorithm matches the performance of Pseudo, the online algorithm with the best competitive ratio known so far. It is also an interesting question whether it holds that Any Fit algorithms cannot be better than 3-competitive for two colors.

References

1. Balogh, J., Békési, J., Dósa, G., Epstein, L., Kellerer, H., Tuza, Z.: Online results for black and white bin packing. Theory Comput. Syst. 1–19 (2014)
2. Balogh, J., Békési, J., Dósa, G., Kellerer, H., Tuza, Z.: Black and white bin packing. In: Erlebach, T., Persiano, G. (eds.) WAOA 2012. LNCS, vol. 7846, pp. 131–144. Springer, Heidelberg (2013)
3. Balogh, J., Békési, J., Galambos, G.: New lower bounds for certain classes of bin packing algorithms. Theor. Comput. Sci. **440–441**, 1–13 (2012)
4. Chrobak, M., Sgall, J., Woeginger, G.J.: Two-bounded-space bin packing revisited. In: Demetrescu, C., Halldórsson, M.M. (eds.) ESA 2011. LNCS, vol. 6942, pp. 263–274. Springer, Heidelberg (2011)
5. Coffman Jr., E., Csirik, J., Galambos, G., Martello, S., Vigo, D.: Bin packing approximation algorithms: survey and classification. In: Pardalos, P.M., Du, D.-Z., Graham, R.L. (eds.) Handbook of Combinatorial Optimization, pp. 455–531. Springer, New York (2013)
6. Csirik, J., Johnson, D.S.: Bounded space on-line bin packing: best is better than first. Algorithmica **31**(2), 115–138 (2001)
7. Dósa, G., Epstein, L.: Colorful bin packing. In: Ravi, R., Gørtz, I.L. (eds.) SWAT 2014. LNCS, vol. 8503, pp. 170–181. Springer, Heidelberg (2014)
8. Dósa, G., Sgall, J.: First Fit bin packing: a tight analysis. In: 30th International Symposium on Theoretical Aspects of Computer Science (STACS), vol. 20 of Leibniz International Proceedings in Informatics (LIPIcs), pp. 538–549. Dagstuhl, Germany (2013)
9. Dósa, G., Sgall, J.: Optimal analysis of best fit bin packing. In: Esparza, J., Fraigniaud, P., Husfeldt, T., Koutsoupias, E. (eds.) ICALP 2014. LNCS, vol. 8572, pp. 429–441. Springer, Heidelberg (2014)
10. Johnson, D.: Near-optimal bin packing algorithms, project MAC. Massachusetts Institute of Technology, Cambridge (1973)
11. Lee, C.C., Lee, D.T.: A simple on-line bin-packing algorithm. J. ACM **32**, 562–572 (1985)
12. Seiden, S.S.: On the online bin packing problem. J. ACM **49**, 640–671 (2002)
13. Ullman, J.: The performance of a memory allocation algorithm. Technical report 100 (1971)

Improved Bound for Online
Square-into-Square Packing

Brian Brubach[✉]

Department of Computer Science, University of Maryland,
College Park, MD, USA
bbrubach@cs.umd.edu

Abstract. We show a new algorithm and improved bound for the online square-into-square packing problem using a hybrid shelf-packing approach. This 2-dimensional packing problem involves packing an online sequence of squares into a unit square container without any two squares overlapping. We seek the largest area α such that any set of squares with total area at most α can be packed. We show an algorithm that packs any online sequence of squares with total area $\alpha \leq 2/5$, improving upon recent results of 11/32 [3] and 3/8 [8]. Our approach allows all squares smaller than a chosen maximum height h to be packed into the same fixed height shelf. We combine this with the introduction of variable height shelves for squares of height larger than h. Some of these techniques can be extended to the more general problems of rectangle packing with rotation and bin packing.

Keywords: Packing · Online problems · Packing squares · Packing rectangles

1 Introduction

In packing problems, we wish to place a set of objects into a container such that no two objects overlap. These problems have been studied extensively and have numerous applications. However, even common one-dimensional versions of packing problems, such as the Knapsack problem, are NP-hard. For such difficult problems, it is often important to know whether it is even feasible to pack a given set into a particular container. In the two-dimensional case, it is worth noting that merely checking whether a given set of squares can be packed into a unit square was shown to be NP-hard by Leung et al. [4].

To address this fundamental feasibility question, the square-into-square packing problem asks, "What is the largest area α such that any set of squares with total area α can be packed into a unit square without overlapping?" It is trivial to show that an upper bound is $\alpha \leq 1/2$. Two squares of height $1/2 + \epsilon$ cannot be packed into a unit square container. In addition, Moon and Moser [1] showed

---

This work has been supported partially by NSF award CCF-1218620. I also want to thank my mentor Prof. Gandhi for all his support.

E. Bampis and O. Svensson (Eds.): WAOA 2014, LNCS 8952, pp. 47–58, 2015.
DOI: 10.1007/978-3-319-18263-6_5

in 1967 that the bound of $1/2$ is tight in the offline case. Squares can be sorted in decreasing order and packed left-to-right into horizontal "shelves" starting along the bottom of the container. The height of each shelf is set by the largest object in the shelf and when a shelf fills, a new one is opened directly above it.

In the online version of the problem, we have no knowledge of the full set of squares to be packed. Squares are received one at a time and each must be packed before seeing the next square. Once a square is packed, it cannot be moved. Thus, the successful offline approach cannot be used as it requires sorting the set. Although various techniques have been employed to achieve lower bounds for the online case, the most recent work revisits the idea of shelves in a novel way. The current best lower bound for the online version is $\alpha \geq 11/32$ by Fekete and Hoffmann [3] in 2013. They took a dynamic, multi-directional shelf-packing approach that allocates perpendicular shelves within other larger shelves.

1.1   Related Work

Offline Square Packing. Early related work involved packing a set of objects into the smallest possible rectangle container. Moser [5] posed this question in 1966: "What is the smallest number A such that any family of objects with total area at most 1 can be packed into a rectangle of area A?" Since then, there have been many results for the offline packing of squares into rectangle containers.

In 1967, Moon and Moser [1] showed that any set of squares with total area 1 can be packed into a square of height $\sqrt{2}$. This established $A \leq 2$ or in the terms of our problem, $\alpha \geq 1/2$. This result was followed by several improvements on the value of A using rectangular containers. The current best upper bound is 1.3999 by Hougardy [6] in 2011.

Online Square Packing. In 1997, Januszewski and Lassak [7] considered the online variant in many dimensions. For two-dimensional square-into-square packing, their work showed a bound of $\alpha \geq 5/16$ by recursively dividing a unit square container into rectangles of aspect ratio $\sqrt{2}$. In 2008, Han et al. [2] used a similar approach to improve the lower bound to $1/3$. Januszewski and Lassak [7] also considered the general problem of packing d-dimensional cubes into a unit cube and for for $d \geq 5$, showed a tight bound of $2 \left(\frac{1}{2} \right)^d$. For $d = 3$ or 4, they showed cubes of total volume $3/2(1/2)^d$ could be packed.

Fekete and Hoffmann [3] provided a new approach in 2013 which uses multi-directional shelves (horizontal and vertical) that are allocated dynamically within other larger shelves. Using this technique, they were able to improve the lower bound further to $11/32$. More recently, the author of this paper improved the bound to $3/8$ in an unpublished paper [8]. That paper used a multi-directional shelf approach similar to [3] as well as some new techniques which will be included in this paper.

1.2   Our Contributions

We show a new lower bound of $\alpha \geq 2/5$ for online square-into-square packing, improving upon the previous results. Our algorithm combines dynamic, multi-directional *fixed* height shelves with *variable* height shelves. Fixed height shelves are assigned a height at the moment they are opened. Variable height shelves are not assigned a height until they are filled, at which point their height is considered to be the height of the tallest square in the shelf.

One of the key challenges in this problem is packing squares of greatly varying heights. In particular, if squares differ in height by a factor greater than 2, it is difficult to pack them together without creating a lot of wasted space. However, our new approach to multi-directional fixed height shelves allows us to pack all squares with height at most h (in our application $h = 1/6$) together into the same fixed height shelves of height h with very little expense of wasted space. Similarly, we show how to use variable height shelves to pack all squares with height greater than h together.

We introduce new criteria for determining the dimensions of vertical shelves which are dynamically allocated within the horizontal shelves. This allows us to completely avoid the use of *buffer regions* such as those added to the ends of shelves in [3] and minimize wasted space. Our use of variable shelves shared by all larger squares handles many such squares and with fewer special cases. These results may be of independent interest for other 2-dimensional packing problems. Our technique for multi-directional shelf packing can be used for rectangles (if rotation is allowed) as well as squares.

Another interesting future direction for this work would be bin packing problems. Many algorithms for these problems take a natural approach of segregating squares of different sizes into separate bins and maintaining several open bins, each devoted to a different size class. Combining all or most sizes of squares or rectangles into the same bin could be useful in versions of this problem with special restrictions. For instance, there may be a cost associated with switching back-and-forth between different bins or we may require a parallel algorithm in which multiple bins are packed simultaneously.

Finally, we note that our ratio of lower to upper bounds $\left(\frac{2/5}{1/2}\right)$ is the first 2-dimensional result which is tighter than the bounds shown in [7] for the 3 and 4-dimensional versions of this problem. This combined with the aforementioned tight results for dimension at least 5, suggests that new improvements may be possible in the 3 and 4-dimensional cases as well.

1.3   Outline

In Sect. 2, we discuss preliminaries including terminology and notation. In Sect. 3, we present our algorithm for the online packing of squares into a unit square container. In Sect. 4, we analyze our algorithm and show that it successfully packs any online series of squares with total area at most 2/5.

2   Preliminaries

2.1   Terminology

We define a *shelf* S as a subrectangle in the container with height h and length ℓ. For *variable* height shelves, h is equal to the height of the largest square in the shelf. For *fixed* height shelves, the height is determined in advance along with a packing ratio r, $0 < r < 1$. The *packing ratio* is the ratio of the smallest possible height to the largest possible height of squares that can be packed into S. Any square packed into fixed shelf S must have height k, $h \geq k > hr$.

When *packing* a shelf S, squares are added side-by-side to S. In our algorithm, we also pack small vertical fixed shelves into larger horizontal fixed shelves. These vertical shelves are also packed side-by-side with squares and other vertical shelves as seen in Fig. 1.

Shelf Packing

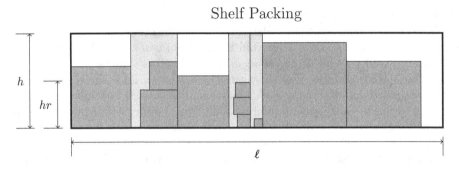

Fig. 1. Illustration of shelf packing with $r = 1/2$ for the horizontal shelf. Vertical shelves are designated with a *light gray* background and have different packing ratios.

In addition, shelves may be considered *open* or *closed*. A shelf S is initially considered open. As squares are added to S, we may receive a square Q with height h_Q, $h \geq h_Q > hr$, such that packing Q into S would exceed the length of S. At this point, we say that S is closed. The new square Q is then packed into some other shelf which is open. When analyzing shelves, we refer to the *used length* l_u of a shelf to describe the length of that shelf which is occupied by squares or vertical shelves.

In the analysis, we will *assign* fractions of the input area to regions or shelves within the container. Typically, the area assigned to a region represents squares which have been packed into that region. However, area assigned to a region may also come from a square packed into some other region subject to the following commonsense rule: A single square may have parts of its area assigned to different regions as long as the sum of those parts is at most the area of the square itself.

We use the term *density* to describe the ratio of the assigned area of a region to the total area of that region. In this paper, we will often count the total area packed into the container in the following way. Let there be a region R with available area A. We show that most of R has been assigned area to a density of $1/2$ with the exception of some small portion of wasted space with area at most W and a density of 0. We then calculate the total area assigned to R as $A/2 - W/2$. We call this $W/2$ value the *waste*.

2.2   Size Classes of Squares

We divide possible input squares into four classes based on height:

- **Large:** height $> \frac{1}{3}$ (also $< 2/3$ due to our input having area at most $2/5$)
- **Medium:** height $\leq \frac{1}{3}$ and $> \frac{1}{6}$
- **Small:** height $\leq \frac{1}{6}$ and $> \frac{1}{12}$
- **Very Small:** height $\leq \frac{1}{12}$

We also refer to small squares as class c_0 and subdivide the very small squares into subclasses c_i, $i \geq 1$. Squares in c_i are packed into shelves with max height h_i and packing ratio r_i. They have height k_i, $h_i \geq k_i > h_i r_i$. We use the notation c_{j+} to refer to all c_i, $i \geq j$.

In our algorithm, we assign ratios as follows: $r_0 = 0.5$, $r_1 = 0.71$, $r_2 = 0.65$, and $r_{3+} = 0.58$. To account for all small and very small squares with height $\leq \frac{1}{6}$, we let $h_0 = \frac{1}{6}$ and for all $i \geq 1$, we set $h_i = h_{i-1} r_{i-1}$. In Sect. 4.2, we will show that these ratios ensure closed vertical shelves for very small squares will have a density greater than 0.5, which is the packing ratio for small squares.

3   Algorithm

For each square, we pack it according to a subroutine based on its size. Very small squares are a special case. They are packed based on their subclass c_i, $i \geq 1$. For each subclass, we maintain exactly one open vertical shelf at any given time. When we receive a very small square, we attempt to add it to the appropriate vertical shelf for its height class. If this new square does not fit, we close that shelf and open a new one. The new vertical shelf itself is packed into the container as if it is a small square. We will show in the analysis that these new vertical shelves can be treated the same as small squares. As such, we only discuss large, medium, and small squares in the description of our algorithm (Fig. 2).

Small: We first alternate packing small squares from left-to-right into the top and bottom shelves of the initial packing region. Each time, we choose the shelf with the shortest used length. Eventually, these shelves will fill up. The first square we receive which doesn't fit into these shelves is packed into shelf M_0 in the main packing region. The second such square is packed into M_1. This can be seen in part (B) of Fig. 2.

After those first two small squares have been packed into the main region, all future small squares are packed left-to-right into the main region according to the following rule. We start packing in the shelf M_1 up to threshold T_1. Let M_i be the current shelf and T_j be the current threshold. When we receive a small square Q that would cause the used length of M_i to exceed threshold T_j, we pack Q into the shelf $M_{(i+1) \bmod 4}$ and that shelf becomes the current shelf. Each time we return to M_0 we increment the threshold to T_{j+1}. Each threshold T_j is at a distance of $j/6$ from the left side of the container.

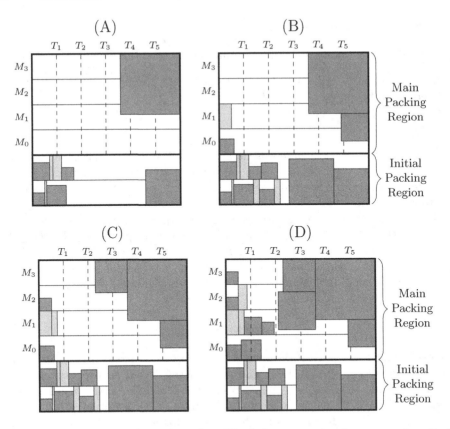

Fig. 2. Illustration of packing algorithm. The *light gray rectangles* represent vertical shelves for very small squares packed into the shelves for small squares. The small shelves in the main region are labeled M_0 through M_3 with thresholds labelled T_1 through T_5. (A) shows a large square in the *upper right corner*, a medium square in the *lower right*, and a set of small and very small squares in the *lower left*. (B) shows the first two "small" squares added to the main region on the left as well as a medium square added to the variable shelf on the *right*. (C) and (D) illustrate the continuation of packing into the main region.

Medium: Medium squares are first packed from right-to-left into the initial region. When we receive medium squares which don't fit into this region, we pack them into main packing region.

In the main region, medium squares will be packed together with large squares in vertical variable shelves from top-to-bottom, starting in the upper right corner. Each time we close one of these shelves, the new one is opened immediately to the left of the previous shelf. We continue adding shelves from right-to-left until the remainder of the input is small enough that no more medium or large squares can be received.

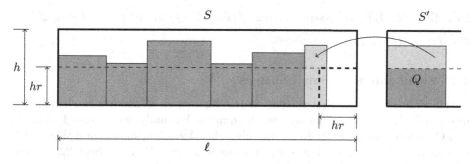

Fig. 3. Illustration of Lemma 1 and Corollary 2 for the case when $r = 1/2$. The upper portion of the square Q is assigned to S, while the lower portion is assigned to S'.

Large: Except in one special case, large squares are packed together with medium squares in the main region as described above. The special case is when we receive a third large square (there can be at most three since the input is at most 2/5). In this case, the third large square is packed into the initial region as if it is a medium square.

4 Analysis

We will show that any input which cannot be packed by our algorithm must have total area greater than 2/5. In Sect. 4.1, we cover basic shelf packing and how we assign area to fixed shelves. In Sect. 4.2, we focus on vertical shelves, showing why they can be treated as small squares (given a sacrifice of 0.235/12 waste due to maintaining one open vertical shelf for each size class). In Sect. 4.3, we bound the waste in the initial region under different circumstances. In Sect. 4.4, we bound the density of variable shelves for medium/large squares and the waste due to small shelves in the main region. In Sect. 4.5, we show that $\alpha \geq 2/5$ for the online square-into-square packing problem.

4.1 Shelf Packing

In our analysis, it is important to determine the area assigned to open and closed shelves. The foundation for many of our lemmas is a generalization of a lemma due to Moon and Moser [1]. The results in this section and Sect. 4.2 can be extended to the setting of rectangles with rotation at the expense of some additional wasted space. We show this in the appendix.

Lemma 1. *Let S be a shelf with height h, length ℓ and ratio r, $0 < r < 1$, that is packed with a set P of squares with height $\leq h$ and $> hr$. Let Q be the first square with height h_Q, $h \geq h_Q > hr$, that does not fit into S. The total area of all the squares packed into S plus the area of Q is greater than $\ell hr - (hr)^2 + h_Q hr$.*

Proof. See Appendix. ☐

Corollary 2. *We can assign an area of $\ell hr - (hr)^2$ to every closed shelf S.*

Proof. See Appendix.   □

4.2   Small and Very Small Shelves

In this section, we show that vertical shelves for very small squares can be packed into small shelves as if they are small squares. Formally, we extend Lemma 1 and Corollary 2 to small shelves containing closed vertical shelves and show that we can bound the waste due to open vertical shelves. Refer to Sect. 2.2 for an overview of how we subdivide the small and very small classes.

For Lemma 1, the used length of a shelf must have density equal to the packing ratio r. Since $r_0 = \frac{1}{2}$, any used section of a small shelf with only c_0 squares satisfies this trivially. We need vertical shelves to supply the same guarantee. Note that vertical shelves within small shelves have a length equal to h_0 (the height of small shelves), which is at least twice the height of any very small shelf.

Lemma 3. *Let S be a shelf with height h, length ℓ and packing ratio r, $0 < r < 1$. If $\ell \geq 2h$, we can choose a value for r, such that the area assigned to S is at least $\frac{\ell h}{2}$ when S is closed.*

Proof. See Appendix.   □

Summary of Heights, Ratios, and Packing Densities:

Height	Ratio	Packing Density
$h_0 = 1/6$	$r_0 = 0.5$	> 0.5
$h_1 \approx 0.08333$	$r_1 = 0.71$	$0.71^2 > 0.5$
$h_2 \approx 0.05917$	$r_2 = 0.65$	$r - \frac{hr^2}{\ell} = 0.65 - \frac{0.05917*0.65^2}{0.25} > 0.5$
$h_3 \approx 0.03846$	$r_{3+} = 0.58$	$r - \frac{hr^2}{\ell} > 0.58 - \frac{0.03846*0.58^2}{0.25} > 0.5$

For simplicity of analysis, we've assigned a ratio of 0.58 for all c_{3+}. We can do this because the packing density only increases with shorter heights as long as the ratio and length remain the same. In short, for $i \geq 3$, we have $h_i = h_{i-1}r_{i-1}$, $r_i = 0.58$, and density greater than 0.5.

Lemma 4. *The waste due to open vertical shelves in the entire container is at most $\frac{0.235}{12}$ and subtracting this number from the sum of all assigned areas allows us to consider all vertical shelves closed.*

Proof. See Appendix.   □

Lemma 5. *Lemma 1 and Corollary 2 can be extended to small horizontal shelves with vertical shelves packed into them at the expense of $\frac{0.235}{12}$ waste subtracted from the total area assigned to the whole container.*

Proof. See Appendix.   □

4.3 Waste in the Initial Packing Region

Our algorithm packs this region from left-to-right with small and very small squares and right-to-left with medium squares and possibly one large square. In this section, we define the *empty length* E of the initial packing region as the distance between the rightmost small or very small square and the leftmost medium or large square. In other words, the distance between these two groups.

While used portions of these shelves will have a density of $1/2$, there may be waste due to unevenly packed small shelves or the empty length. We start by analyzing the waste due to unevenly packing the two small shelves in alternating fashion. Then, we address the waste due to the empty length. Finally, we show that this waste is reduced once we pack small squares into the main region.

Lemma 6. *The waste due to unevenly packed small shelves is at most $1/72$.*

Proof. See Appendix. □

Lemma 7. *The waste due to empty length E is at most $E/6$.*

Proof. See Appendix. □

Lemma 8. *After receiving two small squares which do not fit in the initial region, the waste in the initial region is at most $1/72$.*

Proof. See Appendix. □

4.4 The Main Packing Region

In this section, we first consider the section B in the main region which is occupied by variable shelves for medium and large squares. Then, we consider the area C occupied by fixed shelves for small and very small squares.

Lemma 9. *Let B be the rectangular section in the main packing region containing all closed variable height shelves. Let B also include the open variable shelf if that shelf contains one large square or at least two medium squares. The density of B is at least $1/2$.*

Proof. See Appendix. □

Lemma 10. *Let C be the rectangular section in the main packing region containing all small squares. If the small shelf with the longest used length is one of the top two shelves (M_2 or M_3) the waste in C is at most $5/144$. Otherwise, the waste in C is at most $9/144$. The remaining portion of C has a density of $1/2$.*

Proof. See Appendix. □

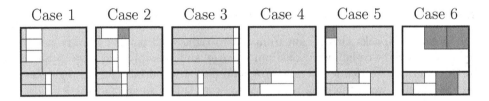

Fig. 4. Illustrations of each case in Theorem 11. *Light gray* sections have been assigned enough area to have a density of 1/2. *Dark gray squares* represent medium squares in Cases 2 and 5 and large squares in Case 6. The *white* sections are wasted space.

4.5 Improved Bound for Online Square-into-Square Packing

Theorem 11. *Any set of squares with total area at most 2/5, which is received in an online fashion, can be packed into a unit square container.*

Proof. Assume for contradiction that the total area of the input is at most 2/5 and some square Q does not fit. We divide the proof into cases based on which size classes of squares have been packed into the main region. As before, we use the term small to refer to both small and very small squares.

In cases 1 and 2, the main region has small squares as well as medium and/or large squares. In case 3, it has only small squares. In cases 4 and 5, the main region has medium squares and may have large squares. In case 6, it has only large squares. We will show that each case can only occur if the input is greater than 2/5.

Case 1: The main region contains small squares as well as medium and/or large squares and the most recent variable height shelf contains at least one large square or at least two medium squares.

Here, we consider the intersection of the most recent variable shelf and the fixed height shelf with the longest used length. By Lemmas 9 and 10, the main region has waste at most 9/144 due only to small shelves. By Lemmas 4 and 8, the waste in the initial region is at most $0.235/12 + 1/72$. Then the total area is

$$1/2 - 9/144 - 0.235/12 - 1/72 > 2/5$$

Case 2: The main region contains small squares as well as medium and/or large squares and the most recent variable height shelf contains exactly one medium square.

Again, we consider the intersection of the most recent variable shelf and the fixed height shelf with the longest used length. Let S be the intersecting variable shelf. By Lemma 9, the area to the right of S has no waste. Note that since S has exactly one medium square, it must intersect one of the top two small shelves. Then by Lemma 10, the area to the left of S has waste at most 5/144. We now consider the waste in S itself. Let M with height h_M be the medium square in S. The waste due to S is $(h_M \cdot 2/3)/2 - h_M^2$. As in the previous case, by Lemmas 4 and 8, the waste in the initial region is at most $0.235/12 + 1/72$.

$$1/2 - 5/144 - 0.235/12 - 1/72 - (h_M \cdot 2/3)/2 + h_M^2 > 2/5$$

Case 3: There are no medium or large squares in the main region.

As in previous cases, the initial region will account for an area of $1/6 - 0.235/12 - 1/72$. In the main region, we can apply Corollary 2 to the bottom three shelves and Lemma 1 to the topmost shelf. By design each time some square closes a shelf, there is room for that square in the shelf above it until we receive some square which would close the top shelf. Since Corollary 2 assigns an area less than Lemma 1, we can apply it to all four shelves to account for a total area of $4(\ell h r - (hr)^2) = 4(1 \cdot 1/6 \cdot 1/2 - 1/12^2) = 1/3 - 1/36$.

$$1/6 - 0.235/12 - 1/72 + 1/3 - 1/36 > 2/5$$

Case 4: The main region contains at least one medium square, may contain large squares, and contains no small squares. The final variable shelf contains at least one large square or at least two medium squares.

In this case, we consider an input which causes the variable shelves to extend beyond the left edge of the container. By Lemma 9, the area occupied by these variable shelves (which includes the entire main region) has a density of $1/2$. However, unlike previous cases, there may be empty space in the initial region of at most $1/3$ since that is the height of the biggest possible medium square. By Lemma 7, the waste due to that empty space is at most $1/3 \cdot 1/6 = 1/18$. So the waste in the initial region is at most $0.235/12 + 1/72 + 1/18$.

$$1/2 - 0.235/12 - 1/72 - 1/18 > 2/5$$

Case 5: The main region contains at least one medium square, may contain large squares, and contains no small squares. The final variable shelf contains exactly one medium square.

Again, we consider an input which causes the variable shelves to extend beyond the left edge of the container. Let S be the open variable shelf containing a single medium square M with height h_M. By Lemma 9, the area to the right of S has no waste. As before, the waste due to S is $(h_M \cdot 2/3)/2 - h_M^2$. Since M was packed into the main region, the empty length in the initial region is at most h_M. Then by Lemma 7, the waste due to that empty space is at most $h_M \cdot 1/6 = h_M/6$. So the waste in the initial region is at most $0.235/12 + 1/72 + h_M/6$.

$$1/2 - 0.235/12 - 1/72 - h_M/6 - (h_M \cdot 2/3)/2 + h_M^2 > 2/5$$

Case 6: Three large squares are received before any small or medium squares are packed into the main region. This is the special case in which we need to pack a large square into the initial region.

Let Q be the third large square with height h_Q. Suppose there isn't room in the initial region for Q or some small/medium square received after Q. The first two large squares represent an area of at least $2/9$ and the area of Q is $h_Q{}^2 > 1/9$. As in case 3, the initial region can be assigned an area of at least $1/6$ minus waste. The empty length can be at most h_Q, otherwise Q would fit.

$$2/9 + h_Q{}^2 + 1/6 - h_Q/6 - 0.235/12 - 1/72 > 2/5 \qquad \square$$

References

1. Moon, J., Moser, L.: Some packing and covering theorems. Colloq. Mathematicum **17**, 103–110 (1967)
2. Han, X., Iwama, K., Zhang, G.: Online removable square packing. Theory Comput. Syst. **43**(1), 38–55 (2008)
3. Fekete, S., Hoffmann, H.: Online square-into-square packing. In: Proceedings of the 16th International Workshop on Approximation Algorithms for Combinatorial Optimization Problems, pp. 126–141 (2013)
4. Leung, J.Y.-T., Tam, T.W., Wong, C.S., Young, G.H., Chin, F.Y.L.: Packing squares into a square. J. Parallel Distrib. Comput. **10**(3), 271–275 (1990)
5. Moser, L.: Poorly formulated unsolved problems of combinatorial geometry. Mimeographed (1966)
6. Hougardy, S.: On packing squares into a rectangle. Comput. Geom.: Theory Appl. **44**(8), 456–463 (2011)
7. Januszewski, J., Lassak, M.: On-line packing sequences of cubes in the unit cube. Geometriae Dedicata **67**(3), 285–293 (1997)
8. Brubach, B.: Improved online square-into-square packing (2014). http://arxiv.org/abs/1401.5583

Improved Approximation Algorithm for Fault-Tolerant Facility Placement

Bartosz Rybicki$^{(\boxtimes)}$ and Jaroslaw Byrka

Institute of Computer Science, University of Wrocław, Wrocław, Poland
{bry,jby}@ii.uni.wroc.pl

Abstract. We consider the Fault-Tolerant Facility Placement problem ($FTFP$), which is a generalization of the classical Uncapacitated Facility Location problem (UFL). In the $FTFP$ problem we have a set of clients C and a set of facilities F. Each facility $i \in F$ can be opened many times. For each opening of facility i we pay $f_i \geq 0$. Our goal is to connect each client $j \in C$ with $r_j \geq 1$ open facilities in a way that minimizes the total cost of open facilities and established connections.

In a series of recent papers $FTFP$ was essentially reduced to Fault-Tolerant Facility Location problem ($FTFL$) and then to UFL showing it could be approximated with ratio 1.575. In this paper we show that $FTFP$ can actually be approximated even better. We consider approximation ratio as a function of $r = min_{j \in C} \, r_j$ (minimum requirement of a client). With increasing r the approximation ratio of our algorithm λ_r converges to one. Furthermore, for $r > 1$ the value of λ_r is less than 1.463 (hardness of approximation of UFL). We also show a lower bound of 1.278 for the approximability of the $FTFL$ for arbitrary r. Already for $r > 3$ we obtain that $FTFP$ can be approximated with ratio 1.275, showing that under standard complexity theoretic assumptions $FTFP$ is strictly better approximable than $FTFL$.

1 Introduction

In the Fault-Tolerant Facility Placement problem, we are given a set F of locations where facilities may be opened (each $i \in F$ costs $f_i > 0$ and can be opened many times) and a set C of clients. Each $j \in C$ has connection requirement $r_j > 0$. Our goal is to open a subset of facilities (possibly many copies of some facilities) and connect each client j with r_j open facilities, such that the total cost of connections and opened facilities is as small as possible. In this paper we consider the metric version of the problem where the connections between elements of the set $C \cup F$ satisfy the triangle inequality.

It is easy to see that the classical UFL problem is a special case of $FTFP$ with all $r_j = 1$. On the other hand, if no facility can be open more than once, then the problem becomes the Fault-Tolerant Facility Location problem (FTFL), in which the demands cannot exceed the number of facilities.

B. Rybicki—Research supported by NCN 2012/07/N/ST6/03068 grant.

E. Bampis and O. Svensson (Eds.): WAOA 2014, LNCS 8952, pp. 59–70, 2015.
DOI: 10.1007/978-3-319-18263-6_6

Facility location problems are typically APX-hard and there exist constant factor approximation algorithms assuming metric connection costs. Shmoys, Tardos and Aardal [6] gave the first constant factor 3.16-approximation algorithm for uncapacitated facility location problem based on LP-rounding. Later Chudak and Shmoys [16] obtained a $(1+\frac{2}{e})$-approximation by marginal-preserving randomized rounding of facility openings, which has became standard for facility location problems. The long line of results for UFL includes a primal-dual algorithm JMS [2], which was then combined with a scaled version of [16] in a work of Byrka and Aardal [5]. The currently best known ratio of 1.488 was obtained by Shi Li [1] by further randomizing the algorithm from [5]. The best known lower bound for UFL is 1.463, see the paper of Guha and Khuller [13]. Many techniques developed for UFL can be directly applied to $FTFP$ which was shown in [18].

First constant factor approximation algorithm for the closely related $FTFL$ problem was given by Guha, Meyerson and Munagala [14]. Next Swamy and Shmoys improved the ratio to 2.076, see [12]. More recently Byrka, Srinivasan and Swamy [11] improved the ratio to 1.725 using dependent rounding [10] and laminar clustering. Moreover it is shown in [12] that JMS algorithm can be adapted to $FTFL$ with uniform requirements of clients.

$FTFP$ was first studied by Xu and Shen [19] and next by Yan and Chrobak who first obtained a 3.16-approximation algorithm [8], and later improved the ratio to 1.575 [18].

1.1   Our Contribution

We extend the work of Yan and Chrobak [18] and propose an algorithm with approximation ratio being a decreasing function of the minimal requirement $r = min_{j \in C} \, r_j$. Our solution benefits from requirements of clients being bigger than one. Instead of considering a client $j \in C$ as r_j distinct clients with unit demand we derive benefits from this multiplicity and use Poisson distribution to estimate the expected number of useful facilities which will be open in a set of a particular volume. We consider both cases: uniform and non-uniform requirements of clients, and obtain the following approximation ratios:

r	1	2	3	4	5	6	7	8	9	10
Non-uniform	1.515	1.439	1.338	1.275	1.234	1.207	1.187	1.171	1.159	1.149
Uniform	1.488	1.410	1.329	1.272	1.234	1.207	1.187	1.171	1.159	1.149

We also prove a lower bound of 1.278 on the approximability of Fault-Tolerant Facility Location (where at most one facility may be opened in each location) for arbitrarily large $r > 1$ (the previous lower bound for $FTFL$ is equal to 1.463 [13] and holds only for $r = 1$).

Observation 1. *Lower bound for FTFL, of value 1.278, is bigger than approximation ratio λ_r for $r \geq 4$. Moreover for $r \geq 2$ FTFP is easier than UFL.*

Note that λ_r for $r = 4$ (in both uniform and non-uniform case) is bounded by 1.275, which is smaller than our lower bound for $FTFL$.

2   The LP Formulation

Consider the following standard LP relaxation of $FTFP$.

$$min \sum_{i \in F} \sum_{j \in C} c_{ij} x_{ij} + \sum_{i \in F} y_i f_i \tag{1}$$

$$\sum_{i \in F} x_{ij} \geq r_j \quad \forall_{j \in C} \tag{2}$$

$$y_i - x_{ij} \geq 0 \quad \forall_{i \in F, j \in C} \tag{3}$$

$$x_{ij}, y_i \geq 0 \quad \forall_{i \in F, j \in C} \tag{4}$$

An optimal solution of the above LP is denoted by a pair (x^*, y^*). Using these variables we express the total facility cost as $F^* = \sum_{i \in F} f_i y_i^*$ and the connection cost of each client $j \in C$ as $C_j^* = \sum_{i \in F} c_{ij} x_{ij}^*$. Summing over all clients gives the total connection cost $C^* = \sum_{j \in C} C_j^*$ of the LP solution. The cost of (x^*, y^*) denoted by $LP^* = F^* + C^*$ is a lower bound on the cost of an optimal integral solution denoted by OPT.

We say that a solution is complete if for each client $j \in C$ and each facility $i \in F$ we have $x_{ij}^* \in \{0, y_i^*\}$. Detailed description of a technique called *facility splitting*, which yields complete solutions, can be found in [4]. The splitting algorithm takes as input a solution of the LP and outputs a complete solution of the same cost to a larger, but equivalent instance of the problem. For clarity of a presentation, throughout the paper we simply assume that all fractional solutions are complete.

Definition 1. *The volume of a set $F' \subseteq F$, denoted by $vol(F')$ is the sum of facility openings in this set, i.e., $vol(F') = \sum_{i \in F'} y_i$.*

One of the problems with input instances is possibly non-polynomial demand of some clients. In [18] we can find an elegant reduction of such instance to instances with requirements bounded by $|F|$. In Sect. 5 we give an algorithm which generalizes this reduction. Our algorithm also reduces the input instance to an instance with polynomial demands of clients, but we also care not to reduce the requirements of clients too much.

3   Algorithm for $FTFP$

The following algorithm $A(\gamma)$ is parametrized by a real constant $\gamma \in (1, 3)$.

Our final *Algorithm* 1 is as follows: run algorithm $A(\gamma_l)$ for each choice of $\gamma_l = 1 + 2 \cdot \frac{n-l}{n}$, where $l = 1, 2, \ldots n - 1$. Select the best of the obtained solutions. Note that $n - 1$ is the number of different values of γ, each of them we use as

---

Algorithm 1. $A(\gamma)$

1: formulate and solve the LP (1)-(4), get an optimal solution (x^*, y^*);
2: scale up facility opening by γ ($\bar{y}_i = \gamma \cdot y_i^*$), then recompute values of x_{ij} to obtain a minimum cost solution $(\bar{x}, \bar{y})$;
3: compute clustering for all clients;
4: round facility opening variables using dependent rounding;
5: connect each client j with r_j closest open facilities;

---

a parameter of algorithm $A(\gamma)$. In the computation of approximation ratios we use n equal 1000, but we will describe our results for a general n.

Scaling facility opening is an idea from [5], it decreases average connection cost of each client, but increases total cost of opening facilities and multiply it by $\gamma > 1$. In *FTFP* we can open more than one facility in one location, so scaling does not cause problems with opening more than one facility in one place. The version of clustering which we use is very close to the one described in [16]. To round facility opening variables we use the randomized algorithm from [10], called dependent rounding. Each step of the algorithm $A(\gamma)$ is carefully described in the following sections.

3.1 Scaling

Let F_j denote the set of facilities with a positive flow from a client $j \in C$, i.e., facilities i with $x_{ij}^* > 0$ in the optimal LP solution.

Let $\gamma_l > 1$. Suppose that all facilities are sorted in an order of non-decreasing distances from a client $j \in C$. Scaling all y^* variables by γ_l divides the set of facilities F_j into two disjoint subsets (we can assume that opening of each facility is ϵ by facility splitting technique): the set of close facilities of a client j, denoted by $F_j^{C_l}$, such that $vol(F_j^{C_l}) = r_j$; and the distant facilities, denoted by $F_j^{D_l} = F_j \setminus F_j^{C_l}$, note that $vol(F_j^{D_l}) = r_j(\gamma_l - 1)$. Certainly for each $i_1 \in F_j^{C_l}$ and $i_2 \in F_j^{D_l}$ we have $c_{i_1 j} \leq c_{i_2 j}$.

By $D_{av}^{C_l}(j), D_{av}^{D_l}(j)$ and $D_{av}(j)$ we denote the average distances to close, distant and all facilities in set F_j, respectively. More formally:

$$D_{av}^{C_l}(j) = \frac{\sum_{i \in F_j^{C_l}} c_{ij} \bar{x}_{ij}}{vol(F_j^{C_l})}, \quad D_{av}^{D_l}(j) = \frac{\sum_{i \in F_j^{D_l}} c_{ij} \bar{x}_{ij}}{vol(F_j^{D_l})}$$

By $D_{max}^l(j)$ we denote the maximal distance to a facility in $F_j^{C_l}$, and by $c_l(j)$ we denote the average distance to $F_j^l = F_j^{C_l} \setminus F_j^{C_{l-1}}$ for $n > l \geq 1$, $F_j^n = F_j \setminus F_j^{C_{n-1}}$ and $F_j^0 = \emptyset$. (see Fig. 1)

3.2 Clustering

Definition 2. *The radius of a set F' for a client j, where $F' \subseteq F$ and $j \in C$, is $max_{i \in F'} c_{ij}$. Assume that $vol(F') \geq r$. By $B(j, F', r)$ we denote the subset of F' of volume r which has the smallest radius.*

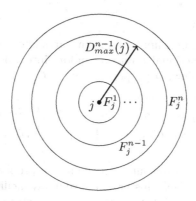

Fig. 1. Figure shows partition of facilities in set F_j.

Each client $j \in C$ initially has a cluster proposition $CP(j) = B(j, F, r_j) = F_j^C$, whose radius is $q_j = D_{max}^C(j)$. In the following algorithm the cluster proposition of a client j changes, but the radius never increases. The above described

---

Algorithm 2. Clustering

1: **for all** $j \in C$ **do**
2:      $q_j := D_{max}^C(j)$
3: **end for**
4: **while** there is a client with positive requirement **do**
5:      select a client $j \in C$ with $r_j > 0$ that minimizes q_j and set $r_j := 0$
6:      **for all** $j' \in N(j) = \{j'' \in C \mid CP(j) \cap CP(j'') \neq \emptyset \wedge r_{j''} > 0\}$ **do**
7:          $r_{j'} := max(0, r_{j'} - \lceil vol(CP(j) \cap CP(j')) \rceil)$
8:          $CP(j') := B(j', CP(j') \setminus CP(j), r_{j'})$
9:      **end for**
10:     create $C(j) = \{j\} \cup N(j) \cup CP(j)$ and call j the center of cluster $C(j)$;
11: **end while**

---

procedure is a variant of the method described in [16]. It is well known that output of the procedure has two important properties. First: each facility is clustered by at most one client. Second: the distance from a client to any of his cluster centers is not too big.

Lemma 1. *Distance from any client $j \in C$ to any close facility of $j' \in C$ such that $j \in C(j')$ is bounded by $3 \cdot D_{max}^C(j)$.*

Proof. Suppose that $j' \in C$ and $j \in C_C(j') = C(j') \cap C$. From the fact that $j \in C_C(j')$ follows that $q_{j'} \leq q_j$, which is equivalent with $D_{max}^C(j') \leq D_{max}^C(j)$. The definition of $CP(\cdot)$ assures that the distance from j (j') to any facility in $CP(j)$ ($CP(j')$) can be bounded by $D_{max}^C(j)$ ($D_{max}^C(j')$). Consider $i' \in CP(j) \cap CP(j')$ and any $i \in CP(j')$. Distance from j to i is $c_{i'j} + c_{i'j'} + c_{ij'} \leq D_{max}^C(j) + 2 \cdot D_{max}^C(j') \leq 3 \cdot D_{max}^C(j)$. $\square$

3.3 Facility Opening

A randomized procedure deciding whether a particular facility should be open or not transforms the fractional $\bar{y}$ into a random integral $\hat{y}$. We would like the procedure to have the following properties, where $C_F(j) = C(j) \cap F$:

- *Marginal distribution:* $Pr[\hat{y}_i = 1] = \bar{y}_i$
- *Sum-preservation:* $\sum_{i \in C_F(j)} \hat{y}_i \in \{\lfloor vol(C_F(j)) \rfloor, \lceil vol(C_F(j)) \rceil\}$
- *Negative correlation:* $\forall S \subseteq C_F(j) \forall b \in \{0,1\} Pr[\bigwedge_{i \in S}(\hat{y}_i = b)] \leq \prod_{i \in S} Pr[\hat{y}_i = b].$

One method which gives an output with the above properties is the dependent rounding (DR) from [10]. Each cluster can have many facilities open fractionally. We first apply DR to each $C_F(j)$, where j is the center of a cluster. Then the remaining fractional facility openings are rounded by DR in an arbitrary order.

4 Analysis

To bound the expected connection cost of an algorithm $A(\gamma)$, we need to first analyse the number of facilities which will be opened in a set of a particular volume. Suppose that facilities are opened independently and that in the limit case all facilities are opened very little in the fractional solution, then the number of eventually open facilities from a set has the Poisson distribution. By the negative correlation this distribution can be used to derive the following lower bound on the number of useful opened facilities from the considered set.

Observation 2. *The expected number of possible connections with set D of volume $\Lambda = vol(D)$, when the requirement is k, is $h(\Lambda, k) \geq \sum_{i=1}^{k-1} iP_\Lambda(X = i) + kP_\Lambda(X \geq k)$. Where $P_\Lambda(X = i) = \frac{\Lambda^i e^{-\Lambda}}{i!}$ is the probability of opening exactly i facilities in a set of volume Λ, if opened independently (Poisson distribution).*

Lemma 2. *Suppose that $\gamma = \gamma_k$. Consider a client $j \in C$ which is not a center of any cluster. The expected connection cost of client j is at most*

$$E[C_j] \leq \sum_{l=1}^{n-1} c_l(j) \cdot \frac{e_1^{k,l}(j)}{r_j} + \frac{e_3^k(j)}{r_j} \cdot 3D_{max}^k(j)$$

where $e_1^{k,l}(j)$ is expected number of open facilities in set F_j^l, in which opening of each facility is scaled by γ_k; $e_3^k(j)$ is r_j decreased by expected number of open facilities in set F_j, in which opening of each facility is scaled by γ_k (or zero if number of open facilities in F_j is bigger than r_j).

Proof. The value of $e_1^{k,l}(j)$ is the expected number of open facilities in the set $F_j^l = F_j^{C_l} \setminus F_j^{C_{l-1}}$, when all fractional openings of facilities are scaled up by γ_k. Connection cost of a client j with an open facility in this set is $c_l(j)$. The expected number of connections which j has to establish with close facilities of his cluster centers is $e_3^k(j)$ - his requirement reduced by the number of facilities opened in F_j. Lemma (1) bounds the distance to close facilities of cluster centers of j. $\square$

4.1   Factor Revealing LP

Consider running an algorithm $A(\gamma_l)$ for $\gamma_l = 1 + 2 \cdot \frac{n-l}{n}$ where $l = 1, 2, \ldots n - 1$. Observe that the following linear program, called FRLP, covers all that executions. Value of the objective function is an upper bound on the approximation ratio of the best of the obtained solutions.

$$max\ \lambda_r \tag{5}$$

$$\gamma_k f + \sum_{l=1}^{n-1} c_l \cdot \frac{e_1^{k,l}}{r} + \frac{e_3^k}{r} \cdot 3 \cdot c_{k+1} \geq \lambda_r\ \forall_{k<n} \tag{6}$$

$$\sum_{l=1}^{n} vol(F^l) \cdot c_l = c \tag{7}$$

$$0 \leq c_i \leq c_{i+1} \leq 1\ \forall_{i<n} \tag{8}$$

$$f + c = 1 \tag{9}$$

$$f, c \geq 0 \tag{10}$$

The above LP encodes the cost of solutions obtained in executions of an algorithm A for different values of the scaling parameter γ_k for $k = 1, 2, \ldots n-1$. Adversary has the freedom to choose the distances from client j to groups of facilities and the relation between values of f and c in the optimal solution, which have to sum up to one, and (both) be non-negative. We consider all facilities in the order of a non-decreasing distance from the client j, so the average distances to consecutive groups of facilities have to be non-decreasing, see constraint (8). We divide facilities into sets F_j^l, for $1 \leq l \leq n$. In each set F_j^l the adversary may choose the distance from client j to the open facility in F_j^l, which is the worst for our algorithm and equals $c_l(j)$. Equality (7) shows that the sum of average distances, each weighted by the volume of facilities at such distance, has to sum up to the total connection cost in the optimal solution. The crucial inequality (6) encodes the expected cost of an algorithm $A(\gamma_k)$ and it is used as an upper bound for the approximation ratio. Client in inequality (6) is a client with minimum requirement r, $e_1^{k,l} = h(\gamma_k \cdot vol(F^{C_l}, r)) - h(\gamma_k \cdot vol(F^{C_{l-1}}), r)$ is expected number of open facilities in set F^l and $e_3^k = r - h(\gamma_k \cdot r, r)$. Correctness of this inequality follows from Lemmas (1), (2) and $D_{max}^l \leq c_{l+1}$. If $r = 1$ then instead of $Algorithm$ 1 we use method from [18]. To improve the approximation ratio from 1.575 to 1.515 we run the algorithm from [18] for a number of values of the scaling parameter $\gamma_l = 1 + 2 \cdot \frac{n-l}{n}$, where $l = 1, 2, \ldots n - 1$. It can be analyzed by FRLP. The computed values of λ_r, for $r = 1, 2, \ldots 10$, are in the following table (Fig. 2):

r	1	2	3	4	5	6	7	8	9	10
λ_r	1.515	1.439	1.338	1.275	1.234	1.207	1.187	1.171	1.159	1.149

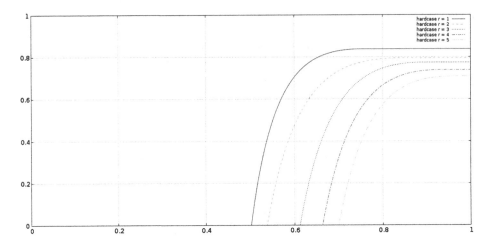

Fig. 2. The profiles of distances in tight instances for *Algorithm* 1 for *FTFP* (in a general, non-uniform case) for $1 \leq r \leq 5$, extracted from the *FRLP* solutions. The x-axis encodes the volume of a set of facilities closest to a client and the y-axis is the distance to the farthest facility in this set.

4.2   Uniform Requirement

As it was shown in [12] the JMS algorithm can be modified to work with *FTFL* with uniform requirements of clients, and the approximation ratio remains the same. In consequence it also works for *FTFP* with uniform requirements of clients. We can add one more constraint $1.11f + 1.78c \geq \lambda_r$ to the *FRLP* in Sect. 4.1 which encodes that we additionally run the (modified) JMS algorithm[1]. Such FRLP for $r = 1$ is a dual of the LP from [1], probabilities of particular algorithms in Shi Li paper are dual values of constraints in FRLP. As you can see in the following table, adding the JMS algorithm makes difference only for small values of r.

r	1	2	3	4	5	6	7	8	9	10
non-uniform	1.515	1.439	1.338	1.275	1.234	1.207	1.187	1.171	1.159	1.149
uniform	1.488	1.410	1.329	1.272	1.234	1.207	1.187	1.171	1.159	1.149

5   Factor λ_r Is a Decreasing Function of r

Lemma 3. *Function* $f(r) = \frac{r - h((1+\epsilon)r, r)}{r}$ *converges to 0 when* $r \mapsto \infty$.

---

[1] An algorithm for UFL is called (a,b)-approximation if the cost of returned solution is upper bounded by $a \cdot F^* + b \cdot C^*$, where F^* and C^* are, respectively, the costs of establishing connections and opening facilities in an optimal solution.

Theorem 1. *For each choice of $\epsilon > 0$ there exists r_0 such that for each $r \geq r_0$ inequality $\lambda_r \leq 1 + \epsilon$ holds.*

The above theorem easy follows from the following lemmas, because the approximation ratio of *Algorithm* 1 is always upper bounded by the approximation ratio of $A(\gamma)$ for each choice of γ.

Consider an instance $\mathcal{I}$ and a client $j \in C$. Lemma (2) implies that the expected connection cost of j in algorithm $A(\gamma)$ is $E[C_j] \leq \sum_{l=1}^{n} c_l(j) \cdot \frac{e_1^{k,l}(j)}{r_j} + \frac{e_3^k(j)}{r_j} \cdot 3D_{max}(j)$. Note that the inequality $\sum_{l=1}^{n} c_l(j) \cdot \frac{e_1^{k,l}(j)}{r_j} \leq C_j^*$ holds, because in the solution (x^*, y^*) client j fractionally uses the same facilities, but with smaller opening values. Therefore, in the expectation he pays not less for connection than in our scaled up solution. Notice that, for a particular choice of $\gamma = 1 + \epsilon$, the value of the expression $3(1 + \frac{1}{\epsilon})$ is a constant. From [10] we know that the following inequality holds.

$$\frac{e_3^k(j)}{r_j} \cdot 3D_{max}(j) \leq f(r) \cdot 3(1 + \frac{1}{\epsilon})C_j^*$$

Observation 3. *For $\gamma = 1 + \epsilon$ the approximation factor for connection cost of the solution produced by $A(\gamma)$ depends only on $f(r)$, where r is the minimum requirement in the considered instance.*

Li Yan showed [7] a result similar to the below lemma, but the result is weaker: he shows that the limit is 1 only for a fixed number of facilities.

Lemma 4. *For each $\epsilon > 0, \gamma = 1 + \epsilon$, there exists r_0 such that for each $r \geq r_0$, approximation ratio of an algorithm $A(\gamma)$ is bounded by $1 + \epsilon$.*

Proof. Lemma 3 and Observation 3 imply that for each choice of ϵ there exists r_0 such that for each instance with minimum requirement $r \geq r_0$ approximation ratio of $A(1 + \epsilon)$ is bounded by $1 + \epsilon$. □

5.1 Dealing with Large Requirements r_j

Yan and Chrobak proved the following theorem

Theorem 2 *(from [18]).* *Suppose that there is a polynomial-time algorithm $\mathcal{A}$ that, for any instance of $FTFP$ with maximum demand bounded by $|F|$, computes an integral solution that approximates the fractional optimum of this instance within factor $\rho > 1$. Then there is a ρ-approximation algorithm $\mathcal{A}'$ for $FTFP$.*

The main result of this section is an extension of Theorem 2 which exploits our Theorem 1. Consider an instance $\mathcal{I}$ for which the approximation ratio of *Algorithm* 1 is almost one, see Theorem 1. As it was mentioned in Sect. 2 we can assume that the optimal solution (x^*, y^*) to the LP (1) - (4) for an instance $\mathcal{I}$ is complete, so for each $i \in F$ and $j \in C$ we have $x_{ij}^* \in \{0, y_i^*\}$.

From optimality of this solution, we can assume that $\sum_{i\in F} x^*_{ij} = r_j$ for all $j \in C$. We split solution (x^*, y^*) into two parts $(x^*, y^*) = (\hat{x}, \hat{y}) + (\dot{x}, \dot{y})$, where

$$\hat{y}_i = max\{\lfloor y^*_i - \bar{r}\rfloor, 0\}, \quad \hat{x}_{ij} = max\{\lfloor x^*_{ij} - \bar{r}\rfloor, 0\} \;\; \forall j \in C, i \in F$$

$$\dot{y}_i = y^*_i - \hat{y}_i, \quad \dot{x}_{ij} = x^*_{ij} - \hat{x}_{ij} \;\; \forall j \in C, i \in F$$

where $1 \leq \bar{r} \leq min_{j\in C}\, r_j$. Now we will construct two instances $\dot{\mathcal{I}}$ and $\hat{\mathcal{I}}$ of $FTFP$ with the same parameters as $\mathcal{I}$, except requirements. Demand of each client j is $\hat{r}_j = \sum_{i\in F} \hat{x}_{ij}$ in the instance $\hat{\mathcal{I}}$ and $\dot{r}_j = \sum_{i\in F} \dot{x}_{ij}$ in $\dot{\mathcal{I}}$.

Lemma 5. *(i) $(\hat{x}, \hat{y})$ is a feasible integral solution to instance $\hat{\mathcal{I}}$*
(ii) $(\dot{x}, \dot{y})$ is a feasible fractional solution to instance $\dot{\mathcal{I}}$
(iii) $(\hat{x}, \hat{y})$ and $(\dot{x}, \dot{y})$ are optimal solutions to $\hat{\mathcal{I}}$ and $\dot{\mathcal{I}}$
(iv) $\forall_{j\in C}\, (\bar{r} + 1) \cdot |F| \geq \dot{r}_j \geq \bar{r}$

Proof. (i) For a feasibility of $(\hat{x}, \hat{y})$, we need to show that all constraints of LP (1)–(4) are satisfied. For each $j \in C$ we have that $\hat{r}_j = \sum_{i\in F} \hat{x}_{ij}$, so (2) holds. Solution (x^*, y^*) is complete, so $x^*_{ij} \in \{0, y^*_i\}$. If $x^*_{ij} = 0$ then $\bar{x}_{ij} = 0 \leq \bar{y}_i$, otherwise $x^*_{ij} = y^*_i > 0$ in that case we have that $\hat{x}_{ij} = \hat{y}_i$. In consequence constraint (3) is satisfied.

(ii) In the case of $(\dot{x}, \dot{y})$ also all inequalities hold. Constraint (2) is satisfied, because $\dot{r}_j = \sum_{i\in F} \dot{x}_{ij}$. Note that both $\dot{x}_{ij}$ and $\dot{y}_i$ are non-negative. We need to show that $\dot{x}_{ij} \leq \dot{y}_i$ which is equivalent with $y^*_i - max\{\lfloor y^*_i - \bar{r}\rfloor, 0\} \geq x^*_{ij} - max\{\lfloor x^*_{ij} - \bar{r}\rfloor, 0\}$. If $x^*_{ij} = 0$ then we have $y^*_i \geq max\{\lfloor y^*_i - \bar{r}\rfloor, 0\}$ which holds. In the other case we have $x^*_{ij} = y^*_i > 0$. With that assumption we trivially obtain the following equality $y^*_i - max\{\lfloor y^*_i - \bar{r}\rfloor, 0\} = x^*_{ij} - max\{\lfloor x^*_{ij} - \bar{r}\rfloor, 0\}$.

(iii) Suppose that at least one of $(\hat{x}, \hat{y})$ and $(\dot{x}, \dot{y})$ is not an optimal solution to $\hat{\mathcal{I}}$ and $\dot{\mathcal{I}}$, respectively. In that situation we are able to obtain solution to instance $\mathcal{I}$ with a smaller cost than $cost(x^*, y^*)$, which is a contradiction.

(iv) To prove $\dot{r}_j \leq (\bar{r} + 1) \cdot |F|$ we have to show that the following inequality holds, where $F' = \{i \in F \mid x^*_{ij} \geq \bar{r} + 1\}$.

$$r_j - \sum_{i\in F'}(x^*_{ij} - (\bar{r} + 1)) \leq (\bar{r} + 1)|F| \iff \sum_{i\in F\setminus F'} x^*_{ij} \leq (\bar{r} + 1)|F \setminus F'|$$

To finish the proof of the lemma we should prove the following inequalities

$$r_j - \sum_{i\in F} max\{\lfloor x^*_{ij} - \bar{r}\rfloor, 0\} \geq r_j - \sum_{i\in F} max\{x^*_{ij} - \bar{r}, 0\} \geq \bar{r}$$

Let $F' = \{i \in F \mid x^*_{ij} > \bar{r}\}$. Using F' we can rewrite the above inequality as $r_j - \sum_{i\in F'}(x^*_{ij} - \bar{r}) \geq \bar{r}$. Consider two cases: $|F'| = 0$ and $|F'| \geq 1$. The first is trivial because $r_j \geq \bar{r}$ holds. In the second case $r_j - \sum_{i\in F'}(x^*_{ij} - \bar{r}) \geq r_j + \bar{r} - \sum_{i\in F} x^*_{ij} \geq \bar{r}$, which trivially holds, because $r_j = \sum_{i\in F} x^*_{ij}$. $\square$

Corollary 1. *For an instance $\mathcal{I}$, with requirement r, for which the approximation ratio of Algorithm 1 is λ_r, we can obtain two other instances: integral $\hat{\mathcal{I}}$ and fractional $\dot{\mathcal{I}}$. Instance $\dot{\mathcal{I}}$ has polynomial demands and a minimum requirement $1 \leq \bar{r} \leq \min_{j \in C} r_j$. The approximation ratio λ and a running time of Algorithm 1 depends on value of $\bar{r}$, which can be arbitrarily selected from $[1, r]$. Sum of the integral solution $\mathcal{S}$ and the optimal integral solution for $\hat{\mathcal{I}}$, is a feasible integral solution for $\mathcal{I}$ with the approximation ratio λ.*

6   Lower Bound for FTFL

We give a reduction from the Set Cover problem. Consider an instance of Set Cover defined as $X = \{x_1, x_2, \dots x_n\}$, and $S = \{S_1, S_2, \dots S_m\}$ such that $S_i \subseteq X$ for each $i \in \{1, 2, \dots m\}$. We would like to find a cover $C \subseteq S$ such that $|C| = k$ is minimized. In our reduction we assume that we know k (we can run algorithm for each value of $1 \leq k \leq m$).

Theorem 3. *If for any $r > 1$ there is a polynomial time algorithm with an approximation factor smaller than 1.278 for the Fault-Tolerant Facility Location problem for instances with minimal requirement r, then $NP = P$.*

The main idea of the proof, which you can find in the full version of the paper [15], is the same as in [13]. We use an algorithm for $FTFL$ to obtain partial covers for the Set Cover instance (X, S). We show that the partial cover cannot be too big in each step, because then it would contradict the Dinur and Steuer result [3]. They proved that approximation algorithm for the Set Cover with ratio $c \cdot ln|X|$ where $c < 1$, implies that $NP = P$ (Fig. 3).

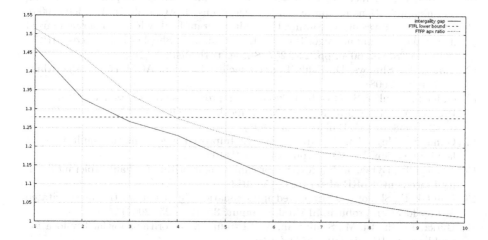

Fig. 3. The figure shows three quantities as a function of $r = 1, 2, \dots 10$: the lower bound for $FTFL$, approximation ratio (in the general, non-uniform case) of our algorithm for $FTFP$, and a lower bound on the integrality gap of the LP (1)–(4). The integrality gap results are also true for the standard LP for $FTFL$, for details see the full version of the paper [15]

7 Open Problems

Is it possible to apply techniques similar to presented in this paper to $FTFL$? Is $FTFL$ getting any easier with increasing value of r? It would also be interesting to derive a lower bound on the approximability of $FTFP$ as a function of $r > 1$.

References

1. Li, S.: A 1.488 approximation algorithm for the uncapacitated facility location problem. Inf. Comput. **222**, 45–58 (2013)
2. Jain, K., Mahdian, M., Saberi, A.: A new greedy approach for facility location problems. In: STOC, pp. 731–740 (2002)
3. Dinur, I., Steurer, D.: Analytical approach to parallel repetition. In: STOC, pp. 624–633 (2014)
4. Sviridenko, M.: An improved approximation algorithm for the metric uncapacitated facility location problem. In: Cook, W.J., Schulz, A.S. (eds.) IPCO 2002. LNCS, vol. 2337, pp. 240–257. Springer, Heidelberg (2002)
5. Byrka, J., Aardal, K.: An optimal bifactor approximation algorithm for the metric uncapacitated facility location problem. SIAM J. Comput. **39**(6), 2212–2231 (2010)
6. Shmoys, D., Tardos, E., Aardal, K.: Approximation algorithms for facility location problems (Extended abstract). In: STOC, pp. 265–274 (1997)
7. Yan, L.: Approximation algorithms for the Fault-Tolerant facility placement problem, Ph.D. Thesis
8. Yan, L., Chrobak, M.: Approximation algorithms for the Fault-Tolerant facility placement problem. Inf. Process. Lett. **111**(11), 545–549 (2011)
9. Feige, U.: A threshold of ln n for approximating set-cover. In: 28th ACM Symposium on Theory of Computing, pp. 314–318 (1996)
10. Gandhi, R., Khuller, S., Parthasarathy, S., Srinivasan, A.: Dependent rounding and its applications to approximation algorithms. J. ACM **53**(3), 324–360 (2006)
11. Byrka, J., Srinivasan, A., Swamy, C.: Fault-Tolerant facility location: a randomized dependent lp-rounding algorithm. In: Eisenbrand, F., Shepherd, F.B. (eds.) IPCO 2010. LNCS, vol. 6080, pp. 244–257. Springer, Heidelberg (2010)
12. Swamy, C., Shmoys, D.: Fault-Tolerant facility location. ACM Trans. Algorithms **4**(4), 1–27 (2008)
13. Guha, S., Khuller, S.: Greedy strikes back: improved facility location algorithms. In: Proceedings of the 9th ACM-SIAM Symposium on Discrete Algorithms (SODA), pp. 228–248. SIAM, Philadelphia (1998)
14. Guha, S., Meyerson, A., Munagala, K.: Improved algorithms for fault tolerant facility location. In: SODA, pp. 636–641 (2001)
15. Rybicki, B., Byrka, J.: Improved approximation algorithm for Fault-Tolerant Facility Placement. CoRR abs/1311.6615 (2013)
16. Chudak, F., Shmoys, D.: Improved approximation algorithms for the uncapacitated facility location problem. SIAM J. Comput. **33**(1), 1–25 (2003)
17. Byrka, J., Ghodsi, M., Srinivasan, A.: LP-rounding algorithms for facility-location problems. CoRR abs/1007.3611 (2010)
18. Yan, L., Chrobak, M.: LP-rounding Algorithms for the Fault-Tolerant Facility Placement Problem. CoRR abs/1205.1281 (2012)
19. Xu, S., Shen, H.: The Fault-Tolerant facility allocation problem. In: Dong, Y., Du, D.-Z., Ibarra, O. (eds.) ISAAC 2009. LNCS, vol. 5878, pp. 689–698. Springer, Heidelberg (2009)

The Submodular Facility Location Problem and the Submodular Joint Replenishment Problem

Sin-Shuen Cheung[(✉)]

School of Operations Research and Information Engineering,
Cornell University, Ithaca, NY 14853, USA
sc2392@cornell.edu

Abstract. In this paper we consider natural generalizations of the facility location problem and the joint replenishment problem in which the opening cost of facilities and the ordering cost over the planning horizon is characterized by a submodular set function in the *oracle model*. Specifically, we can access the function only through a blackbox that returns the value of the function for a given set. We prove information theoretic lower bounds that these two problems cannot be approximated by any polynomial-time algorithm better than a ratio of $O(\sqrt{n}/\log^2 n)$. Moreover, we give $O(\sqrt{n} \cdot \log n)$-approximation algorithms for the two problems.

1 Introduction

A submodular set function $f : F \to \mathbb{R}$ is a function that satisfies the constraint $f(S) + f(T) \geq f(S \cap T) + f(S \cup T)$ for any subsets $S, T \subseteq F$. In the past decades, a vast amount of research has been done on optimization problems involving submodular functions. In this paper, we study two variations of the facility location problem and the joint replenishment problem.

1.1 Facility Location Problem with Submodular Opening Costs

The facility location problem with submodular opening costs (FLS problem) we study is the following model: Given a set of facilities F and a set of clients D, a subset $R \subseteq F$ of facilities can be opened with an *opening cost* of $f(R)$, where $f : 2^F \to \mathbb{R}_+$ is a nonnegative monotone submodular function. Each client $j \in D$ can be connected to an opened facility $i \in F$ with a *connection cost* of c_{ij}. We assume the set of connection costs $\{c_{ij}\}_{i \in F, j \in D}$ is metric. The goal is to find a set of facilities to open such that every client is connected to some opened facility and the total cost is minimized.

Over the past decade, a tremendous amount of research has been done on the facility location problem. Since the first constant factor approximation algorithm by Shmoys et al. [1], many approximation algorithms were developed using various techniques, including LP rounding algorithms, randomized rounding algorithms,

© Springer International Publishing Switzerland 2015
E. Bampis and O. Svensson (Eds.): WAOA 2014, LNCS 8952, pp. 71–82, 2015.
DOI: 10.1007/978-3-319-18263-6_7

primal-dual schema, local search and greedy algorithms (see [2–7]). To the best of our knowledge, the best approximation algorithm known is due to Li [8] and it has an approximation ratio of 1.488. On the hardness side, Guha and Khuller [9] proved that the problem cannot be approximated better than a factor of 1.463 unless NP $\subseteq$ DTIME($n^{O(\log \log n)}$).

There is other work on the facility location problem that involve submodular functions. Svitkina and Tardos [10] studied a variant in which the facility opening costs are specified by submodular functions $g_i : 2^D \rightarrow \mathbb{R}$ for each facility $i \in F$. If a subset of clients $D(i) \subseteq D$ is connected to facility i, then the opening cost for facility i is $g_i(D(i))$. They give an $O(\log n)$-approximation algorithm for this problem, and a constant factor approximation algorithm for the case where the cost functions $\{g_i\}_{i \in F}$ are restricted to be hierarchical functions. Hayrapetyan et al. [11] develop a constant factor approximation algorithm for a variant of the facility location problem in which a subset of clients $D' \subseteq D$ is not connected to opened facilities at a cost of $h(D')$, where the *penalty function* $h : 2^D \rightarrow \mathbb{R}_{\geq 0}$ is a submodular function. Chudak and Nagano [12] give an approximation algorithm for this problem using the Lovász extension and non-smooth convex minimization techniques and Du et al. [13] give a primal-dual algorithm.

1.2   Joint Replenishment Problem with Submodular Ordering Costs

We now define the joint replenishment problem with submodular ordering costs (JRS problem). Given a planning horizon $\mathcal{T} = \{1, 2, \ldots, T\}$ and a set of types of items S, for each time period t in the planning horizon, there are demands $\{d_{i,t}\}_{i \in S}$ to be fulfilled by orders of item $i \in S$ from periods no later than t. Items ordered in period t are allowed fulfill demands of a later period t' at a holding cost of $h_{t,t'}$ per unit. Furthermore, we assume the holding costs to be linear, namely $h_{t,t'} = \sum_{r=t}^{t'-1} h_{r,r+1}$. The ordering cost is a submodular set function $f : 2^{\mathcal{T}} \rightarrow \mathbb{R}_+$; specifically if $R \subseteq \mathcal{T}$ is the set of periods in which orders are placed, then the ordering cost is $f(R)$.

The classical joint replenishment problem is one of the most studied models in the operations research literature and it captures the trade–off between holding costs and ordering costs. In our setting, the ordering cost only depends on whether or not orders are placed on each period, but not how many items are ordered. Other work that incorporates submodular ordering costs in the joint replenishment problems include [14], in which the authors studied a different model where the ordering cost for each period is a submodular function of the types of items that are ordered.

1.3   Our Results

In this paper, we show that no polynomial-time algorithm can approximate FLS problem and JRS problem better than a factor of $O(\sqrt{n}/\log^2 n)$ in the oracle model. Inspired by the work of Svitkina and Fleischer [15] and Iwata and Nagano [16], we construct two families of instances that are hard to distinguish in polynomial time, yet their optimal values differ by a ratio of $O(\sqrt{n}/\log^2 n)$. Additionally, we give $O(\sqrt{n} \cdot \log n)$-approximation algorithms for both problems by utilizing a fundamental result on approximating monotone submodular functions by Goemans et al. [17]. Their algorithm is based on ellipsoidal approximations to a convex body and it makes only polynomially many oracle queries to derive an approximation of a monotone submodular function. In our approximate algorithm for the facility location problem with submodular opening costs, we substitute the submodular opening cost function $f : 2^F \to \mathbb{R}_{\geq 0}$ by the $O(\sqrt{n} \cdot \log n)$-approximation of the function computed by the algorithm of Goemans et al. [17], then approximately solve the resulting problem.

The rest of the paper proceeds as follows. In Sect. 2, we prove the hardness result. Then we design and analyze $O(\sqrt{n} \cdot \log n)$-approximation algorithms for the problems in Sect. 3. In Sect. 4, we conclude and discuss future research.

2   Lower Bounds

In this section, we show that the FLS and JRS problems cannot be approximated better than a factor of $O(\sqrt{n}/\log^2 n)$ in polynomial time in the oracle model. The proofs for FLS and JRS problems follow the same approach. We design two instances of the problem that are hard to distinguish in polynomial time while their optimal values are far apart. We elaborate the proof for the FLS problem first, and then make an analogous proof for the JRS problem.

2.1   FLS Problem

In order to design two instances of the FLS problem polynomial-time indistinguishable, we shall defined the instances as follows.

For the FLS problem, let the two instances have the same set of facilities F and the same set of clients D. We denote $n = |F|$ and $m = |D|$. Let $f_1, f_2 : 2^F \to \mathbb{R}_{\geq 0}$ denote the submodular opening costs for the two instances. Furthermore, we let each client have at least $\lfloor \frac{n}{m} \rfloor$ co-located facilities. Moreover, any two clients are located far away such that it is always suboptimal to connect a client $j \in D$ to a facility $i \in F$ that is not co-located to j. Formally, let the set of facilities F be the union of m disjoint subsets of facilities $\{F_l\}_{l=1}^m$ such that $\lfloor \frac{n}{m} \rfloor \leq |F_l| \leq \lceil \frac{n}{m} \rceil$ and the facilities in F_l are co-located with the same client. We let the connection cost of any client and facility that are not co-located be $O(n)$. Figure 1 shows how we locate the facilities and clients.

Construction of f_1 and f_2. Following the approach of Iwata and Nagano [16], we construct the two submodular functions as follows:

$$f_1(T) = \min\{\mu, |T|\} \quad \text{for } T \subseteq F,$$
$$f_2(T) = \min\{\mu, |T \backslash R| + \min\{72 \log^2 n, |T \cap R|\}\} \quad \text{for } T \subseteq F,$$

where parameter μ and subset $R \subseteq F$ are to be determined. It is not hard to check that both f_1 and f_2 are submodular functions and that $f_2(T) \leq f_1(T)$ for all $T \subseteq F$.

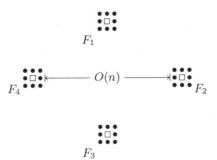

Fig. 1. A demonstration of facility and client locations, where • denotes a facility and □ denotes a client.

Let $p \in [0,1]$. The set $R \subseteq F$ is randomly generated in the following way. *For every $i \in F$, include i in R with probability p independently.* Let parameter μ be the expected size of the set R. Then it follows that

$$\mu := \mathbb{E}[|R|] = pn.$$

The probability that F_l does not contain any facility in R is $\mathbb{P}[F_l \cap R = \emptyset] = (1-p)^{|F_l|}$ for all $l \in \{1, \ldots, m\}$. Then the probability that every set F_l contains at least one facility in R can be lower–bounded as follows

$$\mathbb{P}\left[\bigcap_{l=1}^{m}(F_l \cap R \neq \emptyset)\right] = 1 - \mathbb{P}\left[\bigcup_{l=1}^{m}(F_l \cap R = \emptyset)\right]$$

$$\geq 1 - \sum_{l=1}^{m}(1-p)^{|F_l|}$$

$$\geq 1 - \sum_{l=1}^{m}(1-p)^{\lfloor \frac{n}{m} \rfloor}$$

$$\geq 1 - me^{-p\lfloor \frac{n}{m} \rfloor}.$$

The next lemma follows easily.

Lemma 1. *If $p \geq \ln(2m)\left(\lfloor \frac{n}{m} \rfloor\right)^{-1}$, then $\mathbb{P}[\bigcap_{l=1}^{m}(F_l \cap R \neq \emptyset)] \geq \frac{1}{2}$.*

Let us introduce the following probabilistic result.

Lemma 2 (Chernoff Bounds, Lemma 15 [16]). *Let $\{x_i\}_{i=1}^{k}$ be independently and identically distributed Bernoulli random variables with $\mathbb{P}[x_i = 1] = p$ and $\mathbb{P}[x_i = 0] = 1 - p$ for $i = 1, \ldots, k$. Then,*

$$\mathbb{P}\left[\sum_{i=1}^{k} x_i \geq \alpha\right] \leq k^{-\alpha}$$

holds for any $\alpha \geq 8kp$.

Let us now give a lemma which is the key to our result.

Lemma 3. *For any fixed set $X \subset F$, for $p = \log n / \sqrt{n}$, the probability that $f_1(X) \neq f_2(X)$ is at most $n^{-\omega(1)}$.*

Proof. Note that $p = \log n / \sqrt{n}$ implies that $\mu = \sqrt{n} \log n$. Let us first consider the case where $|X| \geq 9\mu$. If $|R| \leq 8\mu$, then $|X \backslash R| \geq \mu$, which implies $f_1(X) = f_2(X)$. Thus we only need to upper bound $\mathbb{P}\left[|R| \geq 8\mu\right]$. By Lemma 2, $\mathbb{P}\left[|R| \geq 8\mu\right] \leq n^{-8\mu} = n^{-\omega(1)}$. For the case where $|X| \leq 9\mu$,

$$\begin{aligned} \mathbb{P}\left[f_1(X) \neq f_2(X) \,\big|\, |X| \leq 9\mu\right] &\leq \mathbb{P}\left[|X \cap R| \geq 72 \log^2 n\right] \\ &= \mathbb{P}\left[|X \cap R| \geq 8(9\mu)p\right] \\ &\leq (9\mu)^{-72\mu p} \\ &= n^{-\omega(1)}, \end{aligned}$$

where the second inequality is also obtained by applying Lemma 2. $\quad\square$

We then obtain the following theorem.

Theorem 1. *Suppose that each set in $\{F_l\}_{l=1}^m$ has at least 1 facility in R. Let $p = \log n / \sqrt{n}$ and $m = \lfloor \sqrt{n} \frac{\log n}{\log 2n} \rfloor$, and let T' denote the set that contains exactly one facility from each set $R \cap F_l$ for $l = 1, \dots, m$. Then, $f_1(T') = \Omega(\sqrt{n})$ and $f_2(T') = O(\log^2 n)$.*

Proof. If $F_l \cap R \neq \emptyset$ for all $l = 1, \dots, m$, then the set T' so–defined contains m elements and $T' \subseteq R$. By definition, $f_1(T') = \min\{\sqrt{n} \cdot \log n, m\} = \Omega(\sqrt{n})$ and $f_2(T') = 72 \log^2 n$. $\quad\square$

By a similar argument as in Theorem 3.4 of [15], we can show that no polynomial-time algorithm can approximate the facility location problem with submodular opening costs better than $O(\sqrt{n} / \log^2 n)$ in the oracle model. We now give the main result.

Theorem 2. *In the oracle model, no algorithm that uses at most a polynomial number of oracle calls can approximate the facility location problem with submodular opening costs within a factor of $o(\sqrt{n} / \log^2 n)$, where n is the number of facilities.*

Proof. From Lemma 3, any algorithm that makes a polynomial number of calls to the value oracle has probability at most $n^{-\omega(1)}$ of distinguishing f_1 and f_2. By Theorem 1 and Lemma 1, with probability $\frac{1}{2}$ there is a set $T' \subseteq F$ such that for each client there is a co-located facility in T' and $|T'| = m$. The set T' so–defined satisfies that $f_1(T') = \Omega(\sqrt{n})$ and $f_2(T') = 72 \log^2 n$. We let the connection cost between any two clients be large enough, say $O(n)$, for both instances. Therefore, any solution that pays any connection cost would be suboptimal for both instances. Notice that if we choose T' to be the set of facilities to open for both instances, then the total cost for the instance with opening cost function f_2 is at most $72 \log^2 n$, however for the other instance, the total cost is at least $\Omega(\sqrt{n})$. To conclude, no algorithm can distinguish the two instances with a polynomial number of calls to the value oracle, and thus no algorithm can approximate the problem better than a factor $O(\sqrt{n}/\log^2 n)$ with a polynomial number of calls to the value oracle. □

2.2   JRS Problem

We now state the lower bounds on approximating the JRS problem.

Theorem 3. *In the oracle model, no algorithm that uses at most a polynomial number of oracle calls can approximate JRS problem within a factor of $o(\sqrt{n}/\log^2 n)$, where n is the size of the planning horizon.*

Sketch of proof. The construction of the proof is analogous to that of Theorem 2 for FLS problem. Let $n = |\mathcal{T}|$ and let m be the number of demands throughout the planning horizon. We consider the case where we have only one item type. Let the potential orders and the demands be as shown in Fig. 2. Specifically, we split the planning horizon into m segments, each of which has $\lfloor\frac{n}{m}\rfloor$ or $\lceil\frac{n}{m}\rceil$ time periods followed by a unit demand. Let F_i denote a set of $\lfloor\frac{n}{m}\rfloor$ or $\lceil\frac{n}{m}\rceil$ consecutive periods which form a segment in the planning horizon, for $i = 1, \ldots, m$. For periods within some F_i, there is no holding cost. For two periods that are not in the same F_i, let the holding cost be $\Omega(n)$. Then we construct two submodular functions f_1 and f_2 as in Sect. 3.1, except they are defined for subset $R \subseteq \mathcal{T}$ instead. Hence if we choose $m = \lceil\sqrt{n}\frac{\log n}{\log 2n}\rceil$ and $p = \frac{\log n}{\sqrt{n}}$ then in expectation the optimal costs of the two instances are $\Omega(n)$ and $O(\log^2 n)$. On the other hand, from Lemma 3, the probability for distinguishing the two instances is as low as $n^{-\omega(n)}$. We complete the proof by following the same argument as in the proof of Theorem 2.

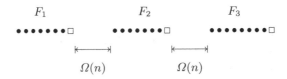

Fig. 2. A demonstration of potential orders and demands, where ● denotes potential order and □ denotes a demand. The $\Omega(n)$ in distance between a demand and the next potential order indicate an $\Omega(n)$ holding cost.

3 $O(\sqrt{n} \cdot \log n)$-Approximation Algorithms

In this section we present a unified approach to generate $O(\sqrt{n} \cdot \log n)$-approximation algorithms for both the FLS and JRS problems. Without loss of generality, we assume the instances have the property that $\min\{c_{ij} \neq 0 : i \in F, j \in D\} = 1$ and $\min\{h_{t,t'} \neq 0 : t, t' \in \mathcal{T}, t \leq t'\} = 1$.

Our algorithms utilize a recent result by Goemans et al. [17]. In their paper, the authors show that for any monotone submodular function $f : 2^F \to \mathbb{R}$ where $|F| = n$, there is a polynomial time algorithm that finds a set of numbers $\{p_i\}_{i \in F}$ such that the function $\hat{f}(S) = \sqrt{\sum_{i \in S} p_i}$ approximates f in the sense that $f(S) \leq \hat{f}(S) \leq O(\sqrt{n} \cdot \log n) \cdot f(S)$ for all sets $S \subseteq F$. To approximate FLS (or JRS) problem, we consider substituting the opening cost function (or the ordering cost function) f by its approximation $\hat{f}$. Formally, we consider the following problems (FLAS) and (JRAS):

(FLAS) the facility location problem with opening cost function $\hat{f}$ and connection costs $\{c_{ij}\}_{i \in F, j \in D}$;

(JRAS) the joint replenishment problem with ordering cost function $\hat{f}$ and holding costs $\{h_{t,t'}\}_{t,t' \in T}$.

Notice that $f(S) \leq \hat{f}(S) \leq O(\sqrt{n} \cdot \log n) f(S)$ implies that any constant factor approximation algorithm to (FLAS) (or (JRAS)) with opening cost function (or ordering cost function) $\hat{f}$ is an approximation algorithm to the original instance. Now we elaborate the approximation algorithm to FLS problem via approximating (FLAS), and then apply the same framework to JRS problem.

3.1   Approximation Algorithm for the FLS Problem

We first define a family of problems (FL_u) as the classic facility location problems with opening cost $\{p_i\}_{i \in F}$ and connection costs $\{u \cdot c_{i,j}\}_{i \in F, j \in D}$ for positive parameter u. Here $\{p_i\}_{i \in F}$ is the output from the aforementioned approximation algorithm in Goemans et al. [17]. Problem (FL_u) is the classic facility location problem for which constant–factor approximation algorithms are known.

Our algorithm calls a constant–factor approximation algorithm for the facility location problem to solve (FL_u) for a set of values of u, and chooses the one that minimizes $\hat{f}(S) + \sum_{i \in F, j \in D} c_{ij} x_{ij}$. Let (x^*, y^*) denote the optimal solution for (FLAS) and define $u^* = \frac{\sum_{i \in F} p_i y_i^*}{\sum_{i \in F} \sum_{j \in D} c_{ij} x_{ij}^*}$ if $\sum_{i \in F} \sum_{j \in D} c_{ij} x_{ij}^* \neq 0$. Let $(x^{(u)}, y^{(u)})$ denote any α-approximate solution for problem (FL_u). Then

$$\left(\sum_{i \in F} p_i y_i^{(u)} + u \cdot \sum_{i \in F} \sum_{j \in D} c_{ij} x_{ij}^{(u)} \right) \leq \alpha \left(\sum_{i \in F} p_i y_i^* + u \cdot \sum_{i \in F} \sum_{j \in D} c_{ij} x_{ij}^* \right). \tag{1}$$

The following Lemma states that if u^* lies between u and θu for some u and some real number $\theta > 1$, then the approximate solution to (FL_u) is also an approximate solution to (FLAS).

Lemma 4. *If $0 < u \leq u^* \leq \theta u$ for some positive number θ, then*

$$\sqrt{\sum_{i \in F} p_i y_i^{(u)}} + \sum_{i \in F} \sum_{j \in D} c_{ij} x_{ij}^{(u)} \leq \max\{\alpha(\theta+1), \sqrt{2\alpha}\} \left(\sqrt{\sum_{i \in F} p_i y_i^*} + \sum_{i \in F} \sum_{j \in D} c_{ij} x_{ij}^* \right).$$

The proof of Lemma 4 is omitted due to the page limit. In the next lemma, we prove that if $u^* \geq \max\{1, 2^{\lceil \log(\sqrt{\sum_{i \in F} p_i}) \rceil}\}$ or $\sum_{i \in F} \sum_{j \in D} c_{ij} x_{ij}^* = 0$, the approximate solution for $(\text{FL}_{2^{\lceil \log(\sqrt{\sum_{i \in F} p_i}) \rceil}})$ also approximates (FLAS), and if $u^* \leq 1$ then the approximate solution for (FL_1) approximates (FLAS). This result, together with Lemma 4, implies that we can enumerate the value of u from the set $\{1, 2, 4, \ldots, 2^{\lceil \log(\sqrt{\sum_{i \in F} p_i}) \rceil}\}$ and solve (FL_u) approximately so as to have an approximate solution to (FLAS).

Lemma 5. *Let* $u = \max\{1, 2^{\lceil \log(\sqrt{\sum_{i \in F} p_i}) \rceil}\}$. *The following statements are true:*

(i). If $u^* \geq u$, *then*

$$\sqrt{\sum_{i \in F} p_i y_i^{(u)}} + \sum_{i \in F} \sum_{j \in D} c_{ij} x_{ij}^{(u)} \leq (\sqrt{2\alpha} + 2\alpha) \left(\sqrt{\sum_{i \in F} p_i y_i^*} + \sum_{i \in F} \sum_{j \in D} c_{ij} x_{ij}^* \right);$$

(ii). if $\sum_{i \in F} \sum_{j \in D} c_{ij} x_{ij}^* = 0$, *then*

$$\sqrt{\sum_{i \in F} p_i y_i^{(u)}} + \sum_{i \in F} \sum_{j \in D} c_{ij} x_{ij}^{(u)} \leq (\sqrt{\alpha} + \alpha) \left(\sqrt{\sum_{i \in F} p_i y_i^*} + \sum_{i \in F} \sum_{j \in D} c_{ij} x_{ij}^* \right);$$

(iii). if $u^* \leq 1$, *then*

$$\sqrt{\sum_{i \in F} p_i y_i^{(1)}} + \sum_{i \in F} \sum_{j \in D} c_{ij} x_{ij}^{(1)} \leq (2\alpha + \sqrt{2\alpha}) \left(\sqrt{\sum_{i \in F} p_i y_i^*} + \sum_{i \in F} \sum_{j \in D} c_{ij} x_{ij}^* \right).$$

Proof. (i). If $\infty > u^* \geq u$, we have that

$$\sum_{i \in F} p_i y_i^{(u)} + u \cdot \sum_{i \in F} \sum_{j \in D} c_{ij} x_{ij}^{(u)} \leq \alpha \left(\sum_{i \in F} p_i y_i^* + u \cdot \sum_{i \in F} \sum_{j \in D} c_{ij} x_{ij}^* \right)$$

$$\leq \alpha \left(\sum_{i \in F} p_i y_i^* + u^* \cdot \sum_{i \in F} \sum_{j \in D} c_{ij} x_{ij}^* \right)$$

$$= 2\alpha \sum_{i \in F} p_i y_i^*.$$

Then it follows that

$$\sqrt{\sum_{i \in F} p_i y_i^{(u)}} + \sum_{i \in F} \sum_{j \in D} c_{ij} x_{ij}^{(u)} \leq \sqrt{2\alpha \sum_{i \in F} p_i y_i^* + \frac{2\alpha}{u} \cdot \sum_{i \in F} p_i y_i^*}$$

$$\leq \sqrt{2\alpha \sum_{i \in F} p_i y_i^* + 2\alpha \sqrt{\sum_{i \in F} p_i y_i^*}}$$

$$\leq (\sqrt{2\alpha} + 2\alpha) \sqrt{\sum_{i \in F} p_i y_i^*}$$

$$\leq (\sqrt{2\alpha} + 2\alpha) \left(\sqrt{\sum_{i \in F} p_i y_i^*} + \sum_{i \in F} \sum_{j \in D} c_{ij} x_{ij}^* \right).$$

(ii). If $\sum_{i \in F} \sum_{j \in D} c_{ij} x_{ij}^* = 0$, then

$$\sum_{i \in F} p_i y_i^{(u)} + u \cdot \sum_{i \in F} \sum_{j \in D} c_{ij} x_{ij}^{(u)} \le \alpha \left(\sum_{i \in F} p_i y_i^* + u \cdot \sum_{i \in F} \sum_{j \in D} c_{ij} x_{ij}^* \right) = \alpha \sum_{i \in F} p_i y_i^*.$$

By a similar argument as in the proof of (i), we have that

$$\sqrt{\sum_{i \in F} p_i y_i^{(u)}} + \sum_{i \in F} \sum_{j \in D} c_{ij} x_{ij}^{(u)} \le (\sqrt{\alpha} + \alpha) \left(\sqrt{\sum_{i \in F} p_i y_i^*} + \sum_{i \in F} \sum_{j \in D} c_{ij} x_{ij}^* \right).$$

(iii). If $u^* \le 1$, by the assumption that $\min\{c_{ij} \ne 0 : i \in F, j \in D\}$, we must have $\sum_{i \in F} \sum_{j \in D} c_{ij} x_{ij}^* \ge 1$. Then

$$\sum_{i \in F} p_i y_i^{(1)} + \sum_{i \in F} \sum_{j \in D} c_{ij} x_{ij}^{(1)} \le \alpha \left(\sum_{i \in F} p_i y_i^* + \sum_{i \in F} \sum_{j \in D} c_{ij} x_{ij}^* \right) = \alpha(1 + u^*) \cdot \sum_{i \in F} \sum_{j \in D} c_{ij} x_{ij}^*.$$

Therefore

$$\sqrt{\sum_{i \in F} p_i y_i^{(1)}} + \sum_{i \in F} \sum_{j \in D} c_{ij} x_{ij}^{(1)} \le \left(\sqrt{\alpha(1 + u^*)} + \alpha(1 + u^*) \right) \sum_{i \in F} \sum_{j \in D} c_{ij} x_{ij}^*$$

$$\le (2\alpha + \sqrt{2\alpha}) \left(\sqrt{\sum_{i \in F} p_i y_i^*} + \sum_{i \in F} \sum_{j \in D} c_{ij} x_{ij}^* \right).$$

$\square$

Now we give the approximation algorithm for the FLS problem in Algorithm 1.

Algorithm 1. Approximation algorithm for the FLS problem

Compute $\{p_i\}_{i \in F}$ by the procedure from [17]

for $u = 1, 2, 2^2, \ldots, 2^{\lceil \log(\sqrt{\sum_{i \in F} p_i}) \rceil}$ **do**

   Compute an α-approximation solution $(x^{(u)}, y^{(u)})$ for (FL_u)

end for

minimize $\sqrt{\sum_{i \in F} p_i y_i} + \sum_{i \in F} \sum_{j \in D} c_{ij} x_{ij}$ over $\{(x^{(u)}, y^{(u)})$: $u =$ $1, 2, 2^2 \ldots, 2^{\lceil \log(\sqrt{\sum_{i \in F} p_i}) \rceil}\}$ and output the minimizer

---

Theorem 4. *Given a polynomial time subroutine that computes an α-approximate for (FL_u) for some constant α, Algorithm 1 returns a constant factor approximate solution to (FLAS), and hence an $O(\sqrt{n} \cdot \log n)$-approximate solution for the facility location problem with submodular opening costs in polynomial time.*

Proof. We first prove that Algorithm 1 returns a constant factor approximate solution to (FLAS). If (x^*, y^*) satisfies $\sum_{i \in F} \sum_{j \in D} c_{ij} x_{ij}^* = 0$, then by Lemma 5, $(x^{(1)}, y^{(1)})$ is a constant factor approximate solution to (FLAS). Notice that the solution that Algorithm 1 returns is at least as good as $(x^{(1)}, y^{(1)})$, hence the output is a constant factor approximate solution to (FLAS). Similarly, if u^* is well–defined and $u^* \geq \max\{1, 2^{\lceil \log(\sqrt{\sum_{i \in F} p_i}) \rceil}\}$, or $u^* \leq 1$, then by Lemma 5, $(x^{(\lceil \log(\sqrt{\sum_{i \in F} p_i}) \rceil)}, y^{(\lceil \log(\sqrt{\sum_{i \in F} p_i}) \rceil)})$ is a constant factor approximate solution to (FLAS). If $u^* \in [1, 2^{\lceil \log(\sqrt{\sum_{i \in F} p_i}) \rceil}]$, then there must be some u that the algorithm visits such that $u \leq u^* \leq 2u$. By Lemma 4, $(x^{(u)}, y^{(u)})$ is a constant factor approximate solution to (FLAS). Combining the above cases, Algorithm 1 returns a constant factor approximate solution to (FLAS). Recall that any constant factor approximate solution to (FLAS) is an $O(\sqrt{n} \cdot \log n)$-approximate solution to the facility location problem with submodular opening costs, hence Algorithm 1 returns an $O(\sqrt{n} \cdot \log n)$-approximate solution to the facility location problem with submodular opening costs.

Regarding the running time, notice that the algorithm calls a polynomial time subroutine to approximate (FL_u) for $\log(\sqrt{\sum_{i \in F} p_i}) + 1$ times, which is still polynomial in the size of the input given that $\{p_i\}_{i \in F}$ are the output of an polynomial time algorithm given by [17]. Therefore, Algorithm 1 is a polynomial-time algorithm.   □

3.2   Approximation Algorithm to JRS Problem

As in the previous section, we consider a family of classic joint replenishment problems (JR_u), which has ordering cost $\sum_{t \in T} p_t y_t$ and holding cost $u \cdot \sum_{i \in S} \sum_{t \in T} \sum_{t':t \leq t'} h_{t,t'} x_{i,t,t'}$ where (x, y) is a feasible assignment to fulfill all demands $\{d_{i,t}\}_{i \in S, t \in T}$. Let (x^*, y^*) be the optimal solution to (JRAS) and define $u^* = \frac{\sum_{t \in T} p_t y_t^*}{u \cdot \sum_{i \in S} \sum_{t \in T} \sum_{t':t \leq t'} h_{t,t'} x_{i,t,t'}^*}$. Then we can prove results analogous to Lemmas 4 and 5. Therefore, by calling a constant–factor approximation algorithm, say the one in [18], to solve (JR_u) for $u = \{1, 2, 2^2, \ldots, 2^{\lceil \log(\sqrt{\sum_{t \in T} p_t}) \rceil}\}$, we can have an approximation algorithm analogous to Algorithm 1. To be specific, the algorithm first computes the set of values $\{p_t\}_{t \in T}$ by the procedure from [17], then finds approximate solution $(x^{(u)}, y^{(u)})$ for (JR_u) by the algorithm from [18] for $u = \{1, 2, 2^2, \ldots, 2^{\lceil \log(\sqrt{\sum_{t \in T} p_t}) \rceil}\}$. Finally, the algorithm outputs the minimizer of $\sqrt{\sum_{t \in T} p_t y_t} + \sum_{i \in S} \sum_{t \in T} \sum_{t':t \leq t'} u \cdot h_{t,t'} x_{i,t,t'}$ over the approximate solutions $(x^{(u)}, y^{(u)})$'s. By arguments similar to those of Theorem 4, one can see that the algorithm halts in polynomial time and it returns a $O(\sqrt{n} \cdot \log n)$-approximate solution to JRS problem.

4   Conclusion

In this paper, we prove that the facility location problem with submodular opening costs and the joint replenishment problem with submodular ordering costs cannot be approximated better than a factor of $O(\sqrt{n}/\log n)$ and give an approximation algorithms that match the lower bound with only polylogarithmic loss. In our algorithms, we first substitute the submodular opening cost function or the submodular ordering cost function by an approximation of it, then solve the approximated problem by doing a line search over the range of the ratio between the opening cost and the connection cost or between the ordering cost and holding cost. It is natural to ask the question that whether we can develop approximation algorithms for these problems by following any of the classic schema for the facility location problem, such as the primal-dual schema.

Acknowledgments. The work was partially supported by grant NSF–CCF1115256. The author also would like to thank David Williamson for his constant help and Chaoxu Tong for inspiring discussions.

References

1. Shmoys, D.B., Tardos, E., Aardal, K.: Approximation algorithms for facility location problems (extended abstract). In: Proceedings of the Twenty-Ninth Annual ACM Symposium on Theory of Computing. STOC'97, pp. 265–274. ACM, New York (1997)
2. Chudak, F.A., Shmoys, D.B.: Improved approximation algorithms for the uncapacitated facility location problem. SIAM J. Comput. **33**(1), 1–25 (2003)
3. Jain, K., Vazirani, V.V.: Approximation algorithms for metric facility location and k-median problems using the primal-dual schema and Lagrangian relaxation. J. ACM (JACM) **48**(2), 274–296 (2001)
4. Charikar, M., Guha, S.: Improved combinatorial algorithms for the facility location and k-median problems. In: 40th Annual IEEE Symposium on Foundations of Computer Science, 1999. FOCS'99, pp. 378–388 (1999)
5. Gupta, A., Tangwongsan, K.: Simpler analyses of local search algorithms for facility location (2008) arXiv preprint arXiv:0809.2554
6. Jain, K., Mahdian, M., Markakis, E., Saberi, A., Vazirani, V.V.: Greedy facility location algorithms analyzed using dual fitting with factor-revealing LP. J. ACM (JACM) **50**(6), 795–824 (2003)
7. Byrka, J., Aardal, K.: An optimal bifactor approximation algorithm for the metric uncapacitated facility location problem. SIAM J. Comput. **39**(6), 2212–2231 (2010)
8. Li, S.: A 1.488 approximation algorithm for the uncapacitated facility location problem. Inf. Comput. **222**, 45–58 (2013)
9. Guha, S., Khuller, S.: Greedy strikes back: improved facility location algorithms. J. Algorithms **31**(1), 228–248 (1999)
10. Svitkina, Z., Tardos, E.: Facility location with hierarchical facility costs. In: Proceedings of the Seventeenth Annual ACM-SIAM Symposium on Discrete Algorithm. SODA'06,pp. 153–161. ACM, New York (2006)
11. Hayrapetyan, A., Swamy, C., Tardos, E.: Network design for information networks. In: Proceedings of the Sixteenth Annual ACM-SIAM Symposium on Discrete Algorithms. SODA'05, pp. 933–942. Society for Industrial and Applied Mathematics, Philadelphia (2005)
12. Chudak, F.A., Nagano, K.: Efficient solutions to relaxations of combinatorial problems with submodular penalties via the Lovász extension and non-smooth convex optimization. In: Proceedings of the Eighteenth Annual ACM-SIAM Symposium on Discrete Algorithms. SODA'07, pp. 79–88. Society for Industrial and Applied Mathematics, Philadelphia (2007)

13. Du, D., Lu, R., Xu, D.: A primal-dual approximation algorithm for the facility location problem with submodular penalties. Algorithmica **63**(1–2), 191–200 (2012)
14. Cheung, M., Elmachtoub, A.N., Levi, R., Shmoys, D.B.: The submodular joint replenishment problem. Unpublished Manuscript (2013)
15. Svitkina, Z., Fleischer, L.: Submodular approximation: sampling-based algorithms and lower bounds. SIAM J. Comput. **40**(6), 1715–1737 (2011)
16. Iwata, S., Nagano, K.: Submodular function minimization under covering constraints. In: 50th Annual IEEE Symposium on Foundations of Computer Science, 2009. FOCS'09, pp. 671–680 (2009)
17. Goemans, M.X., Harvey, N.J., Iwata, S., Mirrokni, V.: Approximating submodular functions everywhere. In: Proceedings of the Twentieth Annual ACM-SIAM Symposium on Discrete Algorithms, pp. 535–544. Society for Industrial and Applied Mathematics, New York (2009)
18. Levi, R., Roundy, R., Shmoys, D., Sviridenko, M.: A constant approximation algorithm for the one-warehouse multiretailer problem. Manag. Sci. **54**(4), 763–776 (2008)

Online Multi-Coloring with Advice

Marie G. Christ, Lene M. Favrholdt, and Kim S. Larsen[✉]

University of Southern Denmark, Odense, Denmark
{christm,lenem,kslarsen}@imada.sdu.dk

Abstract. We consider the problem of online graph multi-coloring with advice. Multi-coloring is often used to model frequency allocation in cellular networks. We give several nearly tight upper and lower bounds for the most standard topologies of cellular networks, paths and hexagonal graphs. For the path, negative results trivially carry over to bipartite graphs, and our positive results are also valid for bipartite graphs. The advice given represents information that is likely to be available, studying for instance the data from earlier similar periods of time.

1 Introduction

We consider the problem of graph *multi-coloring*, where each node may receive multiple requests. Whenever a node is requested, a color must be assigned to the node, and this color must be different from any color previously assigned to that node or to any of its neighbors. The goal is to use as few colors as possible. In the online version, the requests arrive one by one, and each request must be colored without any information about possible future requests. The underlying graph is known to the online algorithm in advance.

The problem is motivated by *frequency allocation* in cellular networks. These networks are formed by a number of base transceiver stations, each of which covers what is referred to as a cell. Due to possible interference, neighboring cells cannot use the same frequencies. In this paper, we use classic terminology and refer to these cells as nodes in a graph where nodes are connected by an edge if they correspond to neighboring cells in the network. Frequencies can then be modeled as colors. Multiple requests for frequencies can occur in one cell and overall bandwidth is a critical resource.

Two basic models dominate in the discussion of cellular networks, the highway and the city model. The former is modeled by linear cellular networks, corresponding to paths, and the latter by hexagonal graphs. We consider the problem of multi-coloring such graphs.

If A is a multi-coloring algorithm, we let $A(I)$ denote the number of colors used by A on the input sequence I. When I is clear from the context, we simply write A instead of $A(I)$. The quality of an online algorithm is often given in terms of the competitive ratio [28,37]. An online multi-coloring algorithm is

---

Supported in part by the Danish Council for Independent Research and the Villum Foundation.

E. Bampis and O. Svensson (Eds.): WAOA 2014, LNCS 8952, pp. 83–94, 2015.
DOI: 10.1007/978-3-319-18263-6_8

c-competitive if there exists a constant α such that for all input sequences I, $A(I) \leq c\ \text{OPT}(I) + \alpha$. The (asymptotic) *competitive ratio* of A is the infimum over all such c. Results that can be established using $\alpha = 0$ are referred to as *strict* (or absolute). Often, it is a little unclear when one refers to an *optimal* online algorithm, whether this means that the solution produced is as good as the one produced offline or that no better online algorithm can exist. For that reason, we may use the term *strictly 1-competitive* to emphasize that an algorithm is as good as an optimal offline algorithm, and *optimal* to mean that no better online algorithm exists under the given conditions. Throughout, we let n denote the number of requests in a given input sequence.

For practical applications, the assumption that absolutely nothing is known about the future is often unrealistic. A way of relaxing this very strict and somewhat unnatural assumption is to give the algorithm some *advice*. A recent trend in the analysis of online algorithms has been to consider advice, formalized under the notion of *advice complexity*, starting in [20].

This realization that input is not arbitrary (uniformly random, for instance) is not new, and work focused on locality of reference in input data has tried to capture this. Early work includes access graph results, starting in [8], and with references to additional related work in [10], but also more distributional models, such as [1], have been developed. In [12] an entirely different concept of accommodating sequences was introduced and further developed in [9,13]. The idea is that for many problems requiring resources, there is a close connection between the resources available and the resources required for an optimal offline algorithm, as when capacity of transportation systems are matched with expected demand. This leans itself closely up against many of the results that we report here, where the advice needed to do better is often some information regarding the resources required by an optimal offline algorithm.

Thus, the results in this paper could have practical applications. The results establish which type of information is useful, how algorithms should be designed to exploit this information, and what the limits are for what can be obtained.

Returning to the *advice complexity modeling*, some problems need very little advice. On the other hand, complete information about the input or the desired output is a trivial upper bound on the amount of advice needed to be optimal. The first approach to formalizing the concept of advice measured the number of bits per request [20]. This model is well suited for some problems where information is tightly coupled with requests and the number of bits needed per request is constant. However, for most problems, we prefer the model where we simply measure the total advice needed throughout the execution of the algorithm. As also discussed in [5,25], this model avoids some modeling issues present in the "per request" modeling, and at the same time makes it possible to derive sublinear advice requirements. Thus, we use the advice model from [25], where the online algorithm has access to an infinite advice tape, written by an offline oracle with infinite computation power. In other words, the online algorithm can ask for the answer to any question and read the answer from the tape. Competitiveness is defined and measured as usual, and the advice complexity is simply the number of bits read from the tape, i.e., the maximum index of the bits read from the advice tape.

As the advice tape is infinite, we need to specify how many bits of advice the algorithm should read and if this knowledge is not implicitly available, it has to be given explicitly in the advice string. For instance, if we want OPT as advice (the number of colors an optimal offline algorithm uses on a given sequence, for instance), then we cannot merely read $\lceil\log(\text{OPT}+1)\rceil$ (all logs in this paper are base 2) bits, since this would require knowing something about the value of OPT. One can use a *self-delimiting encoding* as introduced in [22]. We use the variant from [11], defined as follows: The value of a non-negative integer X is encoded by a bit sequence, partitioned into three consecutive parts. The last part is X written in binary. The middle part gives the number of bits in the last part, written in binary. The first part gives the number of bits in the middle part, written in unary and terminated with a zero. These three parts require $\lceil\log(\lceil\log(X+1)\rceil+1)\rceil+1$, $\lceil\log(\lceil\log(X+1)\rceil+1)\rceil$, and $\lceil\log(X+1)\rceil$ bits, respectively, adding a lower-order term to the number of bits of information required by an algorithm. We define $enc(x)$ to be the minimum number of bits necessary to encode a number x, and note that the encoding above is a (good) upper bound on $enc(x)$.

We now discuss *previous work* on multi-coloring and then state our results. When working with online algorithms, decisions are generally irrevocable, i.e., once a color is assigned to a node, this decision is final. However, in some applications, local changes of colors may be allowed (reassignment of frequencies). This is called *recoloring*. An algorithm is *d-recoloring* if, in the process of treating a request, it may recolor up to a distance d away from the node of the request.

For multi-coloring a path, the algorithm 4-BUCKET is $\frac{4}{3}$-competitive [19], and this is optimal [15]. Even with 0-recoloring allowed (that is, colors at the requested node may be changed), 4-BUCKET is optimal [16]. Furthermore, if 1-recoloring is allowed, the algorithm GREEDYOPT is strictly 1-competitive [16].

For multi-coloring bipartite graphs, the optimal asymptotic competitive ratio lies between $\frac{10}{7}\approx1.428$ and $\frac{18-\sqrt{5}}{11}\approx1.433$ [18].

In [14], it was shown that, for hexagonal graphs, no online algorithm can be better than $\frac{3}{2}$-competitive or have a better strict competitive ratio than 2. They also gave an algorithm, HYBRID, with an asymptotic competitive ratio of approximately 1.9 on hexagonal graphs. On k-colorable graphs, it is strictly $\frac{k+1}{2}$-competitive, and hence, it has an optimal strict competitive ratio on hexagonal graphs. Recoloring was studied in [27]: No d-recoloring algorithm for hexagonal graphs has an asymptotic competitive ratio better than $1+\frac{1}{4(d+1)}$. For $d=0$, the lower bound was improved to $\frac{9}{7}$. In [38], a $\frac{4}{3}$-competitive 2-recoloring algorithm is given. The best known 1-recoloring algorithm for hexagonal graphs is $\frac{33}{24}$-competitive [39]. For the offline problem of multi-coloring hexagonal graphs, no polynomial time algorithm can obtain an absolute approximation ratio better than $\frac{4}{3}$ [32,34,35], unless P = NP.

Many other problems have been considered in the advice model, see e.g., [2,4–7,20,21,23,29,30]; also variants of graph coloring different from ours [3,24, 31,36].

Table 1. Overview of our results. Recall that n denotes the number of requests in the input sequence. We mark the ratios that are strict by "s" and the ones that are asymptotic by "a". Note that a strict lower bound can be larger than an asymptotic upper bound. For each bound, we indicate the number of the theorem proving the result. For readability, many of the bounds stated are weaker than those proven in the paper. Moreover, the upper bounds for the path hold for any bipartite graph. The result of Theorem 3 in the third row of the table is valid only for *neighborhood-based* algorithms, as defined just before Theorem 3 in Sect. 2.

	Ratio	Lower bound	Type	Result	Upper bound	Type	Result
Path	1	$\log n - 2$	s	Theorem 1	$\log n + O(\log\log n)$	s	Theorem 4
	$1 + \frac{1}{2^b}$	$b - 2$	a	Theorem 2	$b + 1 + O(\log\log n)$	s	Theorem 5
	$< \frac{4}{3}$	$\omega(1)$	a	Theorem 3			
Hex	1				$(n+1)\lceil \log n \rceil$	s	Theorem 8
	$< \frac{5}{4}$	$\Omega(n)$	a	Theorem 7			
	$\frac{4}{3}$				$n + 2|V|$	a	Theorem 10
	$\frac{3}{2}$	$\lfloor \frac{n-1}{3} \rfloor$	s	Theorem 6	$\log n + O(\log\log n)$	a	Theorem 9

An overview of *our results* is given in Table 1. For the path, these results are nearly tight, even with upper bounds that also apply to bipartite graphs. For hexagonal graphs, note that with a linear number of advice bits, it is possible to be $\frac{4}{3}$-competitive, and the lower bound for being better than $\frac{5}{4}$-competitive is close to linear. The advice given to the algorithms is essentially (an approximation of) OPT or the maximum number of requests given to any clique in the graph. For the underlying problem of frequency allocation, guessing these values based on previous data may not be unrealistic.

Due to space restrictions, some proofs have been removed or shortened. These can be found in the full version [17].

2   The Path

As explained earlier, we establish all lower bounds for paths, and since a path is bipartite, all these negative results carry over to bipartite graphs. Similarly, all our (constructive) upper bounds are given for bipartite graphs and therefore also apply to paths. We start with three lower bound results.

Theorem 1. *Any strictly 1-competitive online algorithm for multi-coloring paths of at least 10 nodes has advice complexity at least* $\lceil \log(\lfloor \frac{n}{4} \rfloor + 1) \rceil$.

Proof. We let $m = \lfloor \frac{n}{4} \rfloor$ and define a set S of $m + 1$ sequences, all having the same prefix of length $2m$. The set S will have the following property: for no two sequences in S can their prefixes be colored in the same way while ending up using the optimal number of colors on the complete sequence. Starting from one end of the path, we denote the nodes $v_1, v_2, \ldots$.

We define the set S to consist of the sequences $I_0, I_1, \ldots, I_m$, where I_i is defined in the following way. First m requests are given to each of the nodes v_1

and v_4. Then i requests to each of v_2 and v_3. To give all sequences the same length, the sequence ends with $\lceil n - 2m - 2i \rceil$ requests distributed as evenly as possible among v_6, v_8, and v_{10}. Since $\lceil \lceil n - 2m - 2i \rceil / 3 \rceil \leq m$, the optimal number of colors will not be influenced by this part of the sequence.

Note that $\text{OPT}(I_i) = m + i$. In order not to use more than $\text{OPT}(I_i)$ colors for I_i, exactly i of the colors assigned to v_4 have to be different from the colors assigned to v_1. The prefixes of length $2m$ in S are identical, so all information to distinguish between the different sequences must be given as advice. The cardinality of S is $m+1$. To specify one out of $m+1$ possible actions, $\lceil \log(m + 1) \rceil$ bits are necessary. □

For algorithms that are $\frac{9}{8}$-competitive or better, we give the following lower bound.

Theorem 2. *Consider multi-coloring paths of at least 10 nodes. For any $b \geq 3$ and any $(1 + \frac{1}{2^b})$-competitive algorithm, A, there exists an $N \in \mathbb{N}$ such that A has advice complexity at least $b - 2$ on sequences of length at least N.*

Proof. For any $(1 + \frac{1}{2^b})$-competitive algorithm, A, there exists an $\alpha \geq 1$ such that $A(I) \leq (1+\frac{1}{2^b})\text{OPT}(I)+\alpha$, for any input sequence I. We consider sequences of length $n \geq 2^{2b+2}\alpha + 3$.

Let $m = \lfloor \frac{n}{4} \rfloor$ and consider the same set of sequences as in the proof of Theorem 1. Recall that $\text{OPT}(I_i) = m + i$. For the sequence I_i, let x_i denote the number of colors that A uses on v_4 but not on v_1. Then, A uses $m + x_i$ colors in total for v_1 and v_4. On v_3, it can use at most x_i of the colors used at v_1, so the total number of colors used at v_1, v_2, and v_3 is at least $m + 2i - x_i$. Thus, $A(I_i) \geq \max\{m + x_i, m + 2i - x_i\}$.

We will prove that there are $p \geq 2^{b-2}$ sequences $I_{i_1}, I_{i_2}, \ldots, I_{i_p}$ such that, for any pair $i_j \neq i_k$, we have $x_{i_j} \neq x_{j_k}$, or otherwise A would not be $(1 + \frac{1}{2^b})$-competitive. This will immediately imply that A must use at least $b-2$ advice bits.

Let $\varepsilon = \frac{1}{2^b} + \frac{1}{2^{2b}}$. From $A(I_i) \leq (1+\frac{1}{2^b})\text{OPT}(I_i)+\alpha$ and $m \geq 2^{2b}\alpha$, we obtain the inequalities

$$m + x_i \leq (1 + \varepsilon)(m + i) \qquad \text{and} \qquad m + 2i - x_i \leq (1 + \varepsilon)(m + i)$$

which reduce to

$$x_i \leq \varepsilon m + (1 + \varepsilon)i \tag{1}$$

and

$$i \leq \frac{x_i + \varepsilon m}{1 - \varepsilon} \tag{2}$$

Hence, by (1), $x_0 \leq \varepsilon m$. Thus, by (2), we can have $x_i = x_0$ only if $i \leq \frac{2\varepsilon m}{1-\varepsilon}$. Therefore, we let $i_1 = 0$ and $i_2 = \lfloor \frac{2\varepsilon m}{1-\varepsilon} + 1 \rfloor$. In general, we ensure $x_{i_j} \neq x_{i_{j+1}}$ by letting $i_{j+1} = \lfloor \frac{x_{i_j}+\varepsilon m}{1-\varepsilon} + 1 \rfloor$. Thus,

$$i_{j+1} \leq \frac{x_{i_j} + \varepsilon m}{1 - \varepsilon} + 1 \leq \frac{1 + \varepsilon}{1 - \varepsilon} \cdot i_j + \frac{2\varepsilon m}{1 - \varepsilon} + 1,$$

where the second inequality follows from (1). Solving this recurrence relation, we get

$$i_{j+1} \leq \left(\frac{1+\varepsilon}{1-\varepsilon}\right)^j i_1 + \sum_{k=0}^{j-1} \left(\frac{1+\varepsilon}{1-\varepsilon}\right)^k \left(\frac{2\varepsilon m}{1-\varepsilon}+1\right) = \frac{\left(\frac{1+\varepsilon}{1-\varepsilon}\right)^j - 1}{2\varepsilon} (2\varepsilon m + 1 - \varepsilon)$$

We let p equal the largest j for which $i_j \leq m$. Through arithmetic manipulations using the various bounds established above, one can show that $p \geq 2^{b-2}$.   □

For the following theorem, we define the class of *neighborhood-based* algorithms: A multi-coloring algorithm, A, is called neighborhood-based, if there exists a constant d such that, when assigning a color to a request to a node v, A bases its decision only on requests to nodes a distance of at most d away from v. Note that, in particular, a neighborhood-based algorithm cannot base its decision on the current value of OPT.

Using the family of request sequences from the proofs of Theorems 1 and 2, it is fairly easy to establish a lower bound of $\omega(1)$ on the advice complexity for neighborhood-based algorithms that are better than $\frac{4}{3}$-competitive:

Theorem 3. *No neighborhood-based online algorithm for multi-coloring paths with advice complexity $O(1)$ can be better than $\frac{4}{3}$-competitive.*

We now turn to upper bounds. For multi-coloring a path, there is a strictly 1-competitive 1-recoloring algorithm, GREEDYOPT [16]. GREEDYOPT divides the nodes into two sets, *upper* and *lower*, such that every second node belongs to *upper* and the remaining nodes belong to *lower*. The following invariant is maintained: After each request, each node in *lower* uses consecutive colors starting with color 1 and each node in *upper* uses consecutive colors ending with a color no larger than the optimal number of colors for the request sequence seen so far.

The algorithm for paths from [16] is easily generalized to work on bipartite graphs, letting the nodes of one partition, L, belong to *lower* and the nodes of the other partition, U, belong to *upper*. Recoloring is only needed if the number of colors used by an optimal offline algorithm is not known. Hence, using $enc(\text{OPT})$ advice bits, an online algorithm can be strictly 1-competitive, even if recoloring is not allowed. For the resulting algorithm, GREEDYOPTADVICE, we prove the following.

Theorem 4. *Algorithm GREEDYOPTADVICE is correct, strictly 1-competitive, and has advice complexity $enc(\text{OPT})$.*

We now turn to nonoptimal variants of GREEDYOPTADVICE using fewer than $enc(\text{OPT})$ advice bits. We show how to obtain a particular competitive ratio of $1 + \frac{1}{2^b}$, using $b + 1 + O(\log \log \text{OPT})$ bits of advice. Thus, essentially, we are approaching optimality exponentially fast in the number of bits of advice.

Theorem 5. *For any integer $b \geq 1$, there exists a strictly $(1 + \frac{1}{2^{b-1}})$-competitive online algorithm for multi-coloring bipartite graphs with advice complexity $b + enc(a)$, where $a + b$ is the total number of bits in the value OPT.*

Proof. As advice, the algorithm asks for the b high order bits of the value OPT, as well as the number $a = \lceil \log(\text{OPT}+1) \rceil - b$ of low order bits, but not the value of these bits. The algorithm knows b and can just read the first b bits, while a needs to be encoded. Thus, $b + enc(a)$ bits are sufficient to encode the advice.

If OPT contains fewer than b bits, this is detected by a being zero. In this case, some of the b bits may be leading zeros. By Theorem 4, we can then be strictly 1-competitive. Now, assume this is not the case, and let $\text{OPT}_b = \lfloor \frac{\text{OPT}}{2^a} \rfloor$ denote the value represented by the b high order bits. The algorithm computes $m = 2^a \text{OPT}_b + 2^a - 1$ and runs GREEDYOPTADVICE with this m. Since OPT $\leq$ $m \leq \text{OPT} + 2^a - 1$, the algorithm is correct and uses at most $\text{OPT} + 2^a - 1$ colors.

For any number $x \geq 1$, consisting of c bits, with the most significant bit being one, $2^c \leq 2x$. Thus, $2^{b+a} \leq 2\text{OPT}$, so $2^a \leq \frac{2\text{OPT}}{2^b}$. This means that the number of colors used by GREEDYOPTADVICE is less than $\text{OPT} + \frac{2\text{OPT}}{2^b} = (1 + \frac{1}{2^{b-1}})\text{OPT}$, so the algorithm is strictly $(1 + \frac{1}{2^{b-1}})$-competitive. □

Corollary 1. *For any $\varepsilon > 0$, there exists a strictly $(1+\varepsilon)$-competitive deterministic online algorithm for multi-coloring bipartite graphs with advice complexity $O(\log\log \text{OPT})$.*

Proof. Except for the term b, the advice stated in Theorem 5 is $O(\log\log \text{OPT})$ and OPT $\leq n$. Thus, we just need to bound b. For a given ε, choose b large enough such that $\frac{1}{2^{b-1}} \leq \varepsilon$. Using this value for b in Theorem 5, we obtain an algorithm with a strict competitive ratio of at most $1 + \frac{1}{2^{b-1}} \leq 1 + \varepsilon$. Since, for any given ε, b is a constant, the total amount of advice is $O(\log\log \text{OPT})$. □

The Multi-Coloring problem is also considered in the context of request cancellations, i.e., a color already given to a node disappears again. We remark that just by allowing 0-recoloring, where requests at the node where the cancellation takes place may be recolored, we can extend the algorithm GREEDYOPTADVICE, using the same advice, to a strictly 1-competitive algorithm.

3   Hexagonal Graphs

A hexagonal graph is a graph that can be obtained by placing (at most) one node in each cell of a hexagonal grid (such as the one sketched in Fig. 1) and adding an edge between any pair of nodes placed in neighboring cells. Note that any hexagonal graph can be 3-colored. This is easily seen, since it is possible to use the three colors cyclically on the cells of each row of the underlying hexagonal grid, such that no two neighboring cells receive the same color.

As in the previous section, we first focus on lower bound results.

Theorem 6. *Any online algorithm for multi-coloring hexagonal graphs with a strict competitive ratio strictly smaller than $\frac{3}{2}$ has advice complexity at least $\lfloor \frac{n-1}{3} \rfloor$.*

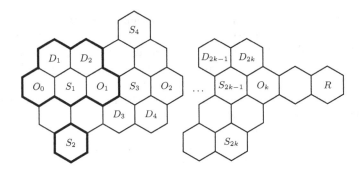

Fig. 1. Hexagonal lower bound construction.

Proof. First, we explain a small part of the construction that we will use in many copies. We consider two sequences with the same prefix of length 2. Both sequences can be colored with two colors, but this requires coloring the two prefixes of length two differently. Consider the left-most part of Fig. 1 (surrounded by thick lines) consisting of the "double" nodes D_1 and D_2, the "outer" nodes O_0 and O_1 and the "single" nodes S_1, and S_2. These nodes form the same type of configuration as the nodes D_3, D_4, O_1, O_2, S_3, and S_4.

First the nodes O_0 and O_1 get one request each. Then, either D_1 and D_2 or S_1 and S_2 receive one request each. The node S_2 is used to get up to the same sequence length in all cases. In order not to use more than two colors, the outer nodes have to use different colors if we later give requests to the two D-nodes. Similarly, the O-nodes should have the same color if we later give a request to the S-node in between them. Since the prefix of length two is $\langle O_0, O_1 \rangle$ for both sequences, all information for an algorithm to distinguish between the two sequences must be given as advice.

We can repeat this graph pattern $\lfloor \frac{n-1}{3} \rfloor$ times, as illustrated in Fig. 1 with $k = \lfloor \frac{n-1}{3} \rfloor$, giving the requests to all O-nodes first. This results in a sequence set of size $2^{\lfloor \frac{n-1}{3} \rfloor}$, implying the result. $\qquad \square$

Theorem 7. *Any online algorithm for multi-coloring hexagonal graphs with competitive ratio strictly smaller than $\frac{5}{4}$ has advice complexity $\Omega(n)$.*

Proof. We use the basic construction from Theorem 6. Assume p requests are given to one of the components like this:

First, we give $\frac{p}{4}$ requests to each of O_0 and O_1. Let q, $0 \le q \le \frac{p}{4}$, denote the number of colors used at both nodes. Then following up by giving $\frac{p}{4}$ requests to each S-node results in a minimum of $\frac{3p}{4} - q$ colors used, while giving the requests to the D-nodes instead results in a minimum of $\frac{p}{2} + q$ colors.

Note that $\mathrm{OPT} = \frac{p}{2}$, independent of in which of the two ways the sequence is continued. Thus, for any $\varepsilon > 0$, any $(\frac{5}{4} - \varepsilon)$-competitive algorithm must choose q such that, for some constant α, $\frac{3p}{4} - q \le (\frac{5}{4} - \varepsilon) \frac{p}{2} + \alpha$ and $\frac{p}{2} + q \le (\frac{5}{4} - \varepsilon) \frac{p}{2} + \alpha$. Adding these two inequalities, we obtain $\frac{5p}{4} \le (\frac{5}{4} - \varepsilon)p + 2\alpha$ which is equivalent to $\varepsilon p \le 2\alpha$. Thus, if p is non-constant, no $(\frac{5}{4} - \varepsilon)$-competitive algorithm can use the same value of q for both sequences.

Now assume for the sake of contradiction that for some advice of $g(n) \in o(n)$ bits, we can obtain a ratio of $\frac{5}{4} - \varepsilon$. Let $f(n) = \frac{1}{2}\frac{n}{g(n)}$. Since $g(n) \in o(n)$, $f(n) \in \omega(1)$. The idea is now to repeat the construction as in the proof of Theorem 6 and give $f(n)$ requests to each construction ($f(n)$ has the role of p in the above). Since a pair of neighboring constructions share $f(n)/4$ requests, this results in $\frac{n - f(n)/4}{3f(n)/4} = \frac{4n - f(n)}{3f(n)} \geq \frac{n}{f(n)}$ constructions. We assume without loss of generality that all our divisions result in integers.

In order to be $(\frac{5}{4} - \varepsilon)$-competitive, an online algorithm must, for each two neighboring O-nodes, choose between at least two different values of q. These are independent decisions, and the ratio only ends up strictly better than $\frac{5}{4}$ if the algorithm decides correctly in every subconstruction. Thus, it needs at least $\frac{n}{f(n)}$ bits of advice. However, $\frac{n}{f(n)} = \frac{n}{\frac{1}{2}\frac{n}{g(n)}} = 2g(n) > g(n)$, a contradiction.   □

For upper bounds, we first have the following trivial upper bound on the advice necessary to be optimal, independent of the graph topology:

Theorem 8. *There is a strictly 1-competitive online multi-coloring algorithm with advice complexity $(n + 1) \lceil \log \text{OPT} \rceil$.*

In the following, we will show how two known approximation algorithms can be converted to online algorithms with advice. In the description of the algorithms, we let the *weight* of a clique denote the total number of requests to the nodes of the clique. Note that the only maximal cliques in a hexagonal graph are isolated nodes, edges, or triangles. We let ω denote the maximum weight of any clique in the graph.[1]

A $\frac{3}{2}$-competitive algorithm called the Fixed Preference Allocation algorithm, FPA, was proposed in [26]. In [33], the strategy was simplified and it was noted that the algorithm can be converted to a 1-recoloring online algorithm. We describe the simplified offline algorithm below.

The algorithm uses three color classes, R, G, and B. The color classes represent a partitioning of the nodes in the graph so that no two neighbors are in the same partition. Each of the three color classes has its own set of $\lceil \frac{\omega}{2} \rceil$ colors, and each node in a given color class uses the colors of its color class, starting with the smallest. This set of colors is also referred to as the node's *private* colors. If more than $\lceil \frac{\omega}{2} \rceil$ requests are given to a node, then it borrows colors from the private colors of one of its neighbors, taking the highest available color. R nodes can borrow colors from G nodes, G from B, and B from R. Since $\lceil \frac{\omega}{2} \rceil \leq \lceil \frac{\text{OPT}}{2} \rceil$, we can give $\lceil \frac{\omega}{2} \rceil$ as advice and obtain the following:

Theorem 9. *There is a $\frac{3}{2}$-competitive online algorithm for multi-coloring hexagonal graphs with advice complexity $\text{enc}(\lceil \frac{\text{OPT}}{2} \rceil)$.*

In [32], an algorithm with an approximation ratio of $\frac{4}{3}$ was introduced. This algorithm uses color classes in the same way as FPA, except that the private

---

[1] The Greek letter ω is traditionally used here, so we will also do that. Since there is no argument, this should not give rise to confusion with the $\omega(f)$, stemming from asymptotic notation.

color sets contain only $\lfloor \frac{\omega+1}{3} \rfloor$ colors each. In describing the algorithm, we use the following notation. For any node v, we let n_v denote the number of requests to v. Furthermore, b_v denotes the maximum number of colors that v can borrow, i.e., $b_v = \max\{0, \lfloor \frac{\omega+1}{3} \rfloor - n'_v\}$, where n'_v is the maximum number of requests to any of the neighboring nodes in the color class that v can borrow from.

The algorithm can be seen as working in up to three phases: In the first phase, the algorithm colors $\min\{n_v, \lfloor \frac{\omega+1}{3} \rfloor\}$ requests to each node, v, using the node's private colors. In the second phase, each node v with more than $\lfloor \frac{\omega+1}{3} \rfloor$ requests borrows $\min\{n_v - \lfloor \frac{\omega+1}{3} \rfloor, b_v\}$ colors. Let G_2 be the graph induced by nodes that still have uncolored requests after Phase 2. In [32] it is proven that G_2 is bipartite and that any pair of neighbors in G_2 has a total of at most $\omega - 2\lfloor \frac{\omega+1}{3} \rfloor \leq \lfloor \frac{\omega+1}{3} \rfloor + 1$ uncolored requests after Phase 2. Thus, in the third phase, the remaining requests can be colored with GREEDYOPT (see the path section) using $\lfloor \frac{\omega+1}{3} \rfloor + 1$ additional colors.

We now show how an online algorithm, given the right advice, can behave like the offline $\frac{4}{3}$-approximation algorithm. Note that the three phases of the offline $\frac{4}{3}$-approximation algorithm are characterized by the coloring strategy (using the node's own private colors, borrowing private colors from neighbors, or coloring a bipartite graph). However, when requests arrive online, the nodes may not go from one phase to the next simultaneously.

Theorem 10. *There is a $\frac{4}{3}$-competitive online algorithm for multi-coloring hexagonal graphs with advice complexity at most $n + 2|V|$.*

Proof. Initially, each node is in Phase 1. On a request, the algorithm reads an advice bit and if it is zero, the next color from its private colors is used. If, instead, a one is read, this is treated as a stop bit for Phase 1, and this particular node enters Phase 2.

The algorithm starts with empty private color sets, and adds one color to each set whenever necessary, i.e., whenever a Phase 1 node that has already used all its private colors receives an additional request (this includes the first request to the node). As soon as a node leaves Phase 1, the algorithm knows that this node received $\lfloor \frac{\omega+1}{3} \rfloor$ requests, which is then the final size of each private color set. Knowing the size of the private color sets, the algorithm can calculate the maximum color for the complete coloring of the graph as $m = 4\lfloor \frac{\omega+1}{3} \rfloor + 1$.

In Phase 2, every zero indicates that the algorithm should borrow a color. When another stop bit is received (which could be after no zeros at all if the borrowing phase is empty), it moves to Phase 3. In Phase 3, it reads one bit to decide which partition, upper or lower, of the bipartite graph it is in, and does not need more information after that, since it simply uses the colors $3\lfloor \frac{\omega+1}{3} \rfloor + 1, \ldots, m$, either top-down or bottom-up.

If we allow the algorithm one bit per request, it needs at most two more bits per node, since the stop bits are the only bits that do not immediately tell the algorithm which action to take. Thus, $n + 2|V|$ bits of advice suffice.   □

Acknowledgments. We would like to thank anonymous referees for comments, improving the presentation of our results.

References

1. Albers, S., Favrholdt, L.M., Giel, O.: On paging with locality of reference. J. Comput. Syst. Sci. **70**(2), 145–175 (2005)
2. Barhum, K., Böckenhauer, H.-J., Forišek, M., Gebauer, H., Hromkovič, J., Krug, S., Smula, J., Steffen, B.: On the power of advice and randomization for the disjoint path allocation problem. In: Geffert, V., Preneel, B., Rovan, B., Štuller, J., Tjoa, A.M. (eds.) SOFSEM 2014. LNCS, vol. 8327, pp. 89–101. Springer, Heidelberg (2014)
3. Bianchi, M.P., Böckenhauer, H.-J., Hromkovič, J., Keller, L.: Online coloring of bipartite graphs with and without advice. In: Gudmundsson, J., Mestre, J., Viglas, T. (eds.) COCOON 2012. LNCS, vol. 7434, pp. 519–530. Springer, Heidelberg (2012)
4. Böckenhauer, H.-J., Komm, D., Královič, R., Královič, R.: On the advice complexity of the k-server problem. ICALP. LNCS **6755**, 207–218 (2011)
5. Böckenhauer, H.-J., Komm, D., Královič, R., Královič, R., Mömke, T.: On the advice complexity of online problems. In: Dong, Y., Du, D.-Z., Ibarra, O. (eds.) ISAAC 2009. LNCS, vol. 5878, pp. 331–340. Springer, Heidelberg (2009)
6. Böckenhauer, H.-J., Komm, D., Královič, R., Rossmanith, P.: On the advice complexity of the Knapsack problem. In: Fernández-Baca, D. (ed.) LATIN 2012. LNCS, vol. 7256, pp. 61–72. Springer, Heidelberg (2012)
7. Borodin, A., El-Yaniv, R.: Online Computation and Competitive Analysis. Cambridge University Press, Cambridge (1998)
8. Borodin, A., Irani, S., Raghavan, P., Schieber, B.: Competitive paging with locality of reference. J. Comput. Syst. Sci. **50**(2), 244–258 (1995)
9. Boyar, J., Favrholdt, L.M., Larsen, K.S., Nielsen, M.N.: Extending the accommodating function. Acta Informatica **40**(1), 3–35 (2003)
10. Boyar, J., Gupta, S., Larsen, K.S.: Access graphs results for LRU versus FIFO under relative worst order analysis. In: Fomin, F.V., Kaski, P. (eds.) SWAT 2012. LNCS, vol. 7357, pp. 328–339. Springer, Heidelberg (2012)
11. Boyar, J., Kamali, S., Larsen, K.S., López-Ortiz, A.: Online bin packing with advice. In Thirty-First International Symposium on Theoretical Aspects of Computer Science (STACS), vol. 25 of LIPIcs, pp. 174–186. Schloss Dagstuhl - Leibniz-Zentrum für Informatik GmbH, German (2014)
12. Boyar, J., Larsen, K.S.: The seat reservation problem. Algorithmica **25**(4), 403–417 (1999)
13. Boyar, J., Larsen, K.S., Nielsen, M.N.: The accommodating function: a generalization of the competitive ratio. SIAM J. Comput. **31**(1), 233–258 (2001)
14. Chan, J.W.-T., Chin, F.Y.L., Ye, D., Zhang, Y.: Absolute and asymptotic bounds for online frequency allocation in cellular networks. Algorithmica **58**(2), 498–515 (2010)
15. Chan, J.W.-T., Chin, F.Y.L., Ye, D., Zhang, Y., Zhu, H.: Frequency allocation problems for linear cellular networks. In: Asano, T. (ed.) ISAAC 2006. LNCS, vol. 4288, pp. 61–70. Springer, Heidelberg (2006)
16. Christ, M.G., Favrholdt, L.M., Larsen, K.S.: Online multi-coloring on the path revisited. Acta Informatica **50**(5–6), 343–357 (2013)
17. Christ, M.G., Favrholdt, L.M., Larsen, K.S.: Online multi-coloring with advice. (2014) arXiv:1409.1722 [cs.DS]
18. Chrobak, M., Jez, L., Sgall, J.: Better bounds for incremental frequency allocation in bipartite graphs. Theor. Comput. Sci. **514**, 75–83 (2013)

19. Chrobak, M., Sgall, J.: Three results on frequency assignment in linear cellular networks. Theor. Comput. Sci. **411**(1), 131–137 (2010)
20. Dobrev, S., Královič, R., Pardubská, D.: Measuring the problem-relevant information in input. RAIRO Theor. Inf. Appl. **43**(3), 585–613 (2009)
21. Dorrigiv, R., He, M., Zeh, N.: On the advice complexity of buffer management. In: Chao, K.-M., Hsu, T., Lee, D.-T. (eds.) ISAAC 2012. LNCS, vol. 7676, pp. 136–145. Springer, Heidelberg (2012)
22. Elias, P.: Universal codeword sets and representations of the integers. IEEE Trans. Inf. Theory **21**(2), 194–203 (1975)
23. Emek, Y., Fraigniaud, P., Korman, A., Rosén, A.: Online computation with advice. Theor. Comput. Sci. **412**(24), 2642–2656 (2011)
24. Forišek, M., Keller, L., Steinová, M.: Advice complexity of online coloring for paths. In: Dediu, A.-H., Martín-Vide, C. (eds.) LATA 2012. LNCS, vol. 7183, pp. 228–239. Springer, Heidelberg (2012)
25. Hromkovič, J., Královič, R., Královič, R.: Information complexity of online problems. In: Hliněný, P., Kučera, A. (eds.) MFCS 2010. LNCS, vol. 6281, pp. 24–36. Springer, Heidelberg (2010)
26. Janssen, J., Kilakos, K., Marcotte, O.: Fixed preference channel assignment for cellular telephone systems. IEEE Trans. Veh. Technol. **48**(2), 533–541 (1999)
27. Janssen, J., Krizanc, D., Narayanan, L., Shende, S.M.: Distributed online frequency assignment in cellular networks. J. Algorithms **36**(2), 119–151 (2000)
28. Karlin, A.R., Manasse, M.S., Rudolph, L., Sleator, D.D.: Competitive snoopy caching. Algorithmica **3**, 79–119 (1988)
29. Komm, D., Královič, R.: Advice complexity and barely random algorithms. RAIRO Theor. Inf. Appl **45**(2), 249–267 (2011)
30. Komm, D., Královič, R., Mömke, T.: On the advice complexity of the set cover problem. In: Hirsch, E.A., Karhumäki, J., Lepistö, A., Prilutskii, M. (eds.) CSR 2012. LNCS, vol. 7353, pp. 241–252. Springer, Heidelberg (2012)
31. Bianchi, M.P., Böckenhauer, H.-J., Hromkovič, J., Krug, S., Steffen, B.: On the advice complexity of the online $L(2,1)$-coloring problem on paths and cycles. In: Du, D.-Z., Zhang, G. (eds.) COCOON 2013. LNCS, vol. 7936, pp. 53–64. Springer, Heidelberg (2013)
32. McDiarmid, C., Reed, B.A.: Channel assignment and weighted coloring. Networks **36**(2), 114–117 (2000)
33. Narayanan, L.: Channel Assignment and Graph Multicoloring, pp. 71–94. Wiley, New York (2002)
34. Narayanan, L., Shende, S.M.: Static frequency assignment in cellular networks. Algorithmica **29**(3), 396–409 (2001)
35. Narayanan, L., Shende, S.M.: Corrigendum: static frequency assignment in cellular networks. Algorithmica **32**(4), 679 (2002)
36. Seibert, S., Sprock, A., Unger, W.: Advice complexity of the online coloring problem. In: Spirakis, P.G., Serna, M. (eds.) CIAC 2013. LNCS, vol. 7878, pp. 345–357. Springer, Heidelberg (2013)
37. Sleator, D.D., Tarjan, R.E.: Amortized efficiency of list update and paging rules. Commun. ACM **28**(2), 202–208 (1985)
38. Sparl, P., Zerovnik, J.: 2-local 4/3-competitive algorithm for multicoloring hexagonal graphs. J. Algorithms **55**(1), 29–41 (2005)
39. Witkowski, R., Žerovnik, J.: 1-local 33/24-competitive algorithm for multicoloring hexagonal graphs. In: Frieze, A., Horn, P., Prałat, P. (eds.) WAW 2011. LNCS, vol. 6732, pp. 74–84. Springer, Heidelberg (2011)

Approximating Steiner Trees and Forests with Minimum Number of Steiner Points

Nachshon Cohen and Zeev Nutov[✉]

The Open University of Israel, Raanana, Israel
nachshonc@gmail.com, nutov@openu.ac.il

Abstract. Let R be a finite set of terminals in a metric space (M, d). We consider finding a minimum size set $S \subseteq M$ of additional points such that the unit-disc graph $G[R \cup S]$ of $R \cup S$ satisfies some connectivity properties. In the Steiner Tree with Minimum Number of Steiner Points (ST-MSP) problem $G[R \cup S]$ should be connected. In the more general Steiner Forest with Minimum Number of Steiner Points (SF-MSP) problem we are given a set $D \subseteq R \times R$ of demand pairs and $G[R \cup S]$ should contains a uv-path for every $uv \in D$. Let Δ be the maximum number of points in a unit ball such that the distance between any two of them is larger than 1. It is known that $\Delta = 5$ in $\mathbb{R}^2$. The previous known approximation ratio for ST-MSP was $\lfloor (\Delta+1)/2 \rfloor + 1 + \epsilon$ in an arbitrary normed space [15], and $2.5 + \epsilon$ in the Euclidean space $\mathbb{R}^2$ [5]. Our approximation ratio for ST-MSP is $1 + \ln(\Delta-1) + \epsilon$ in an arbitrary normed space, which in $\mathbb{R}^2$ reduces to $1 + \ln 4 + \epsilon < 2.3863 + \epsilon$. For SF-MSP we give a simple Δ-approximation algorithm, improving the folklore ratio $2(\Delta - 1)$. Finally, we generalize and simplify the Δ-approximation of Calinescu [3] for the 2-Connectivity-MSP problem, where $G[R \cup S]$ should be 2-connected.

Keywords: Wireless network · Unit-disc graph · Steiner tree · Steiner forest · 2-connectivity · Approximation algorithms

1 Introduction

In the Survivable Network problem we are given a graph $G = (V, E)$ with edge-costs (or node-costs) and a set R of terminals, and seek a minimum-cost subgraph H of G that satisfies some prescribed connectivity requirements between the terminals. A fundamental problem of this type is the Steiner Tree problem, where every pair of terminals should be connected in H. In the Steiner Forest problem, we are given a set D of *demand pairs* from R, and H should contains a uv-path for every $uv \in D$. Steiner Tree is a particular case of Steiner Forest, when the graph (D, R) formed by the demand pairs is connected; if also $R = V$, then we get the Minimum Spanning Tree problem. In the k-Connectivity problem, R should be k-connected in H, namely, H should contain k internally-disjoint paths between any pair of nodes in R; if also $R = V$, then we get the k-Connected Subgraph problem. Note that for $k = 1, 2$, whenever only inclusion minimal solution graphs H that contain R are considered, the condition "R is k-connected in H" is equivalent to the condition "H is k-connected"; this is not so for $k \geq 3$.

© Springer International Publishing Switzerland 2015
E. Bampis and O. Svensson (Eds.): WAOA 2014, LNCS 8952, pp. 95–106, 2015.
DOI: 10.1007/978-3-319-18263-6_9

In wireless networks, the range and the location of the transmitters determines the resulting communication network. We consider adding a minimum number of transmitters such that the communication network satisfies some connectivity properties. If the range of the transmitters is fixed, our goal is to add a minimum number of transmitters, and we get the following type of problems.

Definition 1. *Let (M, d) be a metric space and let $V \subseteq M$. The* unit-disk graph *of V, denoted by $G[V]$, has node set V and edge set $\{uv : u, v \in V, d(u, v) \leq 1\}$.*

In Survivable Network with Minimum Number of Steiner Points (SN-MSP) problems we are given a set R of terminals in a metric space (M, d), and seek a minimum size set $S \subseteq M$ of additional points such that $G[R \cup S]$ satisfies some prescribed connectivity requirements between the terminals. In this setting, Steiner Tree is transformed into the following problem.

> Steiner Tree with Minimum Number of Steiner Points (ST-MSP)
> *Instance:* A finite set $R \subseteq M$ of terminals in a metric space (M, d).
> *Objective:* Find a minimum size set $S \subseteq M$ of additional points such that $G[R \cup S]$ is connected.

In the Steiner Forest with Minimum Number of Steiner Points (SF-MSP) problem, we are given a set D of demand pairs from R, and $G[R \cup S]$ should contains a uv-path for every $uv \in D$. In the k-Connectivity-MSP $G[R \cup S]$ should contain k internally-disjoint uv-paths for any $u, v \in R$. Note that 1-Connectivity-MSP is the ST-MSP problem, while 2-Connectivity-MSP is equivalent to the problem of finding a minimum size $S \subseteq M$ such that $G[R \cup S]$ is 2-connected.

As in previous work, we will assume that our metric space is induced by some normed space, allow to place several points at the same location, and assume that the maximum distance between terminals is polynomial in their number.

The Steiner Tree problem was studied extensively (c.f. [2,17,18] and the references therein) and the currently best approximation ratio for it is $\ln 4 + \epsilon$ [2]. Steiner Forest and 2-Connectivity admit ratio 2, c.f. [8] and [12].

We survey some literature on SN-MSP problems. ST-MSP is NP-hard even in $\mathbb{R}^2$, and arises in various wireless network design problems, c.f. [1,3–5,10,11, 14,15] for only a sample of papers in the area, where it is studied both in $\mathbb{R}^2$ and in general metric spaces. In the latter case, the approximation ratio is usually expressed in terms of the following parameter. Let Δ be the maximum number of "independent" points in the unit ball, such that the distance between any two of them is larger than 1. It is known [16] that Δ equals the maximum degree of a minimum-degree Minimum Spanning Tree in the normed space. For Euclidean distances we have $\Delta = 5$ in $\mathbb{R}^2$ and $\Delta = 11$ in $\mathbb{R}^3$, and in $\mathbb{R}^\ell$ Δ is at most the Hadwiger number [16]; hence $\Delta \leq 2^{0.401\ell(1+o(1))}$, by [9].

In finite metric spaces, ST-MSP is equivalent to the variant of the Node-Weighted Steiner Tree problem when terminals have costs 0 and other nodes have cost 1. Klein and Ravi [13] proved that this variant is Set-Cover hard to approximate, and gave an $O(\ln |R|)$-approximation algorithm for general weights. Hence up to constants, even for finite metric spaces, the ratio $O(\ln |R|)$ of [13] is the best possible unless P = NP. Note however, that this does not exclude constant ratios for metric/normed spaces with small Δ, e.g., $\Delta = 5$ in $\mathbb{R}^2$.

Most algorithms for SN-MSP problems applied the following reduction method, by solving the corresponding Survivable Network instance obtained as follows.

Definition 2. *Given a finite set R of points in a metric space (M, d) and an integer $k \geq 1$, the (multi)graph G_R has node set R and k parallel edges between every pair of nodes. The costs of the k edges between u, v are defined as follows. Let $\hat{d}_{uv} = \max\{\lceil d(u, v) \rceil - 1, 0\}$. If $\hat{d}_{uv} > 0$, then all the k edges have cost $\hat{d}_{uv}$. If $\hat{d}_{uv} = 0$, then one edge has cost 0 and the others have cost 1.*

It is easy to see that any solution of cost C to the corresponding Survivable Network instance defines a solution S of size C to the original SN-MSP instance, where every node in S has degree exactly 2; such a solution is called a *bead solution*. Conversely, any bead solution S can be converted into a solution to the Survivable Network instance (in a normed space) of cost at most $|S|$ (c.f. [3, 10]). Due to this bijective correspondence, we simply define a bead solution as a solution to the corresponding Survivable Network instance, and denote the optimal value of a bead solution to an instance I by $\tau = \tau(I)$. If the Survivable Network instance admits a ρ-approximation algorithm, and if for the given SN-MSP instance there exists a bead solution S of size $\leq \alpha$opt, then we get a $\rho\alpha$-approximation algorithm for the SN-MSP instance. Equivalently, for a class $\mathcal{I}$ of SN-MSP instances, define a parameter α by $\alpha = \alpha(\mathcal{I}) = \sup_{I \in \mathcal{I}} \frac{\tau(I)}{\text{opt}(I)}$. Then approximation ratio ρ for Survivable Network instances that correspond to class $\mathcal{I}$ implies approximation ratio $\alpha\rho$ for SN-MSP instances in class $\mathcal{I}$.

Măndoiu and Zelikovsky [14] showed that $\alpha = \Delta - 1$ for ST-MSP. Since the SN instance that corresponds to ST-MSP is the MST problem that can be solved in polynomial time, this gives a $(\Delta - 1)$-approximation algorithm for ST-MSP.

A common method to attack various Steiner Tree problems is by a reduction to the Minimum Connected Spanning Subhypergraph problem. This method was initiated by Zelikovsky [17], and improved in a long series of papers culminating in the paper of Byrka et al. [2]. For ST-MSP in $\mathbb{R}^2$, Chen et al. [5] applied it to get the currently best known ratio $2.5 + \epsilon$. In arbitrary normed spaces, the ratio $\Delta - 1$ of [14] was improved to $\lfloor (\Delta + 1)/2 \rfloor + 1 + \epsilon$ in [15] also using the same method. In this work we use the so called "Relative Greedy Heuristic" due to Zelikovsky [18], and obtain the following result.

Theorem 1. ST-MSP *with constant Δ admits an approximation scheme with ratio $1 + \ln(\Delta - 1) + \epsilon$. In particular, in $\mathbb{R}^2$ the ratio is $1 + \ln 4 + \epsilon < 2.3863 + \epsilon$.*

We note that previous works, as well as Theorem 1, assume that ST-MSP instances with a constant number of terminals can be solved in polynomial time. This condition holds in $\mathbb{R}^2$ if the maximum distance between terminals is polynomial in the number of terminals, see [4, Lemma 11] and the discussion there. If such instances can be approximated within a factor of β, then an additional factor of β is invoked in the ratio of Theorem 1.

In the proof of Theorem 1, most of our effort is spent on bounding the parameter α_k (so called "Steiner ratio") for ST-MSP (see Theorem 3), which is the loss in the approximation ratio as a result of reducing the problem to the Minimum Connected Spanning Subhypergraph problem in a hypergraph of rank k (see Theorem 3). Bounds on α_k have also been computed for several other "Steiner problems", including ST-MSP. However, none of the previous papers succeeded to reveal the somewhat irregular behavior of α_k in the case of ST-MSP. The key difficulty lies in the fact that α_k has a "good" behavior only when $k = \Omega(\Delta)$. This is the reason why in Theorem 1 we require that Δ is a constant, to ensure that our algorithm can be implemented in polynomial time.

We now consider SF-MSP and k-Connectivity-MSP problems. The result in [14] implies that $\alpha = \Delta - 1$ for SF-MSP. Combined with the known 2-approximation for Steiner Forest, this gives ratio $2(\Delta - 1)$ for SF-MSP. Kashyap et al. [11] considered 2-Connectivity-MSP. Their algorithm constructs a 2-Connectivity instance as in Definition 2 and then converts its solution into a bead solution to the 2-Connectivity-MSP instance. Although they analyzed a performance of specific 2-approximation algorithm – the algorithm of Khuller and Raghavachari [12] for 2-Connectivity, they essentially proved that $\alpha = \Delta$ in this case, which implies ratio 2Δ. The analysis of this specific algorithm was recently improved by Calinescu [3], showing that its tight performance is Δ.

We now discuss k-Connectivity-MSP with $k \geq 3$. Bredin et al. [1] considered in $\mathbb{R}^2$ a related problem of adding a minimum size S such that $G[R \cup S]$ is k-connected (note that in k-Connectivity-MSP we require k-connectivity only between terminals). For this problem in $\mathbb{R}^2$, they gave an $O(k^5)$-approximation algorithm, but essentially they implicitly proved that for this class of problems $\alpha = O(\Delta k^3)$. Recently, it was shown in [10] that $\alpha = \Theta(\Delta k^2)$ for a much more general class of Survivable Network problems in any normed space.

Let $\tau^* = \tau^*(I)$ denote the optimal value of a *fractional* bead solution of an SN-MSP instance I, namely, τ^* is the optimum value of a standard cut-LP relaxation for the corresponding Survivable Network instance (see Sect. 4). We observe that if the algorithm we use for the Survivable Network instance computes a solution of cost at most $\rho\tau^*$, then the relevant parameter is the following.

Definition 3. *For a class $\mathcal{I}$ of SN-MSP instances, let* $\alpha^* = \alpha^*(\mathcal{I}) = \sup_{I \in \mathcal{I}} \frac{\tau^*(I)}{\mathsf{opt}(I)}$.

Theorem 2. *For any feasible solution S to SF-MSP there exists a half-integral bead solution of value at most $\Delta|S|/2$; thus $\alpha^* = \Delta/2$ for SF-MSP. Consequently, if Steiner Forest admits a polynomial time algorithm that computes a solution of cost at most $\rho\tau^*$, then SF-MSP admits approximation ratio $\rho \cdot \Delta/2$; thus SF-MSP admits a Δ-approximation algorithm. The same holds for 2-Connectivity-MSP.*

The idea behind Theorem 2 is as follows. From previous work [3,16] we get that for any solution S, $G[R \cup S]$ contains a solution G in which the nodes in S have degree at most Δ. Our main innovation is comparing the optimal solution with a *fractional* (in fact, half-integral) bead solution, rather than an actual bead solution. For 2-Connectivity-MSP this idea appeared implicitly in the paper of Calinescu [3], but our explicit approach is much simpler and more general.

2 Proof of Theorem 1

We consider a generic problem defined in [15], that includes both ST-MSP and the classic Steiner Tree problem.

Generic Steiner Tree
Instance: A (possibly infinite) graph $G = (V, E)$, a finite set $R \subseteq V$ of terminals, and a monotone subadditive cost function c on subgraphs of G.
Objective: Find a minimum-cost connected finite subtree T of G containing R.

Definition 4. *For an instance of* Generic Steiner Tree *and* $2 \le k \le |R|$, *the hypergraph* $\mathcal{H}_k = (R, \mathcal{E}_k)$ *has hyperedge set* $\mathcal{E}_k = \{A \subseteq R : 2 \le |A| \le k\}$. *The cost* $c^*(A)$ *of* $A \in \mathcal{E}_k$ *is the cost of an optimal solution* T_A *to the* Generic Steiner Tree *instance with terminal set* A.

The construction in Definition 4 converts a Generic Steiner Tree instance into a Minimum Connected Spanning Subhypergraph instance in a hypergraph $\mathcal{H}_k$ of rank k. Any solution of cost C to this instance correspond to a solution of value at most C to Generic Steiner Tree instance, by the subadditivity and monotonicity of the cost function in the Generic Steiner Tree problem. This reduction invokes a fee in the approximation ratio, given in the following definition.

Definition 5. *Given an instance* I *of* Generic Steiner Tree *let* $\tau_k(I)$ *denote the minimum cost of a connected spanning sub-hypergraph of* $\mathcal{H}_k$. *The k-ratio for a class* $\mathcal{I}$ *of* Generic Steiner Tree *instances is defined by* $\alpha_k = \sup_{I \in \mathcal{I}} \frac{\tau_k(I)}{\mathrm{opt}(I)}$.

Note that for $\mathcal{I}$ being the class of ST-MSP instances, α_2 is the parameter α defined in the introduction, and that by [14] we have $\alpha_2 = \alpha = \Delta - 1$. We have $\alpha_k = 1$ for instances with $|R| = k$, and in general α_k is monotone decreasing and approaching 1 when k becomes larger.

Particular cases of the following statement can be found in several papers. We failed to find the general version in the literature, and thus provide a proof for completeness of exposition.

Lemma 1. *There exists a polynomial time algorithm that given a hypergraph* $\mathcal{H} = (R, \mathcal{E})$ *with hyper-edge cost* $\{c(A) : A \in \mathcal{E}\}$ *and a spanning tree* T^* *of (edges of size 2 of)* $\mathcal{H}$ *computes a spanning connected sub-hypergraph* T *of* $\mathcal{H}$ *of cost at most* $\tau\left(1 + \ln \frac{c(T^*)}{\tau}\right)$, *where* τ *is the minimum-cost of a connected spanning sub-hypergraph of* $\mathcal{H}$.

Proof. Given a tree $T = (R, F)$ let us say that $A \subseteq R$ overlaps $F' \subseteq F$ if the graph obtained from $T \backslash F'$ by shrinking A into a single node is a tree. Given edge cost $\{c(e) : e \in F\}$ let $F(A)$ be a maximum cost edge set overlapped by A.

Note that $F \backslash F(A)$ is an edge set of a minimum cost spanning tree in the graph obtained from T by shrinking A into a single node; hence for given A, $F(A)$ can be computed in polynomial time. Consider the following algorithm.

Algorithm 1. LOCAL REPLACEMENT ALGORITHM

---

1 **Input:** A hypergraph $\mathcal{H} = (R, \mathcal{E})$ with hyper-edge cost $\{c(A) : A \in \mathcal{E}\}$,
   and a spanning tree $T^* = (R, F^*)$ of (edges of size 2 of) $\mathcal{H}$.
2 **Initialization:** $\mathcal{J} \leftarrow \emptyset$, $F \leftarrow F^*$, $T \leftarrow (R, F)$.
3 **while** $c(F) > 0$ **do**
4   Find $A \in \mathcal{E}$ with $\frac{c(F(A))}{c(A)}$ maximum.
5   **if** $c(F(A)) > c(A)$ **then**
6   Update $T, \mathcal{H}$: remove $F(A)$ and shrink A into a single node.
           $F \leftarrow F \backslash F(A)$ and $\mathcal{J} \leftarrow \mathcal{J} \cup \{A\}$.
7   **Else STOP** and **return** $\mathcal{T} = (R, F \cup \mathcal{J})$.
8 **return** $\mathcal{T} = (R, F \cup \mathcal{J})$

The following statement appeared in [17] (see also [2]); we provide a proof for completeness of exposition.

Claim. Let $T = (R, F)$ be a tree with edge costs $\{c(e) : e \in F\}$ and let $(R, \mathcal{E})$ be a connected hypergraph. Then $\sum_{A \in \mathcal{E}} c(F(A)) \geq c(F)$. Thus there exists $A \in \mathcal{E}$ such that

$$\frac{c(F(A))}{c(A)} \geq \frac{c(F)}{c(\mathcal{E})} \ .$$

Proof. For a node $v \in A$, let C_v be the connected component in $T \backslash F(A)$ that contains v. For an edge $e \in F(A)$ that connects two components C_u, C_v, let $y(e) = uv$ be the replacement edge of e, of cost $c(y(e)) = c(e)$. The graph $T \cup \{y(e)\}$ contains a single cycle and $y(e)$ is the heaviest edge in this cycle, since otherwise $F(A)$ is not minimal. For a hyperedge $A \in \mathcal{E}$ let $y(A) = \cup_{e \in F(A)} y(e)$ be the replacement set of A, and let $y(\mathcal{E}) = \cup_{A \in \mathcal{E}} y(A)$. It is easy to see that $y(A)$ spans A, and $y(\mathcal{E})$ spans R. Consider a MST on $T \cup y(\mathcal{E})$. By the cycle property of a MST, no edge from $y(\mathcal{E})$ would participate in that MST, so $c(T) \leq c(y(\mathcal{E}))$. Finally, $c(y(\mathcal{E})) = \sum_{A \in \mathcal{E}} y(A) = \sum_{A \in \mathcal{E}} c(F(A))$, and the claim follows. □

At every iteration $|F|$ decreases by at least 1, hence the algorithm runs in polynomial time, and clearly it computes a feasible solution. We prove the approximation ratio. Let F_i and $\mathcal{J}_i$ be the set stored in F and $\mathcal{J}$, respectively, at the beginning of iteration $i + 1$, and let A_i be the hyperedge picked at iteration i. Denote $f_i = c(F_i)$ and $s_i = c(A_i)$, and recall that τ denotes the minimum cost of a connected spanning sub-hypergraph of $\mathcal{H}$. At iteration i we remove $F_{i-1}(A_i)$ from F_{i-1} after verifying that $c(F_{i-1}(A_i)) > c(A_i) = s_i$. Hence

$$f_i \leq f_{i-1} - \max\{c(F_{i-1}(A_i)), c(A_i)\} = f_{i-1} - s_i \cdot \max\left\{\frac{c(F_{i-1}(A_i))}{c(A_i)}, 1\right\}$$

By the claim above, $\frac{c(F_{i-1}(A_i))}{c(A_i)} \geq \frac{f_{i-1}}{\tau}$. Thus we have

$$f_i \leq f_{i-1} - s_i \cdot \max\{f_{i-1}/\tau, 1\} \ . \tag{1}$$

The algorithm stops if either $c(F_q) = 0$ or $c(F(A)) \leq c(A)$ at iteration $q+1$. In the latter case, $1 \geq c(F_q)/\tau$ follows by the claim above. In both cases, we have that there exists an index q such that $f_{q-1} > \tau \geq f_q$ holds. Now we use the following statement from [6].

Claim. Let $\tau > 0$ and $f_0, \ldots, f_q$ and $s_1, \ldots, s_q$ be sequences of positive reals satisfying $f_0 > \tau \geq f_q$, such that (1) holds. Then $f_q + \sum_{i=1}^q s_i \leq \tau(1 + \ln(f_0/\tau))$.

Let q be an index such that $f_{q-1} > \tau \geq f_q$ holds. We may assume that $f_0 = c(F^*) > \tau > 0$. Note that $c(\mathcal{J}_q) = \sum_{i=1}^q s_i$ and that $c(F_i) + c(\mathcal{J}_i) \leq c(F_{i-1}) + c(\mathcal{J}_{i-1})$ for any i. Hence from the claim above we conclude that

$$c(T) \leq c(F_q) + c(\mathcal{J}_q) = f_q + \sum_{i=1}^q s_i \leq \tau(1 + \ln(f_0/\tau)) = \tau \left(1 + \ln \frac{c(T^*)}{\tau}\right).$$

□

Corollary 1. *For any constant k,* Generic Steiner Tree *admits an approximation ratio $\alpha_k (1 + \ln \alpha_2)$, provided that for any $A \in \mathcal{E}_k$, the instance with the terminal set A can be solved in polynomial time.*

Proof. By the assumptions, the hypergraph $\mathcal{H}_k$, and the costs $c^*(A)$ with the corresponding trees T_A for $A \in \mathcal{E}_k$, can be computed in polynomial time. We can also compute in polynomial time an optimal spanning tree T^* in $\mathcal{H}_2$; note that $c(T^*) \leq \alpha_2 \cdot \text{opt}$. Then we apply the algorithm in Lemma 1 to compute a sub-hypergraph T of $\mathcal{H}_k$ of c^*-cost at most $\tau \left(1 + \ln \frac{c(T^*)}{\tau}\right)$. Let opt denote the optimal solution value for the Generic Steiner Tree instance. Note that $\text{opt} \leq \tau \leq \alpha_k \text{opt}$. Let $T = \cup_{A \in T} T_A$. Since T is a connected hypergraph, T is a feasible solution to the Generic Steiner Tree instance. We have $c(T) \leq \sum_{A \in T} c(T_A) = c^*(T)$, by the monotonicity and the subadditivity of the c-costs. Thus we have:

$$c(T) \leq c^*(T) \leq \tau \left(1 + \ln \frac{c(T^*)}{\tau}\right) = \tau \left(1 + \ln \frac{c(T^*)/\text{opt}}{\tau/\text{opt}}\right) \leq \alpha_k \text{opt} (1 + \ln \alpha_2).$$

□

Du and Zhang [7] showed that for the classic Steiner Tree problem, $\alpha_k \leq 1 + 1/\lfloor \lg k \rfloor$, where $\lg k = \log_2 k$. In Sect. 3 we prove the following.

Theorem 3. *For* ST-MSP, *$\alpha_k \leq 1 + \frac{2}{\lceil \lg \lfloor k/(\Delta-1) \rfloor \rceil}$ for any integer $k \geq 2\Delta - 2$.*

Note that for an instance I of ST-MSP with Δ independent points on the unit ball we have $\tau_k(I) = \frac{\Delta}{k}$ and $\text{opt}(I) = 1$, which implies $\alpha_k \geq \frac{\tau_k(I)}{\text{opt}(I)} = \frac{\Delta}{k}$. Hence $k > \Delta/2$ is necessary if we want $\alpha_k < 2$.

From Corollary 1 and Theorem 3 we conclude that for any constant $k \geq 2\Delta - 2$, it is possible to compute in polynomial time a solution to an ST-MSP instance of size at most $\alpha_k (1 + \ln(\Delta - 1))$ opt, where α_k is as in Theorem 3. For the metric space $\mathbb{R}^2$, and given a constant $\epsilon > 0$ let $k = 2^{\Theta(1/\epsilon)}$. Then by Theorem 3, $\alpha_k \leq 1 + \epsilon/(1 + \ln 4)$, and the approximation ratio of our algorithm is $1 + \ln 4 + \epsilon$. This completes the proof of Theorem 1.

3   Proof of Theorem 3

For a tree $T = (V, F)$ and $A \subseteq V$ let $T_A = (V_A, F_A)$ be the inclusion minimal subtree of T that contains A. To prove Theorem 3 we prove the following.

Lemma 2. *Let $T = (V, F)$ be a tree of maximum degree $\Delta \geq 2$, let $R \subseteq V$, and let $S = V \backslash R$. Then for any integer $k \geq 2\Delta - 2$ there exists a connected hypergraph $\mathcal{H} = (R, \mathcal{E})$ of rank $\leq k$ such that $\sum_{A \in \mathcal{E}} |V_A \cap S| \leq \left(1 + \frac{2}{\lfloor \lg \lfloor k/(\Delta - 1) \rfloor \rfloor}\right) |S|$.*

To prove Lemma 2 we prove the following.

Lemma 3. *Let $T = (V, F)$ be a tree with edge costs $\{c(e) \geq 1 : e \in F\}$ and let $R \subseteq V$. Then for any integer $p \geq 2$ there exists a connected hypergraph $\mathcal{H} = (R, \mathcal{E})$ of rank $\leq p$ such that $\sum_{A \in \mathcal{E}} c(F_A) + |\mathcal{E}| - 1 \leq \left(1 + \frac{2}{\lfloor \lg p \rfloor}\right) c(T)$.*

Lemma 3 will be proved later. Now we show that it implies Lemma 2. An R-*component* of T is a maximal inclusion subtree of T such that all its leaves are in R but none of its internal nodes is in R. It is easy to see that it is sufficient to prove Lemma 2 for each R-component separately, hence we may assume that R is the set of leaves of T.

If T is a star, then since $k \geq 2\Delta - 2 \geq \Delta$, we let $\mathcal{E}$ to consist of a single hyperedge $A = R$. Then $|V_A \cap S| = 1 = |S|$, and Lemma 2 holds in this case.

Henceforth assume that T is not a star. For $v \in S$ let $R(v)$ be the set of neighbors of v in R, and note that $|R(v)| \leq \Delta - 1$. Let $T' = (V', F') = T \backslash R$ and let $R' = \{v \in S : R(v) \neq \emptyset\}$. Applying Lemma 3 on T' with unit edge-costs and R', we obtain that for $p = \lfloor k/(\Delta - 1) \rfloor$ there exists a connected hypergraph $\mathcal{H}' = (R', \mathcal{E}')$ of rank $\leq p$ such that $\sum_{A' \in \mathcal{E}'} |F'_{A'}| + |\mathcal{E}'| - 1 \leq \left(1 + \frac{2}{\lfloor \lg p \rfloor}\right) |F'|$. Note that $|F'| = |V'| - 1$ and that $|V'_{A'}| = |F'_{A'}| - 1$ for every $A' \in \mathcal{E}'$. Hence

$$\sum_{A' \in \mathcal{E}'} |V'_{A'}| - 1 \leq \left(1 + \frac{2}{\lfloor \lg p \rfloor}\right) (|V'| - 1) \leq \left(1 + \frac{2}{\lfloor \lg p \rfloor}\right) |V'| - 1 .$$

For $A' \in \mathcal{E}'$ let $A = \cup_{v \in A'} R(v)$; then $|A| \leq p(\Delta - 1)$. Let $\mathcal{E} = \{A : A' \in \mathcal{E}'\}$. Then $\mathcal{H} = (R, \mathcal{E})$ is a connected hypergraph of rank $\leq p(\Delta - 1) \leq k$, and

$$\sum_{A \in \mathcal{E}} |V_A \cap S| = \sum_{A' \in \mathcal{E}'} |V'_{A'}| \leq \left(1 + \frac{2}{\lfloor \lg p \rfloor}\right) |V'| = \left(1 + \frac{2}{\lfloor \lg \lfloor k/(\Delta - 1) \rfloor \rfloor}\right) |S| .$$

In the rest of this section we prove Lemma 3, by extending the proof of Du and Zhang [7] of an existence of a connected hypergraph $\mathcal{H} = (R, \mathcal{E})$ of rank $\leq p$ such that $\sum_{A \in \mathcal{E}} c(F_A) \leq \left(1 + \frac{1}{\lfloor \lg p \rfloor}\right) c(T)$. We have an extra term of $|\mathcal{E}| - 1$, and we show that this term can be bounded by $\frac{c(T)}{\lfloor \lg p \rfloor}$.

We start by transforming the tree into a (rooted) binary tree T with edge-costs, which node set is partitioned into a set R of terminals and a set S of non-terminals, such that the following properties hold:

(A) R is the set of leaves of T.

(B) The cost of any edge of T is either 0 or is at least 1, and among the edges that connect a node in $S = V \backslash R$ to its children, at most one has cost 0.

(C) T is a full binary tree, namely, every $v \in S$ has exactly 2 children.

To obtain such a tree, root T at an arbitrary non-leaf node $\hat{s} \in S = V \backslash R$, and apply the following standard reductions.

1. While T has a leaf in S, remove this leaf; hence every leaf of T is in R. Then, for every $v \in R$ that is not a leaf, add to T a new node v' and an edge vv' of cost 0, add v' to R, and move v from R to S. After this step, properties (A) and (B) hold.

2. While there is $v \in S$ that has one child, replace the path P of length 2 that contains v by a single edge of cost $c(P)$, and exclude v from S. After this step, every $v \in S$ has at least 2 children.

3. While there is $v \in S$ that has more than 2 children, do the following. Let u be a child of v such that the cost of the edge vu is at least 1. Add a new node v' and the edge vv' of cost 0, and for every child of u' of v distinct from u replace the edge vu' by the edge vu'. After this step, all the three properties (A), (B), and (C) hold.

Consequently, to prove Lemma 3, it is sufficient to prove the following.

Lemma 4. *Let $T = (V, F)$ be a tree with edge costs $c(e)$ and leaf set R, satisfying (A),(B),(C), Then for any integer $p \geq 2$ there exists a connected hypergraph $\mathcal{H} = (R, \mathcal{E})$ of rank $\leq p$ such that $\sum_{A \in \mathcal{E}} c(F_A) + |\mathcal{E}| - 1 \leq \left(1 + \frac{2}{\lfloor \lg p \rfloor}\right) c(T)$.*

Let $T = (V, F)$ be a rooted tree with leaf set R and let $S = V \backslash R$. For two nodes u, v of T let $P_T(u, v)$ denote the unique path in T between u and v.

Definition 6. *We say that T is* proper *if every node in S has at least 2 children. We say that a mapping $f : S \rightarrow R$ is T-proper if:*
(i) For every $u \in S$, $f(u)$ is a descendant of u.
(ii) The paths $\{P_T(u, f(u)) : u \in S\}$ are edge disjoint.
Given a subtree T' of T with leaf set L' and a proper mapping f, the set of terminal connecting paths *of T' is $\{P_T(u, f(u)) : u \in L' \backslash R\}$. Let $\hat{T}'$ denote the tree obtained from T' by adding to T' all the terminal connecting paths.*

Du and Zhang [7] proved that any proper tree T admits a proper mapping. We prove the following.

Lemma 5. *Let $T = (V, F)$ be a proper tree and let $F_1 \subseteq F$ be such that any $u \in S$ has a child connected to u by an edge in F_1. Then there exists a T-proper mapping f such that for every $u \in S$, the path $P_T(u, f(u))$ contains at least one edge in F_1.*

Proof. The proof is by induction on the height of the tree. Let T be a tree as in the lemma of height h. If $h = 1$, then T has one internal node (the root), say u, and we set $f(u)$ to be the node that is connected to u by an edge in F_1. Suppose that the statement is true for trees with height $h - 1 \geq 1$, and we prove it for trees of height h. Let T' be obtained from T by removing nodes of distance h from the root. By the induction hypothesis, for T' there exists a mapping f' as in the lemma. Let u be an internal node of T. Consider two cases.

Suppose that u is an internal node of T'. If $f'(u)$ is a leaf of T, then define $f(u) = f'(u)$. If $f'(u)$ is an internal of T, then $f'(u)$ is a leaf of T', and all its children in T are leaves. Then we set $f(u)$ to be a child of $f'(u)$ that is connected to $f'(u)$ by an edge in F_1.

Suppose that u is a leaf of T'. Then the children of u in T are leaves, and we set $f(u)$ to be a child of u that is connected to u by an edge in F_1.

It is easy to verify that the obtained mapping f meets the requirements. □

The following statement is implicitly proved by Du and Zhang [7].

Lemma 6 ([7]). *Let T be a proper binary tree with non-negative edge costs and let f be a proper mapping. Then for any integer $p \geq 2$ there exists an edge-disjoint partition $\mathcal{T}$ of T into subtrees such that the following holds:*

(i) *The hypergraph with node set R and hyperedge set $\mathcal{E} = \{\hat{T}' \cap R : T' \in \mathcal{T}\}$ is connected and has rank at most p.*

(ii) *The total number of terminal connecting paths of all subtrees in $\mathcal{T}$ is at least $|\mathcal{T}| - 1$, and their total cost is at most $c(T)/\lfloor \lg p \rfloor$.*

We now finish the proof of Lemma 4, and thus also of Lemma 3. Let $F_1 = \{e \in F : c(e) \geq 1\}$ and let f be a proper mapping as in Lemma 5. Let $\mathcal{T}$ be a partition as in Lemma 6, and let $\mathcal{E}$ be as in Lemma 6(i), so the hypergraph $\mathcal{H} = (R, \mathcal{E})$ is connected and has rank at most p. By Lemma 6(ii), the total number of terminal connecting paths of all subtrees is at least $|\mathcal{T}| - 1 = |\mathcal{E}| - 1$, while their total cost is at most $c(T)/\lfloor \lg p \rfloor$. Every terminal connecting path contains an edge from F_1, by Lemma 5, and thus has cost at least 1. Hence the total cost of all terminal connecting paths is at least $|\mathcal{E}| - 1$. Consequently

$$|\mathcal{E}| - 1 \leq \frac{c(T)}{\lfloor \lg p \rfloor} \ .$$

For $A = \hat{T}' \cap R \in \mathcal{E}$ let $P(T')$ denote the union of the edge sets of the terminal connecting paths of T'. Then $c(F_A) \leq c(\hat{T}') = c(T) + c(P(T'))$, hence

$$\sum_{A \in \mathcal{E}} c(F_A) \leq \sum_{T' \in \mathcal{T}} [c(T') + c(P(T'))] = \sum_{T' \in \mathcal{T}} c(T') + \sum_{T' \in \mathcal{T}} c(P(T')) \leq c(T) + \frac{c(T)}{\lfloor \lg p \rfloor} \ .$$

Summarizing, we have

$$\sum_{A \in \mathcal{E}} c(F_A) + |\mathcal{E}| - 1 \leq c(T) + \frac{c(T)}{\lfloor \lg p \rfloor} + \frac{c(T)}{\lfloor \lg p \rfloor} = \left(1 + \frac{2}{\lfloor \lg p \rfloor}\right) c(T) \ .$$

The proof of Lemma 4, and thus also of Lemma 3 and Theorem 3 is now complete.

4   Proof-sketch of Theorem 2

Definition 7. *For a subset C of nodes of a (multi-)graph $G = (V, E)$ let us use the following notation: $\Gamma_G(C)$ is the set of neighbors of C in G; $\delta_G(C) = \delta_E(C)$ is the set of edges in E with exactly one endnode in C; $E(C)$ is the set of edges in E with both endnodes in C. Given $R \subseteq V$, an R-component of G is a subgraph of G with node set $C \cup \Gamma_G(C)$ and edge set $E(C) \cup \delta_G(C)$, where C is a connected component of $G \backslash R$.*

The following important property of feasible solutions was proved for SF-MSP by Robins and Salowe [16] and for 2-Connectivity-MSP by Calinescu [3].

Lemma 7 ([3,16]). *Let S be an inclusion minimal feasible solution for an instance of SF-MSP or 2-Connectivity-MSP. Then $G[R \cup S]$ contains a subgraph G that satisfies the requirements such that every R-component of G is a tree and such that $\deg_G(v) \leq \Delta$ for every $v \in S$.*

Lemma 8. *Suppose that $G[R \cup S]$ contains a tree T with leaf set R. Let S' be obtained from S by replacing each $v \in S$ by $\deg_T(v)$ copies of v. Then $G[R \cup S']$ contains a simple cycle on $R \cup S'$, called a DFS cycle of T.*

Proof. Traverse the tree T in a DFS order; each time a node $v \in S$ is visited, choose a different copy of v.   □

Given a Steiner Forest instance, we say that a set $A \subset V$ is *deficient* if $|A \cap \{u, v\}| = 1$ for some $uv \in D$. It is easy to see that H is a feasible solution to a Steiner Forest instance iff $\delta_H(A) \geq 1$ for every deficient set A. To formulate a similar condition for 2-Connectivity we need a definition of a *biset*, which is an ordered pair of sets $\mathbb{A} = (A, A^+)$ such that $A \subseteq A^+$; $\Gamma(\mathbb{A}) = A^+ \backslash A$ is the *boundary* of $\mathbb{A}$. Let $\delta_E(\mathbb{A})$ denote the set of edges in E with one end in A and the other in $V \backslash A^+$. It is known that H is a feasible solution to 2-Connectivity iff $x(\delta_E(\mathbb{A})) \geq 2 - |\Gamma(\mathbb{A})|$ for every biset $\mathbb{A}$. The cut-LP relaxations for Steiner Forest and 2-Connectivity minimize $\sum_{e \in E} c_e x_e$ over the polytopes Π_{SF} and Π_{2C}, respectively, defined by:

$$\Pi_{\mathsf{SF}} = \{x \in \mathbb{R}^E : x(\delta_E(A)) \geq 1 \ \forall \ \text{deficient set } A, x_e \geq 0\}$$
$$\Pi_{\mathsf{2C}} = \{x \in \mathbb{R}^E : x(\delta_E(\mathbb{A})) \geq 2 - |\Gamma(\mathbb{A})| \ \forall \ \text{biset } \mathbb{A}, 0 \leq x_e \leq 1\}$$

We will say that a graph with edge capacities x_e is a *fractional bead solution* for SF-MSP or for 2-Connectivity-MSP, if the characteristic vector of the edge-set of the graph belongs to Π_{SF} or to Π_{2C}, respectively.

Lemma 9. *Let G be as in Lemma 7. Then replacing each connected component C of $G \backslash R$ by a DFS cycle on $\Gamma_G(C)$ of capacity $1/2$ results in a fractional half-integral bead solution for SF-MSP or 2-Connectivity-MSP, respectively.*

Proof. For SF-MSP the statement is obvious, while for 2-Connectivity-MSP, due to space limitation, the proof will be provided in the full version.   □

Theorem 2 easily follows from Lemmas 7, 8, and 9.

5   Conclusions

Our main results are a $(1 + \ln(\Delta - 1) + \epsilon)$-approximation scheme for ST-MSP, and a Δ-approximation algorithm for SF-MSP. For ST-MSP in $\mathbb{R}^2$ this improves the ratio $2.5 + \epsilon$ of [5]. For SF-MSP this improves the folklore ratio $2(\Delta - 1)$ that follows from the work of [14]. We believe that the methods presented in this paper will lead to improved approximation algorithms for related problems.

References

1. Bredin, J., Demaine, E., Hajiaghayi, M., Rus, D.: Deploying sensor networks with guaranteed fault tolerance. IEEE/ACM Trans. Netw. **18**(1), 216–228 (2010)
2. Byrka, J., Grandoni, F., Rothvoß, T., Sanità, L.: Steiner tree approximation via iterative randomized rounding. J. ACM **60**(1), 6:1–6:33 (2013)
3. Calinescu, G.: Relay placement for two-connectivity. Networking **2**, 366–377 (2012)
4. Chen, D., Du, D.-Z., Hu, X.-D., Lin, G.-H., Wang, L., Xue, G.: Approximations for Steiner trees with minimum number of Steiner points. Theor. Comput. Sci. **262**(1), 83–99 (2001)
5. Cheng, X., Du, D., Wang, L., Xu, B.: Relay sensor placement in wireless sensor networks. Wirel. Netw. **14**(3), 347–355 (2008)
6. Cohen, N., Nutov, Z.: A $(1 + \ln 2)$-approximation algorithm for minimum-cost 2-edge-connectivity augmentation of trees with constant radius. Theor. Comput. Sci. **67–74**, 489–490 (2013)
7. Du, D.-Z., Zhang, Y.: On better heuristics for Steiner minimum trees. Math. Program. **57**, 193–202 (1992)
8. Goemans, M.X., Williamson, D.P.: A general approximation technique for constrained forest problems. SIAM J. Comput. **24**(2), 296–317 (1995)
9. Kabatjansky, G., Levenstein, V.: Bounds for packing of the sphere and in space. Prob. Inf. Trans. **14**, 117 (1978)
10. Kamma, L., Nutov, Z.: Approximating survivable networks with minimum number of Steiner points. Networks **60**(4), 245–252 (2012)
11. Kashyap, A., Khuller, S., Shayman, M.: Relay placement for fault tolerance in wireless networks in higher dimensions. Comput. Geom. **44**, 206215 (2011)
12. Khuller, S., Raghavachari, B.: Improved approximation algorithms for uniform connectivity problems. J. Algorithms **21**(2), 434–450 (1996)
13. Klein, P., Ravi, R.: A nearly best-possible approximation algorithm for node-weighted Steiner trees. J. Algorithms **19**(1), 104–115 (1995)
14. Măndoiu, I., Zelikovsky, A.: A note on the MST heuristic for bounded edge-length Steiner trees with minimum number of Steiner points. Inf. Process. Lett. **75**(4), 165–167 (2000)
15. Nutov, Z., Yaroshevitch, A.: Wireless network design via 3-decompositions. Inf. Process. Lett. **109**(19), 1136–1140 (2009)
16. Robins, G., Salowe, J.: Low-degree minimum spanning trees. Discrete Comput. Geom. **14**, 151–165 (1995)
17. Zelikovsky, A.: An 11/6-approximation algorithm for the network Steiner problem. Algorithmica **9**, 463–470 (1993)
18. Zelikovsky, A.: Better approximation bounds for the network and Euclidean Steiner tree problems. Technical Report CS-96-06, University of Virginia (1996)

Energy-Efficient Algorithms for Non-preemptive Speed-Scaling

Vincent Cohen-Addad[1]($\boxtimes$), Zhentao Li[1], Claire Mathieu[1], and Ioannis Milis[2]

[1] Département d'Informatique, UMR CNRS 8548, École Normale Supérieure,
Paris, France
{vcohen,zhentao,cmathieu}@di.ens.fr
[2] Department of Informatics, Athens University of Economics and Business,
Athens, Greece
milis@aueb.gr

Abstract. We improve complexity bounds for energy-efficient non-preemptive scheduling problems for both the single processor and multiprocessor cases. As energy conservation has become a major concern, traditional scheduling problems have been revisited in the past few years to take into account the energy consumption [1]. We consider the speed scaling setting introduced by Yao et al. [20] where a set of jobs, each with a release date, deadline and work volume, are to be scheduled on a set of identical processors. The processors may change speed as a function of time and the energy they consume is the αth power of their speed integrated over time. The objective is then to find a feasible non-preemptive schedule which minimizes the total energy used.

We show that for an arbitrarily number of processors and jobs with equal work volumes there is a $2(1+\varepsilon)(5(1+\varepsilon))^{\alpha-1}\tilde{B}_\alpha = O_\alpha(1)$ approximation algorithm, where $\tilde{B}_\alpha$ is the generalized Bell number. This is the first constant factor algorithm for the multi-processor case, and this also extends to arbitrary processor-dependent work volumes, up to losing a factor of $(\frac{(1+r)r}{2})^\alpha$ in the approximation, where r is the maximum ratio between two work volumes. For the single processor case, we introduce a new linear programming formulation of speed scaling, using a new constraint capturing non-preemption, and prove that its integrality gap is at most $12^{\alpha-1}$. With our new constraint we improve on the previously known unbounded integrality gap of at least $\Omega(n^{\alpha-1})$. Finally, we deal with the inapproximabilty of speed scaling and we prove that the multiprocessor case is APX-hard, even in the special case where all release dates and deadlines are equal and r is 4.

1   Introduction

While traditional scheduling problems aim to process jobs as quickly as possible given a variety of side constraints, energy-efficient scheduling aims to also

---

I. Milis—Partially supported by the project THALES-ALGONOW co-financed by the European Union (European Social Fund - ESF) and Greek national funds, through the Operational Program "Education and Lifelong Learning".

© Springer International Publishing Switzerland 2015
E. Bampis and O. Svensson (Eds.): WAOA 2014, LNCS 8952, pp. 107–118, 2015.
DOI: 10.1007/978-3-319-18263-6_10

minimize the energy consumed by the system, typically by changing processor's frequency to scale its speed dynamically, slowing it down at times to conserve energy. Thus, standard scheduling problems must now be revisited to take energy into account, and this has been part of the agenda of the scheduling community for the past few years (see the survey [1] and the references therein).

In minimum energy scheduling problems, introduced by Yao et al. [20], we wish to execute jobs on a single (or a set of) processor(s) so that all jobs complete between their release date and deadline in a way that minimizes the energy consumed. Now, each job has to execute a work volume w and as the processors may change their speed, a job may be completed faster (or slower) than the time w it needs to execute at speed 1. It is observed that a processor running at speed s consumes power at the rate s^α, for a constant $\alpha > 1$ (typical values of α are less than 3) and so a processor running at speeds $s(t)$ during an interval I would consume energy $\int_{t \in I} s(t)^\alpha dt$.

Problem Definition. In this paper we examine a minimum energy scheduling problem, which in its simplest *non preemptive* form can be stated as follows:

NON PREEMPTIVE MINIMUM ENERGY SCHEDULING (α)

Input: A set of m processors $P = \{p_1, p_2, \ldots, p_m\}$; a set of n jobs $J = \{1, \ldots, n\}$, with a life interval $L_j = [r_j, d_j]$ and a work volume w_j for each $j \in J$.

Output: An assignment S of an *execution interval* $[s_j, e_j] \subseteq [r_j, d_j]$ for each job j such that no $m + 1$ execution intervals have a common intersection.

Objective: Minimize $E(S) = \sum_j (e_j - s_j) \left(\dfrac{w_j}{e_j - s_j} \right)^\alpha$

An assignment S of execution intervals for all jobs is called a *schedule*. Equivalently, we could ask an algorithm to also output an assignment of each job to a processor in P but this assignment is obtained greedily from the output above. By convexity of $s \to s^\alpha$, it is more efficient to run a processor at constant speed for the same job. Hence, the energy consumed by a job j with execution interval I is $E(S, j) = |I| \cdot (w_j/|I|)^\alpha$, and we can think of $w_j/|I|$ as the speed given to job j. Clearly, $E(S) = \sum_j E(S, j)$.

For a job j with life interval $[r_j, d_j]$, we say r_j is the *release date* of j and d_j is the *deadline* of j. Hence j must be executed between r_j and d_j. We say that j is *alive* at time t if $t \in L_j$. In the *preemptive* case, we are allowed to stop a job, execute some other job and restart the first job later on. Equivalently, we can think of breaking each job j into as many pieces as we want, that all have the same life interval as j, their total work volume is the work volume of j and they are executed non-preemptively. In the *migratory* version of the problem, stopped jobs can even continue their execution on a different processor.

Related Work. The single processor preemptive problem is polynomial: Yao et al. [20] proposed an elegant greedy algorithm whose optimality was proved by Bansal et al. [11]. On the other hand, the single processor non-preemptive problem is NP-hard (Antoniadis and Huang [5]), even for instances where for any pair of jobs such that $r_j \leq r_{j'}$, it holds that $d_j \geq d_{j'}$; they also proposed a

$(2^{5\alpha-4})$-approximation algorithm for general instances. Moreover, a $(1 + \frac{w_{max}}{w_{min}})^{\alpha}$-approximation algorithm for general instances of this problem was proposed in [7], while [8] introduced a $2^{\alpha-1}(1+\varepsilon)^{\alpha}\tilde{B}_{\alpha}$-approximation which is better for small values of α, where $\tilde{B}_{\alpha} = \sum_{k=0}^{\infty} \frac{k^{\alpha-1}e^{-1}}{k!} < \left(\frac{e^{-0.6+\varepsilon}\alpha}{\ln(\alpha+1)}\right)^{\alpha}$, is the generalized version of the Bell number introduced by [8], for any $\alpha \in \mathbb{R}^{+}$.

The homogeneous multiprocessor preemptive problem remains polynomial when migration of jobs is allowed [2,4]. However, [3] proved that the non-migratory variant of this problem, even for jobs with common release dates and deadlines, is NP-Hard and gave a PTAS for such instances. Greiner et al. [14] proposed a transformation of an optimal solution to general migratory instances to a $B_{\lceil\alpha\rceil}$-approximate solution for non-migratory problem. For the homogeneous multiprocessor non-preemptive problem Bampis et al. [7] proposed a $m^{\alpha-1}(n^{1/m})^{\alpha-1}$-approximation algorithm.

Bampis et al. [8] studied the heterogeneous multiprocessor preemptive problem where every processor i has a different speed-to power function, $s^{\alpha(p_i)}$, and both the life interval and the work of jobs are processor dependent. For the migratory variant they proposed a polynomial in $\frac{1}{\varepsilon}$ algorithm returning a solution within an additive factor of ε far from the optimal solution, and for non-migratory variant an $(((1 + \frac{\varepsilon}{1-\varepsilon})(1 + \frac{2}{n-2}))^{\alpha}\tilde{B}_{\alpha})$-approximation algorithm.

In Table 1 we summarize the results mentioned above and our contribution (in bold). There are also results for special cases of the energy minimization problems when jobs have life intervals of a specific structure (common, agreeable, laminar, purely laminar) or/and equal work volumes [3,5,7,15,20]. Some of the works mentioned above [2,3,11,20] as well as [9,10] study online algorithms for energy minimization problems in the speed scaling setting on a single processor or homogeneous multiprocessors.

Our Results. In Sect. 2, we give a $2(1+\varepsilon)(5(1+\varepsilon))^{\alpha-1}\tilde{B}_{\alpha} = O_{\alpha}(1)$-approximation to the NON PREEMPTIVE MINIMUM ENERGY SCHEDULING (α) problem when all work volumes are equal. This is the first constant factor approximation algorithm for this problem. Our algorithm extends to the case where job volume differ: it provides a $(\frac{5}{2})^{\alpha-1}\tilde{B}_{\alpha}((1+\varepsilon)(1 + \frac{w_{max}}{w_{min}}))^{\alpha}$-approximation to the NON PREEMPTIVE MINIMUM ENERGY SCHEDULING (α) problem where $w_{max} = \max_i w_i$ and $w_{min} = \min_i w_i$ (Theorem 2.1). This is the first multiprocessor algorithm with an approximation factor independent of n and m, improving on the previous approximation of $m^{\alpha-1}(n^{1/m})^{\alpha-1}$ [7]. Up to an additional factor of $(w_{max}/w_{min})^{\alpha}$, our algorithm further extends to the case where the work of jobs w_{ij} depends on the processor i on which j is executed.

In Sect. 3, we prove (Theorem 3.1) that a natural LP relaxation for NON PREEMPTIVE MINIMUM ENERGY SCHEDULING (α) on a single processor has integrality gap at most $12^{\alpha-1}$. Our LP relaxation is obtained from the *compact* LP relaxation in the preemptive setting of [8] (equivalent to their configuration LP) by adding a constraint capturing non-preemption. In general, when faced with hard optimization problems with weak linear programming relaxations, it is a basic approach to find additional constraints that will reduce the integrality gap.

Table 1. Known and our (in bold) results for minimum energy scheduling problems. Problems are denoted by extending the standard three-field notation of Graham et al. [13]. P denotes a homogeneous multiprocessor where all processors obey the same speed-to-power function s^α, while R is used to denote a heterogeneous multiprocessor where each processor has its own speed-to-power function $s^{\alpha(p_i)}$. For both environments the work volume of each job may depend on the processor it is executed and this is indicated by including w_{ij} in the second field.

Problem	Complexity	Approximation ratio
$1\|r_j, d_j, \mathrm{pmtn}\|E$	Polynomial [20]	
$1\|r_j, d_j\|E$	NP-Hard [5]	$2^{5\alpha-4}$ [5]
		$2^{\alpha-1}(1+\varepsilon)^\alpha \tilde{B}_\alpha$ [8]
		$(12(1+\varepsilon))^{\alpha-1}$
		[Theorem 3.1]
$P\|r_j, d_j, \mathrm{pmtn}, \mathrm{mig}\|E$	Polynomial [2,4]	
$P\|r_j = 0, d_j = 1, \mathrm{pmtn}, \mathrm{no\text{-}mig}\|E$	NP-Hard [3]	PTAS [3]
$P\|r_j, d_j, \mathrm{pmtn}, \mathrm{no\text{-}mig}\|E$	NP-Hard	$B_{\lceil\alpha\rceil}$ [14]
$P\|r_j, d_j\|E$	NP-Hard	$m^\alpha(n^{1/m})^{\alpha-1}$ [7]
		$(\frac{5}{2})^{\alpha-1}\tilde{B}_\alpha((1+\varepsilon)(1+\frac{w_{\max}}{w_{\min}}))^\alpha$
		[Theorem 2.1]
$P\|r_j = 0, d_j = 1, w_{i,j}, \mathrm{pmtn}, \mathrm{no\text{-}mig}\|E$	APX-hard	
	[Theorem 4.1]	
$R\|r_{ij}, d_{ij}, w_{ij}, \mathrm{pmtn}, \mathrm{mig}\|E$	Polynomial$(\frac{1}{\varepsilon})$ [8]	$OPT + \varepsilon$ [8]
$R\|r_{ij}, d_{ij}, w_{ij}, \mathrm{pmtn}, \mathrm{no\text{-}mig}\|E$	NP-Hard	$(1+\varepsilon)^\alpha \tilde{B}_\alpha$ [8]

Our constraint closes the previously unbounded integrality gap (Lemma 3.3). This results in the first LP relaxation with a gap independent of n and the work w_j of the jobs for this problem.

In Sect. 4, we deal with the inapproximabilty of speed scaling problems and we prove (Theorem 4.1) that the multiprocessor problem is APX-hard even for jobs with common life intervals and work volumes in $\{1, 3, 4\}$. This is the first APX-hardness result for an energy minimization problem.

Due to space limitations the proofs of the lemmas and some theorems will be given in the full version of the paper [12].

Preliminaries. We define an *independent* set of jobs as a set of jobs whose life intervals do not mutually intersect. Moreover, an independent set of jobs is *good* if the life interval of no job falls between the deadlines of two consecutive jobs of this independent set.

Proposition 1.1. [5] *Let S and S' be two schedules such that job j is executed during intervals I and I' respectively. Then $E(S', j) = (|I|/|I'|)^{\alpha-1}E(S, j)$.*

2   Multiprocessor Scheduling

In this section we present an approximation algorithm for NON-PREEMPTIVE MINIMUM ENERGY SCHEDULING (α) problem and show the following theorem:

Theorem 2.1. *There exists a polynomial-time approximation algorithm for* NON-PREEMPTIVE MINIMUM ENERGY SCHEDULING (α), *with approximation factor* $((\frac{5}{2})^{\alpha-1}\tilde{B}_\alpha((1+\varepsilon)(1+\frac{w_{\max}}{w_{\min}}))^\alpha)$. *When all jobs have the same work volume, this factor becomes* $2(1+\varepsilon)(5(1+\varepsilon))^{\alpha-1}\tilde{B}_\alpha$.

Our algorithm uses a reduction to (a special case of) the non-migratory variant of PREEMPTIVE FULLY HETEROGENEOUS MINIMUM ENERGY SCHEDULING problem, studied by Bampis et al. [8]. In this problem, every processor i has a different speed-to power function, $s^{\alpha(p_i)}$, and both the life interval and the work of jobs are processor dependent.

Theorem 2.2. *[8] There is an approximation algorithm for the* PREEMPTIVE FULLY HETEROGENEOUS MINIMUM ENERGY SCHEDULING *problem without migration with approximation ratio* $(1+\varepsilon)^\alpha \tilde{B}_\alpha$.

Overview. The algorithm proceeds as follows: we consider the life intervals of all the jobs, greedily find m maximal independent sets, and assign the jobs of the ith independent set $\mathcal{J}_i$ to processor p_i. Then we partition time on p_i according to the deadlines of the jobs in $\mathcal{J}_i$, and restrict ourselves to schedules such that no execution interval on p_i overlaps such a deadline. We solve the resulting restricted problem using the algorithm from Bampis et al.'s [8] to obtain a feasible schedule. To analyze its cost, we first show that an optimal solution can be transformed into a solution satisfying our additional constraints and without increasing the cost by too much. We start with an optimal solution and attempt, for each job j in the i-th independent set, to move j to processor p_i and execute it in the middle fifth of its life interval. Its execution interval is then shrunk by a factor of at most 5 and, by Proposition 1.1, its energy consumption is increased by a factor of at most $5^{\alpha-1}$. If we are unable to do so for some j, it is because of some other job j' on processor p_i with a significant overlap with j, and we execute both j and j' during the time of overlap. To guarantee that no execution interval on p_i crosses one of our selected deadlines, we argue that each execution interval crosses at most one such deadline and further modify the schedule, restricting the execution interval to one of the two sides of the deadline, up to shrinking its execution interval by a factor of 2. Finally, the algorithm of Bampis et al. [8] provides an approximation to our constrained problem.

Algorithm and Analysis. To give a detailed description of our algorithm, let $\mathcal{J}_i$ and $\mathcal{J}_{i,l}$ denote subsets of jobs, $\forall i, l$, and I_i^k denote time interval, $\forall i, k$.
ALGORITHM 1

1. $R \leftarrow J$
2. For $i = 1$ to m:
   (a) $\mathcal{J}_i \leftarrow \emptyset$, $k \leftarrow 0$ and $t_0^i \leftarrow 0$.
   (b) While $\exists j \in R$ such that $\{j\} \cup \mathcal{J}_i$ is an independent set
   (c)   Find such a j with d_j minimum and let $\mathcal{J}_i \leftarrow \mathcal{J}_i \cup \{j\}$ and $R \leftarrow R \setminus \{j\}$.
   (d) $k \leftarrow k+1$ and $t_k^i \leftarrow d_j$.
   (e) $t_{k+1}^i \leftarrow +\infty$

3. For every processor p_i, for $k = 1$ to $|\mathcal{J}_i| + 1$, let $I_k^i = [t_{k-1}^i, t_k^i]$.
4. Create an instance of Problem PREEMPTIVE FULLY HETEROGENEOUS MINI-
   MUM ENERGY SCHEDULING as follows:
   (a) For every processor p_i, for every interval I_l^i, create a heterogeneous proces-
       sor (i, l) with $\alpha_{i,l} = \alpha$.
   (b) The new instance has the same set of jobs. For every job $j \in J$ which
       is alive during part or all of some I_l^i, set release date $r_{(i,l)j} = \max(r_j - t_{l-1}^i, 0)$, deadline $d_{(i,l)j} = \min(d_j - t_{l-1}^i, t_l^i - t_{l-1}^i)$, work $w_{(i,l)j} = w_j$.
5. Solve the created problem using the algorithm from [8].
   Let $J_{i,l}$ be the set of jobs scheduled (preemptively) on heterogeneous processor
   (i, l).
6. For each (i, l), reorder the execution intervals inside I_l^i so that the jobs of $J_{i,l}$
   are executed by order of non-decreasing deadline.

For the analysis of ALGORITHM 1, we first give the following Lemma which
has a crucial role in the analysis of the approximation ratio.

Lemma 2.3. *Let $\{\mathcal{J}_1, ..., \mathcal{J}_m\}$ be a subpartition of J such that each $\mathcal{J}_i$ is an
independent set of jobs, and S be a schedule of J. Then there exists a schedule
S' such that for every i all the jobs of $\mathcal{J}_i$ are executed on processor p_i, and whose
cost satisfies $E(S') \le (5/2)^{\alpha-1}(1 + \frac{w_{max}}{w_{min}})^\alpha E(S)$.*

We next show that we can force every job that is executed on processor p_i to be
scheduled during a subinterval of some $[t_{\ell-1}^i, t_\ell^i]$.

Lemma 2.4. *Let $\{\mathcal{J}_1, ..., \mathcal{J}_m\}$ be the subpartition of J found by ALGORITHM 1
and $\mathcal{I}$ be the set of all time intervals found at its Step 3. Let S' be a schedule
such that for every i all the jobs of $\mathcal{J}_i$ are executed on processor p_i. There exists
a schedule S'' such that for each processor p_i and for each job j that is executed
on p_i in S', in S'' j is executed on p_i and the execution interval of j is included
in some I_l^i; moreover, for any j, $E(S'', j) \le 2^{\alpha-1}E(S', j)$.*

Looking at Lemma 2.3, we notice that the jobs whose cost has changed between
S and S' are, in S', now executed on p_i during the life interval of an element
of $\mathcal{J}_i$. Looking at Lemma 2.4, we notice that the jobs whose cost has changed
between S' and S'' are, in S', executed on p_i in an interval that overlaps one of
the t_ℓ^i. Thus for every j we have $E(S, j) = E(S', j)$ or $E(S', j) = E(S'', j)$. Hence
the cost of a job after the two transformations from S to S' to S'' increases by
a factor of at most:

$$\frac{E(S'')}{E(S)} \le \max\left(\left(\frac{5}{2}\right)^{\alpha-1} \cdot \left(1 + \frac{w_{max}}{w_{min}}\right)^\alpha, 2^{\alpha-1}\right) = \left(\frac{5}{2}\right)^{\alpha-1} \cdot \left(1 + \frac{w_{max}}{w_{min}}\right)^\alpha.$$

Finally, observe that the last step of ALGORITHM 1 transforms the schedule
into a non-preemptive schedule that is feasible and has the same cost. Putting
Lemmas 2.3, 2.4 and Theorem 2.2 together, we obtain an approximation ratio of

$$\left(\frac{5}{2}\right)^{\alpha-1} \cdot \left((1 + \varepsilon)\left(1 + \frac{w_{max}}{w_{min}}\right)\right)^\alpha \cdot \tilde{B}_\alpha.$$

By an easy reduction, Theorem 2.1 extends to the case where the work volume of each job, w_{ij}, depends on the processor p_i on which job j is executed, up to losing an additional factor of $(w_{\max}/w_{\min})^\alpha$ in the approximation ratio. Remark, also, that ALGORITHM 1 creates an instance of the PREEMPTIVE FULLY HETEROGENEOUS MINIMUM ENERGY SCHEDULING problem where all the processors have the same α. However, no better approximation ratio is known for this special case.

3 Single Processor Scheduling

In this section, we present a new LP relaxation for NON PREEMPTIVE MINIMUM ENERGY SCHEDULING (α) on a single processor.

Theorem 3.1. *The linear program **LP1** has integrality gap at most $12^{\alpha-1}$.*

Linear Programming Formulation. For each job j and execution interval I, variables $x_{I,j}$ indicates whether j is assigned to I in the schedule. Lemma 3.2, due to Huang and Ott [15], allows us to restrict our attention to execution intervals that begin and end in some polynomial size set T of times. Let $\mathcal{I}$ denote the set of all the intervals with both endpoints at a landmark. Since j must be scheduled, $\sum_I x_{I,j} = 1$. Since at any time t, at most one job is being processed, $\sum_j \sum_{I \ni t} x_{I,j} \leq 1$.

Lemma 3.2 (Discretization of Time). *[15] Let $r_1, \ldots, r_{2n}$ be the release dates and deadlines of jobs. For each $1 \leq i < 2n$, create $n^2(1 + \frac{1}{\varepsilon}) - 1$ equally-spaced "landmarks" in the interval $[r_i, r_{i+1}]$. Let S be a solution of minimal cost such that for each job j and each consecutive landmarks t_i, t_{i+1}, either job j is executed during the whole interval $[t_i, t_{i+1}]$ or not at all. Then $E(S) \leq (1 + \varepsilon)^{\alpha-1} OPT$.*

Linear program LP1

$$\text{minimize } E(\boldsymbol{x}) = \sum_{j \in J} \sum_{I \in \mathcal{I}} x_{I,j} \left(\frac{w_j}{|I|} \right)^\alpha \quad \text{subject to:}$$

$$\text{job constraint} \sum_{I \in \mathcal{I}} x_{I,j} \geq 1 \qquad\qquad \forall j \in J \quad (1)$$

$$\text{processor constraint} \sum_{\substack{j \in J}} \sum_{\substack{I \in \mathcal{I} \\ t \in I}} x_{I,j} \leq 1 \qquad\qquad \forall \text{ landmark } t \quad (2)$$

$$\text{non-preemption} \sum_{\substack{I' \in \mathcal{I} \\ I' \cap I \neq \emptyset}} x_{I',j} + \sum_{\substack{I'' \in \mathcal{I} \\ I \subseteq I'', j' \in J}} x_{I'',j'} \leq 1 \quad \forall I \in \mathcal{I}, \forall j \in J \quad (3)$$

$$x_{I,j} \geq 0 \qquad\qquad \forall j \in J, \forall I \in \mathcal{I}(j) \quad (4)$$

$$x_{I,j} = 0 \qquad\qquad \forall I \notin \mathcal{I}(j) \quad (5)$$

Our linear program contains an additional new constraint (3) "capturing" non-preemption - it is only valid for non preemptive schedules: if job j is scheduled during interval I or a subinterval of I, then no other job can be scheduled during an interval that contains I. This constraint is necessary: without it, the integrality gap is unbounded. To show that the gap without constraint (3) may be large, we first note that for an instance of MINIMUM ENERGY SCHEDULING (α) on a single processor, the ratio between the minimum energy needed when preemption is not allowed and when preemption is allowed can be as last as $\Omega(n^{\alpha-1})$ [7]. Call this ratio the *price of non-preemption*. Figure 1 depicts an instance where this ratio is $\Omega(n^{\alpha-1})$. As a consequence of our constant bound on **LP1**, the gap without constraint (3) is therefore at least as large.

Lemma 3.3. *Without constraint (3), **LP1** has integrality gap at least $\Omega(n^{\alpha-1})$.*

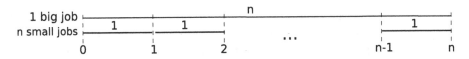

Fig. 1. An instance on which LP1 has an integrality gap of at least $\Omega(n^{\alpha-1})$. The figure shows the life intervals of the $n + 1$ jobs and, above them, their work volume.

The remainder of this section is devoted to proving Theorem 3.1.

Overview. We show that any fractional solution can be transformed into an integral solution without increasing the value of the solution by too much, in three steps. We first divide the time into zones and transform the fractional solution so all (non-zero) fractional execution intervals are inside a zone. Then, each zone is divided into nested subzones and we further transform our fractional solution so that all fractional execution intervals are inside a subzone and the life interval of the corresponding job contains that subzone. Finally, we build a weighted bipartite graph from the transformed fractional solution whose edges represent the possible allocation of execution intervals to subzones. Similarly to [19], we find an integral (weighted) matching in this graph and translate this solution to an integral schedule. We then show that the cost of the integral solution we built is at most $12^{\alpha-1}$ times the cost of the original fractional solution.

Building an Integral Solution from a Fractional Solution. We now detail our transformation of the fractional solution. It consists of three steps. The first step is derived from Antoniadis and Huang's algorithm [5].

Splitting Execution Intervals on Deadlines. Our first transformation turns a fractional solution into a fractional solution where $x_{I,j}$ is 0 for any execution interval I that contains any points in a set of deadlines we pick. The deadlines we pick are the deadlines of a good independent set.

The constraint (3) is crucial for proving the next Lemma.

Lemma 3.4. *Let **LP1** be the LP obtained from an instance of* NON PREEMP-
TIVE MINIMUM ENERGY SCHEDULING *(α) and $\mathcal{J}$ be any good independent set
for this instance. In polynomial time, we can transform any fractional solution
$\boldsymbol{x}$ to **LP1** to a fractional solution $\boldsymbol{y}$ of value at most $2^{\alpha-1}E(\boldsymbol{x})$ where $y_{I,j} = 0$
if I crosses a deadlines of $\mathcal{J}$.*

Further Splits. We now proceed to our second transformation where we further
split the execution intervals of a fractional solution. Now that all execution
intervals (in the support of $\boldsymbol{x}$) lie between two consecutive deadlines (which
we now call a "zone"), we can further partition each zone so the first half is
dedicated to jobs whose life interval ends in that zone and the second half is
dedicated to the others (namely, jobs whose life interval starts in that zone and
jobs whose life interval contains the zone).

Lemma 3.5. *Let **LP1** be the LP obtained from an instance of* NON PREEMP-
TIVE MINIMUM ENERGY SCHEDULING *(α) and $\mathcal{J}$ be any good independent set
for this instance. In polynomial time, we can transform any fractional solution
$\boldsymbol{y}$ where $y_{I,j} = 0$ if I crosses a deadlines of $\mathcal{J}$ to **LP1** to a fractional solution $\boldsymbol{z}$
of value at most $2^{\alpha-1}$ times the value of $\boldsymbol{y}$ where $z_{I,j} > 0$ implies*

1. *$I \subseteq [d_s, d_s + \frac{1}{2^k}(d_e - d_s)] \subseteq L_j$ for some consecutive deadlines d_s, d_e of I and
   $k \geq 1$, or*
2. *$I \subseteq [d_e - \frac{1}{2^k}(d_e - d_s), d_e] \subseteq L_j$ for some consecutive deadlines d_s, d_e of I and
   $k \geq 1$.*

We let $\mathcal{Z}$ consists of all intervals of the form $[d_s, d_s + \frac{1}{2^k}(d_e - d_s)]$ and $[d_e - \frac{1}{2^k}(d_e - d_s), d_e]$ for consecutive deadlines d_s, d_e of I. Though they do not partition the
timeline, we still refer to $\mathcal{Z}$ as *subzones*.

Building a Weighted Bipartite Matching. As a result of Lemma 3.5, for
each $z_{I,j} > 0$, I is contained in some subzone Z and furthermore, the life interval of j contains Z so we can freely shift I to another interval (of the same
length) inside Z. Thus, we will only remember the length of the fractional execution intervals and the subzone Z in which they belong. I.e., we think of $\boldsymbol{z}$ as
a fractional assignment of lengths c_i for each job to Z.

Lemma 3.6. *If for each subzone Z, the lengths c_i assigned to Z and all subzones
included in Z is at most $|Z|$ then there is a feasible schedule where each job is
given their assigned length in Z.*

We now desire an integral assignment of lengths to each Z where the total of all
lengths assigned to Z does not exceed $|Z|$. Note that this constraint is satisfied
by the fractional solution derived from $\boldsymbol{z}$ (as $\boldsymbol{z}$ satisfies constraint (2)).

To obtain such an integral assignment from our fractional assignment derived
from $\boldsymbol{z}$, we build a weighted bipartite graph $G(\boldsymbol{z})$ where assignments correspond
to matchings and the weight of a matching correspond to the energy cost (of the
matching interpreted as a schedule). We will then obtain an integral matching
from the derived fractional matching (whose weight is exactly $E(\boldsymbol{z})$).

We now describe $G(z)$ with bipartition (A, B) and weight $w(e)$ for each edge $e \in E(G)$. We also keep a *length* $\ell(e)$ for each edge e which will be used in the very last step of our proof (but in no way affects the weighted bipartite matching we look for).

- A contains one vertex for each job. I.e., $A = \{a_j \mid j \in J\}$
- B consists of vertices for subzones. However, B may contain more than one vertex for each subzone Z. In fact, the number of subzones it contains is the ceiling of the sum of fractional value of all lengths assigned to Z. I.e.,

$$B = \left\{ b_{Z,i} \mid Z \in \mathcal{Z}, i \in 1, \dots, \left\lceil \sum_{j \in J} \sum_{I \subseteq Z} x_{I,j} \right\rceil \right\}$$

- The edges are constructed as follows. Start with all edges $a_j b_{Z,i}$ for all i if $x_{I,j} > 0$ for some $I \subseteq Z$. We now delete some edges to obtain the edges of $G(z)$ and assign weights and lengths of the remaining edges.
  Sort the lengths assigned to Z by z in decreasing order of length. For each such length c_k for job j of fractional value $z_{I,j}$, set $w(a_j b_{Z,i})$ to $\frac{w_j^\alpha}{c_k^{\alpha-1}}$ where i is the ceiling of the partial sum of all jobs previously considered for Z (i.e., $i = \lceil \sum_{q=1}^{k-1} c_q \rceil$). Set $\ell(a_j b_{Z,i})$ to c_k. Also set $w(a_j b_{Z,i+1})$ to $\frac{w_j^\alpha}{c_k^{\alpha-1}}$ and $\ell(a_j b_{Z,i+1})$ to c_k if adding j to the ceiling of the partial sum increases it by 1. Delete all other edges of the form $w(a_j b_{Z,t})$.

z naturally gives the following fractional matching $M(z)$ of $G(z)$ with total weight $E(z)$: we pick each edge with weight exactly z_k (or z_k split into two as follows if adding z_k increased the ceiling of the partial sum by 1. Whatever we need to add to the partial sum to make it an integer is the fraction we choose of the first edge, and the rest of z_k for the second edge).

To complete the description of our final transformation, we use and the following lemma.

Lemma 3.7. *Let $G(z)$ be the bipartite graph built from a transformed fractional solution z. For any matching M saturating A of $G(z)$, we can obtain a schedule whose energy consumption is at most $3^{\alpha-1}$ times the weight of M.*

To prove the integrality gap of **LP1**, we simply need to apply each Lemma in this section in turn.

Given an optimal fractional solution x to **LP1**, find a good independent set $\mathcal{J}$ (to the instance which generate the LP) and apply Lemma 3.4 to x and I to obtain a fractional solution y of value at most $2^{\alpha-1} E(x)$ where no execution interval crosses a deadline in $\mathcal{J}$.

Then apply Lemma 3.5 to y to obtain z of value at most $2^{\alpha-1} E(y) \leq 4^{\alpha-1} E(x)$ where for any non-zero $z_{I,j}$, I is contained in a "subzone" of the form $[d_s, d_s + \frac{1}{2^k}(d_e - d_s)]$ or $[d_e - \frac{1}{2^k}(d_e - d_s), d_e]$ for some consecutive deadlines d_s, d_e and $k \geq 1$ and furthermore L_j contains this subzone.

Now build $G(z)$ and interpret z as a fractional matching in $G(z)$. It is known that in a weighted bipartite graph, there exists an (integral) matching of same weight as any fractional matching [18,19]. Thus, $G(z)$ has a matching M of same weight as this fractional matching and by Lemma 3.7, we can build a schedule from M whose energy consumption is at most $3^{\alpha-1}E(z) \leq 12^{\alpha-1}E(x)$.

Such a schedule is of course a solution to **LP1** of same value, thus completing the proof.

Algorithm Summary. We can summarize our algorithm for transforming any fractional solution x to an integral solution.

ALGORITHM 2

1. Apply the transformation of Lemma 3.4 and then Lemma 3.5 to the fractional solution to obtain a new fractional solution z.
2. Construct the weighted bipartite graph $G(z) = (A, B)$.
3. Find a minimum weight matching M that matches every node in A.
4. For each edge $e = (a_j, b_{Z,i}) \in M$, schedule job j in the subzone Z with an interval of length $\frac{\ell(e)}{3}$. Use an earliest deadline first schedule for all jobs in Z if Z is in the first half of a zone and a latest release date first schedule if Z is in the second half of a zone.

As a corollary to Theorem 3.1, we obtain the following theorem for non-preemptive scheduling on a single processor.

Theorem 3.8. *There exists a polynomial-time algorithm which computes a $(12(1 + \varepsilon))^{\alpha-1}$-approximation to the* NON PREEMPTIVE MINIMUM ENERGY SCHEDULING (α) *problem on a single processor*

Compared to the previous best approximations of $\min\{2^{\alpha-1}\tilde{B}_\alpha, 2^{5\alpha-4}\}$ [5,8], this is always better than $2^{5\alpha-4}$ and better than $2^{\alpha-1}\tilde{B}_\alpha$ for any $\alpha \geq 25$.

4   Hardness of Approximation

In this section, we sketch the proof of APX-hardness for a variant of NON PRE-EMPTIVE MINIMUM ENERGY SCHEDULING (α) (see [12] for detailed definitions and proofs). In this MINIMUM ENERGY SCHEDULING WITH PROCESSOR DEPEN-DENT WORKS (α) problem, jobs have work volumes w_{ij} depending on which processor they are scheduled on. Our reduction is from MAXIMUM BOUNDED 3-DIMENSIONAL MATCHING; this is the usual 3-Dimensional Matching prob-lem where every element of the ground set appears in at least 1 and at most 3 sets (triples), known to be APX-hard [17] even for instances where the optimal solution has size equal to the cardinality of the ground sets.

Our construction draws some ideas from Azar et al. [6] and Lenstra et al. [16]. We define polynomial time computable mappings between instances of these two problems. Our mapping builds instances of MINIMUM ENERGY SCHEDULING WITH PROCESSOR DEPENDENT WORKS (α) where all jobs have common release dates and deadlines, and are therefore *agreeable*. Moreover, all work w_{ij} are 1, 3 or 4 so, our result shows the problem with constant ratio for which we gave an approximation algorithm in the Sect. 2 is already APX-Hard.

Specifically, our instance has one job for each element of the ground set and one machine for each triple T where jobs for members of T can be scheduled cheaply. We add some "dummy jobs" and "dummy machines" to help us bound the cost of elements we think of as unmatched. Our reduction proves.

Theorem 4.1. MINIMUM ENERGY SCHEDULING WITH PROCESSOR DEPENDENT WORKS *is APX-Hard.*

References

1. Albers, S.: Energy-efficient algorithms. Commun. ACM **53**(5), 86–96 (2010)
2. Albers, S., Antoniadis, A., Greiner, G.: On multi-processor speed scaling with migration: extended abstract. In: SPAA 2011, pp. 279–288 (2011)
3. Albers, S., Müller, F., Schmelzer, S.: Speed scaling on parallel processors. In: SPAA 2007, pp. 289–298 (2007)
4. Angel, E., Bampis, E., Kacem, F., Letsios, D.: Speed scaling on parallel processors with migration. In: Kaklamanis, C., Papatheodorou, T., Spirakis, P.G. (eds.) Euro-Par 2012. LNCS, vol. 7484, pp. 128–140. Springer, Heidelberg (2012)
5. Antoniadis, A., Huang, C.-C.: Non-preemptive speed scaling. J. Sched. **16**(4), 385–394 (2013)
6. Azar, Y., Epstein, L., Richter, Y., Woeginger, G.J.: All-norm approximation algorithms. J. Algorithms **52**(2), 120–133 (2004)
7. Bampis, E., Kononov, A., Letsios, D., Lucarelli, G., Nemparis, I.: From preemptive to non-preemptive speed-scaling scheduling. In: Du, D.-Z., Zhang, G. (eds.) COCOON 2013. LNCS, vol. 7936, pp. 134–146. Springer, Heidelberg (2013)
8. Bampis, E., Kononov, A., Letsios, D., Lucarelli, G., Sviridenko, M.: Energy efficient scheduling and routing via randomized rounding. In: FSTTCS, pp. 449–460 (2013)
9. Bansal, N., Bunde, D.P., Chan, H.-L., Pruhs, K.: Average rate speed scaling. Algorithmica **60**(4), 877–889 (2011)
10. Bansal, N., Chan, H.-L., Katz, D., Pruhs, K.: Improved bounds for speed scaling in devices obeying the cube-root rule. Theory Comput. **8**(1), 209–229 (2012)
11. Bansal, N., Kimbrel, T., Pruhs, K.: Speed scaling to manage energy and temperature. J. ACM **54**(1), 3:1–3:39 (2007)
12. Cohen-Addad, V., Li, Z., Mathieu, C., Milis. I.: Energy-efficient algorithms for non-preemptive speed-scaling. CoRR (2014)
13. Graham, R.L., Lawler, E.L., Lenstra, J.K., Rinnooy Kan, A.H.G.: Optimization and approximation in deterministic sequencing and scheduling: a survey. Ann. Discrete Math. **5**, 287–326 (1979)
14. Greiner, G., Nonner, T., Souza, A.: The bell is ringing in speed-scaled multiprocessor scheduling. Theory Comput. Syst. 1–21 (2013)
15. Huang, C.-C., Ott, S.: New results for non-preemptive speed scaling. Research report, Max-Planck-Institut für Informatik (2013)
16. Lenstra, J.K., Shmoys, D.B., Tardos, É.: Approximation algorithms for scheduling unrelated parallel machines. Math. Program. **46**(3), 259–271 (1990)
17. Petrank, E.: The hardness of approximation: gap location. Comput. Complex. **4**(2), 133–157 (1994)
18. Plummer, M.D., Lovász, L.: Matching Theory. Elsevier Science, New York (1986)
19. Shmoys, D.B., Tardos, É.: An approximation algorithm for the generalized assignment problem. Math. Program. **62**(3), 461–474 (1993)
20. Yao, F., Demers, A., Shenker, S.: A scheduling model for reduced CPU energy. In: FOCS 1995, pp. 374–382 (1995)

Optimal Online and Offline Algorithms for Robot-Assisted Restoration of Barrier Coverage

J. Czyzowicz[1], E. Kranakis[2]($\boxtimes$), D. Krizanc[3], L. Narayanan[4], and J. Opatrny[4]

[1] Département d'informatique, Université du Québec en Outaouais,
Gatineau, QC, Canada
Jurek.Czyzowicz@uqo.ca

[2] School of Computer Science, Carleton University, Ottawa, Canada
kranakis@scs.carleton.ca

[3] Department of Mathematics and Computer Science, Wesleyan University,
Middletown, CT, USA
dkrizanc@wesleyan.edu

[4] Department of Computer Science and Software Engineering, Concordia University,
Montreal, QC, Canada
{lata,opatrny}@cs.concordia.ca

Abstract. Cooperation between mobile robots and wireless sensor networks is a line of research that is currently attracting a lot of attention. In this context, we study the following problem of barrier coverage by stationary wireless sensors that are assisted by a mobile robot with the capacity to move sensors. Assume that n sensors are initially arbitrarily distributed on a line segment barrier. Each sensor is said to cover the portion of the barrier that intersects with its sensing area. Owing to incorrect initial position, or the death of some of the sensors, the barrier is not completely covered by the sensors. We employ a mobile robot to move the sensors to final positions on the barrier such that barrier coverage is guaranteed. We seek algorithms that minimize the length of the robot's trajectory, since this allows the restoration of barrier coverage as soon as possible. We give an optimal linear-time offline algorithm that gives a minimum-length trajectory for a robot that starts at one end of the barrier and achieves the restoration of barrier coverage. We also study two different online models: one in which the online robot does not know the length of the barrier in advance, and the other in which the online robot knows it in advance. For the case when the online robot does not know the length of the barrier, we prove a tight bound of 3/2 on the competitive ratio, and we give a tight lower bound of 5/4 on the competitive ratio in the other case. Thus for each case we give an optimal online algorithm.

1 Introduction

Mobile robots and wireless sensor networks are related areas of research that have largely been studied by different communities of researchers. Recently,

This work was partially supported by NSERC grants.

E. Bampis and O. Svensson (Eds.): WAOA 2014, LNCS 8952, pp. 119–131, 2015.
DOI: 10.1007/978-3-319-18263-6_11

there has been increasing interest in the possibilities uncovered by utilizing *both* technologies [14]: what if mobile robots and wireless sensors could *cooperate* to solve problems and perform tasks? Environments where autonomous networked entities such as robots and sensors cooperate to achieve a common goal are sometimes called *mixed-mode environments* and have been the subject of several recent research events, e.g., [10,18].

In this paper, we study a related mixed-mode problem for barrier coverage. Assume n stationary sensors have initial positions on a line segment barrier. Owing to incorrect initial placement, or the death of some sensors due to battery failure or a disaster, the barrier is not completely covered by the sensors. An illustration is given in Fig. 1(a), where the segment of the barrier covered by a sensor is represented as a box.

The task of the mobile robot is to walk along the barrier segment, pick up and move sensors to final positions such that barrier coverage is restored, i.e., in their final positions, the sensors collectively cover the entire line segment barrier as in Fig. 1(b). Note that the final positions that achieve barrier coverage are not unique. Since sensors may need to be moved in different directions, i.e. left

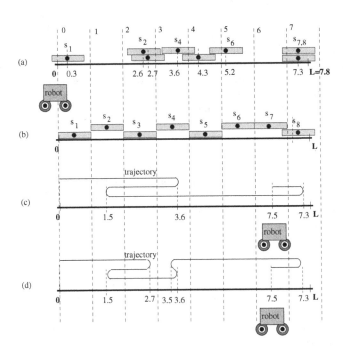

Fig. 1. Robot-assisted restoration of barrier coverage problem with sensor range equal to 0.5: (a) the initial configuration on segment $[0, L]$ with gaps in coverage, (b) a possible solution, (c) and (d) give examples of two trajectories of different length that could be followed by the robot to obtain the final configuration in (b).

or right, to assure coverage, the robot may sometimes need to turn or change direction in order to restore coverage. The robot may decide to resolve the gap as soon as possible, as late as possible, or some time in between. The robot thus follows a certain *trajectory*, which can be specified by the starting point, and a sequence of points where the robot alternately turns left and right before it reaches its termination point. Given the initial configuration of Fig. 1(a), two of the possible trajectories that achieve the same final positions of sensors are shown in Figs. 1(c) and (d). The time needed to restore barrier coverage is clearly related to the length of the robot's trajectory, which in turn depends on the knowledge it has of the initial positions of sensors. The problem we are interested in is finding an *optimal* trajectory for the mobile robot in order to achieve barrier coverage as fast as possible.

Sensor Relocation Model. In the sequel we define the capabilities of the sensors and the robot, as well as the trajectory of the robot.

Sensors. Assume that n *sensors* $s_1, s_2, \ldots, s_n$ are distributed on the line segment $[0, L]$ of length L with endpoints 0 and L in locations $x_1 \leq x_2 \leq \ldots \leq x_n$. The range of all sensors is assumed to be identical, and is equal to a positive real number $r > 0$. Thus sensor s_i in position x_i defines a closed interval $[x_i - r, x_i + r]$ of length $2r$ centered at the current position x_i of the sensor, in which it can detect an intruding object or an event of interest. See Fig. 1(a) for an illustration of a problem instance. We say that the sensor *covers* the closed interval $[x_i - r, x_i + r]$. We assume that the total range of the sensors is sufficient to cover the entire line segment $[0, L]$, i.e., $2rn \geq L$. We define a *gap* to be a closed subinterval G of $[0, L]$ such that no point in G is within the range of a sensor. Clearly, an initial placement of the sensors may have gaps. The sensors provide *complete coverage* of $[0, L]$ if they leave no gaps.

Robot. There is a *mobile robot* that can move the sensors to positions that guarantee coverage of the entire line segment. We assume that the robot can *pick, carry, move* and *drop/deposit* sensors from any initial position to any desired position on the line segment. There is no constraint on the direction and number of turns it can take (left or right) so as to pick and/or drop sensors, and no restriction on where in the line segment it can drop the sensors. We study the case when sensors are small enough and thus the robot can potentially carry all the sensors it needs at the same time.

Robot Trajectory and Length. Our goal is to provide offline and online algorithms so as to minimize the time taken to restore barrier coverage. Assuming constant speed, we measure this by the distance travelled by the robot from its starting position to complete the task of moving sensors to positions which guarantee complete coverage of the barrier. We assume that the mobile robot starts at position 0 and moves to the right. At some point it can turn and move left, then again turn and move right and so on. Thus its trajectory can be specified as a sequence of points on the barrier: $[t_0 = 0, t_1, t_2, \ldots, t_m]$, where the points $t_1, t_3, \ldots$ are the points where the robot turns left, the points $t_2, t_4, \ldots$ are the points where the robot turns right, and finally, the point t_m is the *termination point* of the trajectory. Therefore, $t_i > t_{i-1}$ for all odd i while $t_i < t_{i-1}$ for all even i where $0 < i \le m$, and the robot's trajectory is the sequence of line segments $[0, t_1], [t_1, t_2], \cdots, [t_{m-1}, t_m]$. The length of the trajectory is defined as $\sum_{i=1}^{m} |t_i - t_{i-1}|$.

We seek algorithms that calculate an *optimal* trajectory for the mobile robot that ensure barrier coverage, i.e. a trajectory of smallest possible length. A mobile robot using an *offline* algorithm to calculate its trajectory is assumed to know all the initial positions of sensors before starting its trajectory. On the other hand, a robot using an *online* algorithm knows about sensors only in the parts of the barrier segment where it has already travelled. Specifically, an online robot discovers the presence or absence of a sensor at position x only when reaching x. Therefore, at the start of the algorithm, such a robot has no knowledge about any of the sensors' positions. It can of course remember any sensors that it has seen previously.

Related Work. Barrier coverage using wireless sensors has been the subject of intensive research in the last decade [1,16,19]. Some papers assumed randomized deployment of sensors on the barrier and analyzed the probability of barrier coverage. Other papers have studied the case of relocatable sensors [21,22], which start at arbitrary positions and can move to final positions that achieve barrier coverage. Centralized algorithms for minimizing the maximum and average movement of sensors were studied in [5–7] respectively. Multiple barriers were studied in [2], and distributed algorithms for barrier coverage were given for the first time in [11].

Charikar et al. [4] consider the k-delivery TSP problem for transporting efficiently n identical objects, placed at arbitrary initial locations, to n target locations with a vehicle that can carry at most k objects at a time. Chalopin et al. [3] provide hardness results, exact, approximation, and resource-augmented algorithms for the problem of whether there is a schedule of agents' movements that collaboratively deliver data from specified sources of a network to a central repository. Our problem differs both in being uncapacitated, and in the fact that the locations of the sources and targets are not known in advance.

Online vehicle routing problems and the online travelling salesman problem have been studied previously; see [12] for a survey. Our problem and our conception of online are quite different: the locations the robot needs to deposit sensors are not pre-determined, and we assume an online robot discovers the positions of sensors as it moves along the barrier.

Cooperation between mobile robots and wireless sensors is a relatively new research area and has been explored in several research events in the last couple of years [10,14,17,18]. The authors of [9,20] use information obtained from wireless sensors for the problem of localization of a mobile robot. In [13], mobile robots and stationary sensors cooperate in a target tracking problem: stationary sensors track moving sensors in their sensor range, while mobile robots explore regions not covered by the fixed sensors. A common evaluation platform for mixed-mode environments incorporating both mobile robots and wireless sensor networks is described in [15].

Our Results. We give a linear time offline algorithm that computes an optimal trajectory for a robot starting at an endpoint of the barrier to restore barrier coverage. For the online case, we show that when the robot does not know the length of the barrier and recognizes the end of it only when reaching it, any algorithm must have a competitive ratio of at least 3/2. We give a simple algorithm that matches this bound. When the robot does know the length of the barrier, we show a lower bound of 5/4 on the competitive ratio of any online algorithm for the problem. We then give an adaptive online algorithm whose competitive ratio matches this lower bound.

Due to the page limit, most of the proofs are omitted, they can be found in an extended version of the paper [8].

2   Optimal Offline Algorithm

In our offline algorithm we assume that the robot has global knowledge of the positions of sensors on the line segment, and that during the course of its movement, can pick up and carry as many sensors as necessary and deposit them as required. All sensors have identical range denoted by r, and the robot starts at the endpoint 0 of the interval $[0, L]$, and the number of given sensors is sufficient to cover the given interval.

Obviously, when the barrier does not contain any gap, the trajectory is empty and we consider below instances containing gaps. We begin by establishing the properties of optimal non-empty trajectories of the robot, which are crucial to the development of the algorithm. We say that a solution is *order-preserving* if the final order of the position of the sensors is the same as their initial positions. Secondly, a solution is called *fully stretched* if the robot places all sensors in *attached positions*, i.e., two consecutive sensors encountered by the robot are placed at distance $2r$ and the first sensor is at distance r from 0, except possibly the sensors at or after the termination point t_m as in the example in Fig. 1(b).

Lemma 1. (Order-preserving fully stretched solution) *There exists an optimal trajectory for the robot that produces an order-preserving fully-stretched solution.*

Lemma 2. (Three Visits Lemma) *The trajectory of an optimal algorithm does not contain the same point of the line segment more than three times. Furthermore, the last point of the trajectory can occur in the trajectory at most twice.*

Observe that the above lemmas applies to both offline and online algorithms. Furthermore, once a trajectory is specified, the robot can produce an order-preserving fully stretched solution as discussed below, so it suffices to specify the trajectory of the robot.

Given an optimal trajectory $[t_0, t_1, t_2, t_3 \ldots, t_{m-1}, t_m]$, the robot makes a left turn at $t_1, t_3, \ldots$ and right turns at $t_2, t_4, \ldots$. Therefore, the segments $[t_2, t_1], [t_4, t_3],$ $\ldots, [t_{2i}, t_{2i-1}], \ldots$ of $[0, L]$ are traversed by the robot three times, $1 \le i \le (m-1)/2$, and if m is even, then the segment $[t_m, t_m - 1]$, is traversed twice. Furthermore, all these segments are pairwise disjoint, except possibly for the endpoints of the segments. We call the part of the trajectory $[t_{2i}, t_{2i-1}], 1 \le i \le (m-1)/2$, traversed by the robot three times, a *triple*, t_{2i-1} is called its *left turning point* and

t_{2i} is called its *right turning point*. If m is even, then the segment $[t_m, t_{m-1}]$ is called a *double* and t_{m-1} is called its *left turning point*. Any line segment in the trajectory that is traversed exactly once by the robot is called a *straight line segment* (see Fig. 2). When following a straight line segment the robot necessarily has sufficient supply of sensors and deposits them in attached positions. When following a segment of a triple or a double for the first time, the robot picks all sensors found there and deposits then in attached positions when going back over the segment (see the proof of Lemma 2). Clearly, if two consecutive triples, or a triple and a double share an endpoint, these two moves can be merged into a single triple, or a double. This observation and the preceding lemmas imply the following corollary.

Corollary 1. *There is an optimal order-preserving and fully stretched trajectory of the robot that produces a complete coverage of $[0, L]$ which consists of k consecutive triples and straight line segments for some $k \geq 0$, and ends with a straight line segment or a double (see Fig. 2).*

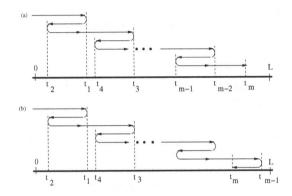

Fig. 2. Two possible shapes of an optimal trajectory.

To construct an optimal trajectory of the robot we need to determine, from the given input instance, the ends of the triples and a double. The following definition of coverage balance is used to determine them.

Definition 1. *The* coverage balance *of sensor s_i at location x_i is defined to be $C_i = (2ri - r) - x_i$, i.e., the difference between the total length that can be covered by sensors $s_1, s_2, \ldots, s_i$ up to the center of s_i, and the distance of s_i to the beginning of the interval.*

Consider the example in Fig. 1. The coverage balance of sensors listed from left to right are $0.2, -1.1, -0.2, -0.1, 0.2, 0.3, -0.8$ and 0.2. Notice that in the two examples of trajectories in this figure each left turn is done at a sensor with negative coverage balance. However, doing a left turn at every sensor with negative coverage balance would be wrong, because it could violate the three visits lemma. Similarly, doing a triple involving many consecutive sensors with negative coverage balance can be sub-optimal as well, as seen in the trajectory of Fig. 1(c). The following lemma specifies all potential left turning points.

Lemma 3. *Let* $([t_0, t_1, t_2, t_3 \ldots, t_{m-1}, t_m])$ *be an optimal trajectory which minimizes the number of triples.*

1. *Every sensor with negative coverage balance is shifted left, and thus its location is in a triple or the double segment.*
2. *No triple segment contains the location of a sensor with nonnegative coverage balance.*
3. *In a double segment the rightmost sensor has negative coverage balance.*
4. *For every triple* $[t_{2i}, t_{2i-1}]$*, the left-turning point* t_{2i-1} *is a location of a sensor* x_j *for some integer* j *such that either* $-2r < C_j < 0$*, or both,* $-2r = C_j$ *and* $x_j = x_{j+1}$*, and the coverage balance of every other sensor located in the interval* $[t_{2i}, t_{2i-1}]$ *is less than or equal to* $-2r$*.*
5. *Let* k *be the smallest integer such that all gaps in* $[0, L]$ *are to the left of* $2rk$*. Then* s_k *is the last sensor to be moved. Let* $c = x_k$ *if* $C_k < 0$*, else* $c = x_k + C_k$*. If the trajectory does not end with a double then* $C_k \geq 0$*,* m *is odd, and the termination point* $t_m = c$*. Otherwise the trajectory ends with a double, and* $t_{m-1} = c$*, i.e., the left-turning point of the double is* c*.*

Thus, by the preceding lemma, the potential left turning points in the example in Fig. 1 are the initial locations of sensors s_3, s_4, and s_7, but not s_2.

Definition 2. *Let* m *be the number of sensors whose coverage balance is either* $-2r < C_j < 0$*, or* $-2r = C_j$ *and* $x_j = x_{j+1}$*. The list* A *of indices of sensors of potential triple delimiters is a list of pairs* $A = [(b_1, a_1), (b_2, a_2), \cdots, (b_m, a_m)]$ *of sensor indices such that*

1. $a_1 < a_2 < \cdots < a_m$ *are the indices of all sensors such that either* $-2r < C_j < 0$*, or* $-2r = C_j$ *and* $x_j = x_{j+1}$
2. b_1 *is the smallest index of a sensor with negative coverage balance, and for* $1 < i \leq m$ *the value of* b_i *is the smallest index larger than* a_{i-1} *with negative coverage balance.*

Lemma 4. *Let* A *be the list of indices of sensors of potential triple delimiters,* m *be the number of pairs in* A*, and* c *be defined as in Lemma 3. There is an optimal, order preserving, fully-stretched trajectory such that for some integer* j*,* $0 \leq j \leq m$*,*

1. *the trajectory contains* j *triples* $[x_{b_i} + C_{b_i}, x_{a_i}]$*,* $1 \leq i \leq j$*,*
2. *If* $j < m$ *then the trajectory ends with a double,* $[x_{b_{j+1}} + C_{b_{j+1}}, c]$*, otherwise the trajectory ends with a straight line and its termination point is* c*.*

The main idea of our offline algorithm is to calculate the list A of potential triple delimiters as defined in Definition 2. Let T_j be a trajectory that uses triples on the first j pairs of A, $0 \leq j \leq m$, and one double if $j < m$. We define the overhead o_j of a trajectory T_j to be the difference between the length of T_j and the straight line trajectory. Clearly,

$$o_j = \begin{cases} c - x_{b_{j+1}} - C_{b_{j+1}} + \sum_{i=1}^{j} 2(x_{a_i} - x_{b_i} - C_{b_i}) \text{ for } 1 < j < m, \\ \sum_{i=1}^{m} 2(x_{a_i} - x_{b_i} - C_{b_i}) \text{ for } j = m \end{cases}$$

The algorithm calculates the overhead of T_j trajectories for $1 \leq j \leq m$ and chooses the trajectory with the minimum overhead. By Lemma 4, the trajectory with the minimum overhead is optimal. Thus a robot finds the coordinates of an optimal trajectory by executing the offline algorithm below.

Offline Algorithm
> **Input:** the length L of the segment, the number n of sensors,
>> their initial locations $x_1 \leq x_2 \leq \ldots \leq x_n$, and their range r;
>
> **Output:** the trajectory points for the robot.

1 Scan $x_1, x_2, \ldots, x_n$ for gaps;
2 if gaps exist **then**
> **2.1 Compute** the smallest integer k such that all gaps are to the left of x_k;
> **2.2 Compute** the sequence $C_i = (2ri - r) - x_i$, $1 \leq i \leq k$;
> **2.3 if** $C_k < 0$ **then** $c \leftarrow x_k$; **else** $c \leftarrow x_k + C_k$;
>> // c is the potential left-turning point of a double.
> **2.4 Scan** the sequence $C_1, C_2, \ldots, C_k$ and
>> **compute** list $< A, B >= [(b_1, a_1), (b_2, a_2), \cdots, (b_m, a_m)]$;
>> // potential triple delimiters
> **2.5** $o_j \leftarrow (c - x_{b_{j+1}} - C_{b_{j+1}}) + \sum_{i=1}^{j} 2(x_{a_i} - x_{b_i} - C_{b_i})$, $1 \leq j \leq m - 1$;
> // T_j overhead
> **2.6** $o_m \leftarrow \sum_{i=1}^{m} 2(x_{a_i} - x_{b_i} - C_{b_i})$; // T_m overhead
> **2.7 Compute** $min\{o_1, o_2, \ldots, o_m\}$; and its index k;
> **2.8 Output** $x_{a_1}, x_{b_1} + C_{b_1}, x_{a_2}, x_{b_2} + C_{b_2}, \ldots, x_{a_k}, x_{b_2} + C_{b_2}$;
>> //the sequence of left/right turning points of the optimal trajectory,
> **2.9 If** $k < m$ **then Output** there is a double from c to $x_{b_{k+1}} + C_{b_{k+1}}$;
> **else the trajectory ends at** c;
> **else** $[0, L]$ is initially completely covered, robot does nothing;

Since algorithm *Offline* calculates the overheads of all trajectories that satisfy Lemma 4 and picks the one with the smallest overhead, the Corollary 1 and Lemma 4 imply that the selected trajectory is optimal. Clearly, all calculations in each step are of $O(n)$ complexity. Thus we have the following theorem.

Theorem 1. *Assume we are given n sensors in the line segment $[0, L]$ and a robot with starting position 0. Algorithm* Offline *computes an optimal trajectory for the robot to follow in $O(n)$ time.*

3   Optimal Online Algorithms

We now consider online algorithms for restoration of barrier coverage by a robot. For the online algorithm we assume that the robot starts at position 0, it can move along the given line segment, but the robot does not know the positions of sensors until it comes upon them. As usual, we define the competitive ratio of an online algorithm as the length of the trajectory of the online algorithm divided by the length of the trajectory of the optimal offline algorithm.

At the outset, observe that on the input instance where the sensors are placed in such a way that there is no gap in the barrier coverage, the offline algorithm produces a trajectory of length 0, while the online algorithm must traverse the entire barrier segment to ensure that the barrier is covered. Thus no online algorithm can have a bounded competitive ratio. To provide a more meaningful comparison of online with offline algorithms, we only consider below input instances where there is a gap in coverage at the very end of the barrier, that is, the point L is uncovered. On such instances, all valid robot trajectories must have length at least $L - r$. We also consider the possibility that L, the length of the barrier, is not known to the robot and the robot will find it out only when reaching the end of the barrier. Since the performance of online algorithms depends on the knowledge of L, we consider the two possibilities separately. We use below the notion of potential left and right turning points as defined in the previous section.

When the value of L is unknown to the robot we show the following result.

Theorem 2. *Assume that the robot does not know the length L of the barrier $[0, L]$. For any $0 < \epsilon \ll 1$, the competitive ratio of any online algorithm is at least $\frac{3}{2} - \epsilon$. Furthermore, there is an online algorithm with competitive ratio at most $\frac{3}{2}$.*

Proof. Assume there is an online algorithm $\mathcal{A}$ with competitive ratio $3/2 - \epsilon$ for some $\epsilon > 0$. We give an adversary argument. Start with an input that has no sensors until position $x = 2ir$ where there are $i > 0$ sensors for some i to be specified later. Clearly there are just enough sensors at x to cover the segment $[0, x]$. Following this, the adversary starts placing sensors in attached position starting at position $x + 2r$. The robot has to make a turn at some point $y \geq x$ to cover the gap in coverage before x. If it does not make a turn before $6x$, the adversary can set $L = 6x$, and the robot must do a double to the beginning, see Fig. 3 (a). The trajectory produced by $\mathcal{A}$ has length at least $2L - r = 12x - r$, while the offline algorithm covers the gap before x with a triple from x using a trajectory of at most $3x - 2r + 5x = 8x - 2r$. This gives the competitive ratio of at least $(12x - r)/(8x - 2r) > 3/2$. If the robot turns at any point y such that $x \leq y < 6x$, then the adversary concludes the barrier at $L = y + r + \delta$, see Fig. 3 (b). Clearly, the trajectory produced by $\mathcal{A}$ has length at least $3y - r + \delta$ while the offline algorithm uses a trajectory of at most $2y + 2\delta$. Thus the competitive ratio of the algorithm is at least $(3y-r+\delta)/(2y+2\delta) \geq 3/2-(r+2\delta)/(2x+2\delta) \geq 3/2-(r+2\delta)/(2ir+2\delta) \geq 3/2-\epsilon$ for sufficiently large i.

To prove the second part of the theorem observe that the algorithm that solves any gap in coverage with a triple from any potential left turning point has competitive ratio at most $3/2$. $\square$

In the remainder of the section, we assume that L, the length of the barrier segment, is known to the online algorithm. We first prove a lower bound on the competitive ratio of any online algorithm for the problem.

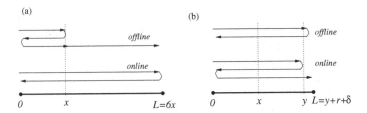

Fig. 3. Competitive ratio for the case when the robot does not know L. In (a) the robot turns at L or $y \geq 6x$, in (b) the robot turns at $y < 6x$.

Theorem 3. *Assume that the online robot knows the length L of the barrier $[0, L]$. For any $0 < \epsilon \ll 1$, the competitive ratio of any online algorithm is at least $\frac{5}{4} - \epsilon$.*

The optimal offline algorithm suggests that if the online robot stops doing triples too soon, it may be beaten by an algorithm that does perhaps just one more triple which avoids the double at the end. However if it keeps doing triples for too long, it may be beaten by an algorithm that does fewer triples. It is natural to ask whether there an optimal fraction of the segment such that the online robot can decide in advance to do triples only until then. We say that an online algorithm has a *fixed switching point* z if it covers each gap before z with a triple, and all gaps after z with the final double. Therefore, the online robot turns left at most once after z, and if it does, it turns at position $L - r$ to do the final double. We show below a tight bound on the competitive ratio of an online algorithm with a fixed switching point.

Theorem 4. *Assume that the robot knows the length L of the barrier $[0, L]$.*

1. *For any $0 < \epsilon \ll 1$, the competitive ratio of any online algorithm with a fixed switching point is at least $\frac{4}{3} - \epsilon$.*
2. *There is an online algorithm with fixed switching point with competitive ratio at most $\frac{4}{3}$.*

Thus, by deciding in advance a switching point at which to stop doing triples, it is impossible to derive an online algorithm that matches the lower bound of Theorem 3. We now specify **AdaptiveOnline**, an online algorithm for a robot which, when starting at location 0, relocates sensors on the segment $[0, L]$ to achieve complete barrier coverage. We calculate an upper bound on the competitive ratio of this online algorithm and prove that it asymptotically matches the lower bound of 5/4 from Theorem 3. Clearly, an online algorithm can calculate the coverage balance of any sensor it encounters. We now describe two functions for the online robot used in the algorithm. The first function is called *walk-in-surplus* and is defined as follows: When at a potential left-turning point (or the start of the barrier) the robot moves right picking up sensors having a positive coverage balance and deposits them $2r$ apart (as the optimal offline algorithm constructing a fully-stretched solution would do), until reaching a point x such that the last sensor it dropped was at location $x - 2r$, and no sensors were encountered in the interval $[x - 2r, x]$. Observe that at such a position x, the robot knows that x is a potential right-turning point. The function then returns the value x. The second function is called *walk-in-deficit*: When first time at a potential right-turning point, robot moves right picking up sensors with negative coverage balance on its way until it reaches a sensor with negative coverage balance greater than $-2r$, or balance exactly $-2r$ and collocated with the next sensor. Thus, this is a potential left turning point y; the function then returns the value y. The functions a *triple*, and a *double* behave the same way as in the offline algorithm. The main challenge for the online algorithm is to determine, when it reaches a potential left turning point, whether to do a triple at this point, or to switch to solving the remaining segment as part of the final double.

We specify our adaptive online algorithm as a recursive procedure **AdaptiveOnline(t,L,r)** in which $[t, L]$ is the subinterval on which the robot has not yet travelled, and barrier coverage remains to be achieved, and r is the range of sensors. To calculate the coverage of $[0, L]$ we execute **AdaptiveOnline(0,L,r)**. We assume that there is a gap at position 0; if not, we simply execute the walk-in-surplus function until reaching a potential right turning-point x and then call **AdaptiveOnline** on the segment $[x - r, L]$. It is clear that the initial part of the trajectory executed until x is optimal, and cannot increase the competitive ratio on the entire input. We give the pseudocode for the algorithm below.

Algorithm. AdaptiveOnline (t, L, r);
  Input: t, L, the subinterval being solved, with a gap at t and r is the range of sensors
  Output: the moves of the robot;
  Variables:: x; // the current position
        T; // current trajectory length
        γ_i; // ratio trajectory/distance at left-turning point in iteration i
        β_i; // ratio trajectory/distance at right-turning point in iteration i
  functions: walk-in-surplus, walk in deficit, triple and double
  $x \leftarrow t + r$; $T \leftarrow 0$; $i \leftarrow 0$; // initialization of variables
  repeat
        $i \leftarrow i + 1$; // iteration of loop
        $b_i \leftarrow x$; // potential right-turning point
        $\beta_i \leftarrow (T + r)/b_i$ // ratio at start of possible triple
        $a_i \leftarrow$ walk-in-deficit; // potential left-turning point
        $T \leftarrow T + r + 3(a_i - b_i)$; // trajectory if triple is done
        $\gamma_i \leftarrow T/a_i$; // ratio at end of possible triple
        if $\gamma_i a_i - a_i > L - t$ **break** // exit the loop
        else
              do a triple on segment $[b_i, a_i]$,
              $x \leftarrow$ walk-in-surplus; // gap starting at $x - r$
              $T \leftarrow T + (x - r - a_i)$; //update trajectory until start of gap
              If $x < L$ and $T/(x - r) \leq 2.5$ **then** AdaptiveOnLine(x-r,L,r);
  until $x = L$;
  if (L not reached) **then**
        do a double (to $L - r$ and back to b_i);
        $T \leftarrow T + (L - a_i) - (a_i - b_i)$;

The key idea is as follows: First, the online robot keeps track of the ratio between its trajectory so far versus the distance it has covered. If it discovers that this ratio is less than 5/2, then it "forgets about" the segment covered so far (it will be shown that it has achieved a competitive ratio of at most $5/4$ for this part), and restarts its computations. The ratio between its trajectory and distance travelled so far is computed only at potential left and right turning points. Secondly, when it reaches a potential left-turning point, the online robot calculates the cost of the triple: the difference between its trajectory if it executes the triple and the distance covered so far. If this difference is too high, it decides to stop doing triples, and finish by doing a double.

Observe that before making a recursive call, at least one gap is covered by the robot. Since the number of gaps is finite, the algorithm terminates. It is also clear that **AdaptiveOnLine** constructs a trajectory that results in barrier coverage. It remains only to analyze the competitive ratio of the trajectory length. Let $T_A(I)$ and $T_o(I)$ be the lengths of the trajectories of the algorithm **AdaptiveOnline** and the optimal offline algorithm on an input instance I respectively. We prove a bound of $5/4$ on $max_I\{T_A(I)/T_o(I)\}$, thereby matching the lower bound of Theorem 3.

Fix an input instance I. Observe that the algorithm **AdaptiveOnline** partitions the segment $[0, L]$ into sub-segments that are solved in each recursive call of the algorithm. We call each of these sub-segments an *epoch*; let n be the number of epochs, such that while traversing epoch j, there is no recursive call. Let T_j be the length of the the trajectory of the online robot in epoch j, and let O_j be the length of the trajectory of the optimal offline robot in the same epoch. Every epoch starts with a gap, and in every epoch except possibly the last, the mobile robot does triples from the first encountered left-turning point to cover gaps. It can be shown that in each epoch the competitive ratio is at most $5/4$. Thus we get the following theorem.

Theorem 5. AdaptiveOnline *is an online algorithm for barrier coverage of a line segment of known length and has competitive ratio at most $5/4$, and is therefore optimal.*

References

1. Balister, P., Bollobas, B., Sarkar, A., Kumar, S.: Reliable density estimates for coverage and connectivity in thin strips of finite length. In: Proceedings of MobiCom 2007, pp. 75–86 (2007)
2. Bhattacharya, B., Burmester, M., Hu, Y., Kranakis, E., Shi, Q., Wiese, A.: Optimal movement of mobile sensors for barrier coverage of a planar region. Theor. Comput. Sci. **410**(52), 5515–5528 (2009)
3. Chalopin, J., Das, S., Mihalák, M., Penna, P., Widmayer, P.: Data delivery by energy-constrained mobile agents. In: Flocchini, P., Gao, J., Kranakis, E., der Heide, F.M. (eds.) ALGOSENSORS 2013. LNCS, vol. 8243, pp. 111–122. Springer, Heidelberg (2014)
4. Charikar, M., Khuller, S., Raghavachari, B.: Algorithms for capacitated vehicle routing. SIAM J. Comput. **31**(3), 665–682 (2001)
5. Chen, D.Z., Gu, Y., Li, J., Wang, H.: Algorithms on minimizing the maximum sensor movement for barrier coverage of a linear domain. In: Fomin, F.V., Kaski, P. (eds.) SWAT 2012. LNCS, vol. 7357, pp. 177–188. Springer, Heidelberg (2012)

6. Czyzowicz, J., Kranakis, E., Krizanc, D., Lambadaris, I., Narayanan, L., Opatrny, J., Stacho, L., Urrutia, J., Yazdani, M.: On minimizing the maximum sensor movement for barrier coverage of a line segment. In: Ruiz, P.M., Garcia-Luna-Aceves, J.J. (eds.) ADHOC-NOW 2009. LNCS, vol. 5793, pp. 194–212. Springer, Heidelberg (2009)

7. Czyzowicz, J., Kranakis, E., Krizanc, D., Lambadaris, I., Narayanan, L., Opatrny, J., Stacho, L., Urrutia, J., Yazdani, M.: On minimizing the sum of sensor movements for barrier coverage of a line segment. In: Nikolaidis, I., Wu, K. (eds.) ADHOC-NOW 2010. LNCS, vol. 6288, pp. 29–42. Springer, Heidelberg (2010)

8. Czyzowicz, J., Kranakis, E., Krizanc, D., Narayanan, L., Opatrny, J.: Optimal online and offline algorithms for robot-assisted restoration of barrier coverage, 20 pages. arXiv, http://arxiv.org/abs/1410.6726 (2014)

9. Djugash, J., Singh, S., Kantor, G.A., Zhang, W.: Range-only SLAM for robots operating cooperatively with sensor networks. In: Proceedings of IEEE International Conference on Robotics and Automation, pp. 2078–2084, May 2006

10. GKmM summer school 2012 website. http://www.gkmm.tu-darmstadt.de/summerschool/. Accessed 25 Jan 2014

11. Hesari, M.E., Kranakis, E., Krizanc, D., Morales-Ponce, O., Narayanan, L., Opatrny, J., Shende, S.: Distributed algorithms for barrier coverage using relocatable sensors. In: Proceedings of PODC 2013, pp. 383–392 (2013)

12. Jaillet, P., Wagner, M.R.: Online vehicle routing problems: a survey. The Vehicle Routing Problem: Latest Advances and New Challenges, pp. 221–237. Springer, New York (2008)

13. Jung, B., Sukhatme, G.: Cooperative tracking using mobile robots and environment-embedded networked sensors. In: International Symposium on Computational Intelligence in Robotics and Automation, pp. 206–211 (2001)

14. Koubaa, A., Khelil, A. (eds.) Cooperative Robots and Sensor Networks. Studies in Computational Intelligence. Springer, Heidelberg (2013)

15. Kropff, M., Reinl, C., Listmann, K., Petersen, K., Radkhah, K., Shaikh, F.K., Herzog, A., Strobel, A., Jacobi, D., von Stryk, O.: MM-ulator: towards a common evaluation platform for mixed mode environments. In: Carpin, S., Noda, I., Pagello, E., Reggiani, M., von Stryk, O. (eds.) SIMPAR 2008. LNCS (LNAI), vol. 5325, pp. 41–52. Springer, Heidelberg (2008)

16. Kumar, S., Lai, T.H., Arora, A.: Barrier coverage with wireless sensors. In: Proceedings of MobiCom 2005, pp. 284–298 (2005)

17. Robosense 2012 workshop website. www.robosense.org. Accessed 25 Jan 2014

18. Robosense 2013 workshop website. http://www.coins-lab.org/events/RoboSense13/. Accessed 25 Jan 2014

19. Saipulla, A., Westphal, C., Liu, B., Wang, J.: Barrier coverage of line-based deployed wireless sensor networks. In: Proceedings of IEEE INFOCOM 2009, pp. 127–135 (2009)

20. Seow, C.K., Seah, W.K.G., Liu, Z.: Hybrid mobile wireless sensor network cooperative localization. In: Proceedings of IEEE 22nd International Symposium on Intelligent Control, pp. 29–34 (2007)

21. Shi, W., Corriveau, J.-P.: A comprehensive review of sensor relocation. In: Proceedings of IEEE/ACM International Conference on Green Computing and Communications, pp. 780–785 (2010)

22. Teng, J., Bolbrock, T., Cao, G., La Porta, T.: Sensor relocation with mobile sensors: design, implementation, and evaluation. In: Proceedings of IEEE MASS, pp. 1–9 (2007)

Linear-Time Approximation Algorithms for Unit Disk Graphs

Guilherme D. da Fonseca[1], Vinícius G. Pereira de Sá[2]([✉]),
and Celina M.H. de Figueiredo[2]

[1] Université Montpellier 2, Montpellier, France
[2] Universidade Federal do Rio de Janeiro, Rio de Janeiro, Brazil
vigusmao@dcc.ufrj.br

Abstract. Numerous approximation algorithms for unit disk graphs have been proposed in the literature, exhibiting sharp trade-offs between running times and approximation ratios. We propose a method to obtain linear-time approximation algorithms for unit disk graph problems. Our method yields linear-time $(4 + \varepsilon)$-approximations to the maximum-weight independent set and the minimum dominating set, bringing dramatic performance improvements when compared to previous algorithms that achieve the same approximation factors. Furthermore, we present an alternative linear-time approximation scheme for the minimum vertex cover, which could be obtained by an indirect application of our method.

1 Introduction

A *unit disk graph* is the intersection graph of n unit disks in the plane. Unit disk graphs are often represented using the coordinates of the disk centers instead of explicit adjacency information. In this geometric setting, two vertices are adjacent if the corresponding points (the disk centers) are within Euclidean distance at most 2 from one another.

Owing to their applicability in wireless networks [10,13], numerous approximation algorithms for unit disk graphs have been proposed in the literature. Such approximations are either *graph-based* algorithms, when they receive as input solely the adjacency representation of the graph, or *geometric* algorithms, when the input consists of a geometric representation of the graph. While the m edges of a graph can be obtained from the vertices' coordinates in $O(n+m)$ time under the real-RAM model with floor function and constant-time hashing [3], obtaining a geometric representation of a given unit disk graph is NP-hard [4].

Linear- and near-linear-time approximation algorithms are an active topic of research, even for problems that can be solved exactly in polynomial time, such as maximum flow and matching (see [5] for references). We note that, when the goal is to design $O(n)$-time algorithms, the geometric representation is required, since the number m of edges in a unit disk graph can be as high as $\Theta(n^2)$.

The *shifting strategy* [7] gave rise to geometric PTASs for several problems for unit disk graphs [8]. Essentially, the shifting strategy reduces the original problem to a set of subproblems of constant diameter. Such reduction takes $O(n)$

© Springer International Publishing Switzerland 2015
E. Bampis and O. Svensson (Eds.): WAOA 2014, LNCS 8952, pp. 132–143, 2015.
DOI: 10.1007/978-3-319-18263-6_12

time and yields a $(1 + \varepsilon)$-approximation to the original problem, given the exact solutions to the subproblems. However, the running times of the PTASs are polynomials of high degree because each subproblem is solved exactly by exploiting the fact that the point set has constant diameter. Graph-based PTASs for these problems are also known [13]. While they do not use the shifting strategy, their running times are even higher than those of their geometric counterparts.

The minimum dominating set problem (MDSP) admits some PTASs [8,13], the fastest of which is geometric and provides a 4-approximation in roughly $O(n^{10})$ time. Such high running times have motivated the study of faster constant-factor approximation algorithms. Examples of graph-based algorithms include a 44/9-approximation that runs in $O(n + m)$ time and a 43/9-approximation that runs in $O(n^2 m)$ time [5]. Among the geometric algorithms, we cite the original 5-approximation, which can be implemented in $O(n)$ time if the floor function and constant-time hashing are available [10]; a 44/9-approximation that uses local improvements and runs in $O(n \log n)$ time [5]; a 4-approximation that uses grids and runs in $O(n^8 \log n)$ time [6]; and a recent 4-approximation that uses hexagonal grids and runs in $O(n^6 \log n)$ time [9].

The maximum-weight independent set problem (MWISP) also admits some PTASs, the fastest of which attains a $(1+\varepsilon)$-approximation in $O(n^{4\lceil 2/\varepsilon\sqrt{3}\rceil})$ time [8,12,13]. A 5-approximation can be obtained in $O(n \log n)$ time by a greedy approach that considers the vertices in decreasing order of weights. In contrast, for the unweighted version, a greedy approach that considers the vertices from left to right [10] can be implemented to give a 3-approximation in $O(n)$ time with floor function and constant-time hashing.

Some efficient PTASs for unit disk graph problems are also known, as the one given in [11] for the minimum vertex cover problem (MVCP).

Our results. We introduce a method to obtain linear-time approximation algorithms for problems on unit disk graphs and other geometric intersection graphs (Sect. 2). Our method is based on approximating the input point set, which can be arbitrarily dense, by a *sparse* set of points, that is, a set of points such that any sufficiently small square contains at most a constant number of points.

To convert the general idea into efficient algorithms, we need to investigate the fundamental question of how well a sparse point set—generated using only local information—can approximate a denser one for each considered problem. Although our algorithms share the same basic idea, their analyses differ significantly. For example, the MWISP analysis applies the Four-Color Theorem for planar graphs [2], while the MDSP analysis applies packing arguments.

By using our method, we obtained linear-time $(4 + \varepsilon)$-approximation algorithms for the MWISP (Sect. 3) and the MDSP (Sect. 4). The proposed algorithms provide significant improvements when compared not only to existing linear-time algorithms, but also to sub-quartic-time algorithms (see Table 1 in Sect. 6). We have also included (Sect. 5) a linear-time $(1 + \varepsilon)$-approximation obtained independently for the MVCP, illustrating an indirect application of our method. Open problems and lower bounds to the approximation ratios of our algorithms are also discussed in Sect. 6.

2   Our Method

The shifting strategy [7] is the main idea behind the existing geometric PTASs for problems on unit disk graphs such as the minimum dominating set, maximum independent set, and minimum vertex cover [8]. Generally, the shifting strategy reduces the original problem with n points to a set of subproblems whose inputs have constant diameter and the sum of the input sizes is $O(n)$. Such reduction is based on partitioning the points according to a number of iteratively shifted grids and takes $O(n)$ time (by using the floor function and constant-time hashing). Exploiting the inputs' constant diameter, each subproblem is solved exactly in polynomial time. The solutions to the subproblems are then combined appropriately (normally in $O(n)$ time) to yield feasible solutions to the original problem, the best of which is returned. The high complexities of these geometric PTASs are due to the exact algorithms that are employed to solve each subproblem.

We propose a method that is based on the shifting strategy. It presents, however, a crucial difference. Rather than obtaining exact, costly solutions for the subproblems, we solve each subproblem *approximately*. To do that, we employ the coresets paradigm [1], where only a subset with a constant number of input points is considered. For a problem whose input is a set P of n points, our method can be briefly described as follows:

1. Apply the shifting strategy to construct a set of r subproblems with inputs $P_1, \ldots, P_r$ such that $\sum_{i=1}^{r} |P_i| = O(n)$ and $\mathrm{diam}(P_i) = O(1)$ for all i.
2. For each subproblem instance P_i, obtain a coreset $Q_i \subseteq P_i$ with $|Q_i| = O(1)$, such that the optimal solution for instance Q_i is an α-approximation to the optimal solution for instance P_i.
3. Solve the problem exactly for each Q_i.
4. Combine the solutions into an $(\alpha+\varepsilon)$-approximation for the original problem.

Coresets for different problems must be devised appropriately. For the MWISP, we create a grid with cells of diameter 0.29 and consider only one point of maximum weight inside each cell. For the MDSP, we create a grid with cells of diameter 0.24 and consider only the (at most four) points, inside each cell, with minimum or maximum coordinate in either dimension (breaking ties arbitrarily). Finally, we solve the MVCP by breaking each subproblem into two cases. In the first one, the number of input points is already bounded by a constant. In the second one, we use the same coreset as in the MWISP.

We assume a real-RAM computation model with floor function and constant-time hashing (as in [3]), so we can partition the input points into grid cells efficiently, yielding an overall $O(n)$ running time for our method. Without these operations, the running time of our algorithms becomes $O(n \log n)$. We also assume that ε is constant. Otherwise, the running time becomes $2^{O(1/\varepsilon^2)}n$ for the WIS and the DS on UDGs, and $2^{O(1/\varepsilon^3)}n$ for the VC.

3   Maximum-Weight Independent Set

In this section, we show how to obtain a linear-time $(4 + \varepsilon)$-approximation to the MWISP. We start by presenting a 4-approximation for point sets of constant diameter, and then we use the shifting strategy to obtain the desired $(4 + \varepsilon)$-approximation.

Given a point p and a set S of points, let $w(p)$ denote the weight of p, and let $w(S) = \sum_{p \in S} w(p)$. We say two or more points are *independent* if their minimum distance is strictly greater than 2.

Theorem 1. *Given a set P of n points with real weights as input, with $\mathrm{diam}(P) = O(1)$, the MWISP can be 4-approximated in $O(n)$ time in the real-RAM.*

Proof. Our algorithm proceeds as follows. First, we find the points of P with minimum or maximum coordinates in either dimension. That defines a bounding box of constant size for P. Within this bounding box, we create a grid with cells of diameter $\gamma = 0.29$ (any value $\gamma < (2 - \sqrt{2})/2$ suffices). Note that the number of grid cells is constant, and therefore the points of P can be partitioned among the grid cells in $O(n)$ time (even without using the floor function or hashing). Then, we build the subset $Q \subseteq P$ as follows. For each non-empty grid cell C, we add to Q a point of maximum weight in $P \cap C$. Afterwards, we determine the maximum-weight independent set I^* of Q. Since $|Q| = O(1)$, this can be done in constant time. We return the solution I^*.

Next, we show that I^* is indeed a 4-approximation. We argue that, given an independent set $I \subseteq P$, there is an independent set $I' \subseteq Q$ with $4\, w(I') \geq w(I)$. Given a point $p \in P$, let $q(p)$ denote the point from Q that is contained in the same grid cell as p. Consider the set $S = \{q(p) : p \in I\}$. Note that $w(q(p)) \geq w(p)$ and $w(S) \geq w(I)$. The set S may not be independent, but since I is independent, the minimum distance in S is at least $2 - 2\gamma = 1.42 > \sqrt{2}$. We claim that the unit disk graph formed by S is a planar graph. To prove the claim, we show that a planar drawing can be obtained by connecting the points of S within distance at most 2 by straight line segments. Given a pair of points p_1, p_2 with distance $\|p_1 p_2\| \leq 2$, the Pythagorean Theorem shows that a unit disk centered within distance greater than $\sqrt{2}$ from both p_1 and p_2 cannot intersect the segment $p_1 p_2$. By the Four-Color Theorem [2], S admits a partition into four independent sets $S_1, \ldots, S_4$. The set I' of maximum weight among $S_1, \ldots, S_4$ must have weight at least $w(I)/4$.[1]

Since I^* is the maximum-weight independent set of Q, we have that I^* is a 4-approximation for the MWISP.   □

The following theorem uses the shifting strategy to obtain a $(4 + \varepsilon)$-approximation for point sets of arbitrary diameter. The proof uses the ideas from [8], presented in a different manner and including details about an efficient implementation of the strategy.

Theorem 2. *Given a set P of n points in the plane as input, the MWISP can be $(4 + \varepsilon)$-approximated in $O(n)$ time on a real-RAM with constant-time hashing and the floor function. Without these operations, it can be done in $O(n \log n)$ time.*

---

[1] Note that the Four-Color Theorem is only used in the argument, and no coloring is ever computed by the algorithm.

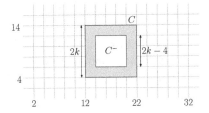

Fig. 1. Cell contraction on a grid rooted at $(2, 4)$ with $k = 5$

Proof. Let k be the smallest integer such that

$$\left(\frac{k-2}{k}\right)^2 \geq \frac{4}{4+\varepsilon}. \tag{1}$$

Throughout this proof, we consider grids with square cells of side $2k$. We say a grid is *rooted at* a point (x, y) if there is a grid cell with corner at (x, y). Given a cell C, the square region $C^- \subset C$, called the *contraction* of C, is formed by removing from C the points within distance at most 2 from the boundary of C. Figure 1 illustrates these concepts.

The algorithm proceeds as follows. For i, j from 0 to $k - 1$, we create a grid with cells of side $2k$ rooted at $(2i, 2j)$. For each cell C in the grid, we run the MWISP 4-approximation algorithm from Theorem 1 with point set $P \cap C^-$, obtaining a solution $I_{i,j}(C)$. Then, the independent set $I_{i,j}$ is constructed as the union of the independent sets $I_{i,j}(C)$ for all grid cells C. We return the maximum-weight set $I_{i,j}$ that is found, call it I^*.

To implement the algorithm efficiently, we create a subgrid of subcells of side 2, assigning each point to the subcell that contains it. In order to partition the n points into subcells, we use the floor function and constant-time hashing, taking $O(n)$ time. If these operations are not available, we determine the connected components of the graph (using the Delaunay triangulation, for example) and for each component we partition the points into subcells by sorting them by x coordinate, separating them into columns, and then sorting the points inside each column by y coordinate. The non-empty subcells are stored in a balanced binary search tree. This process takes $O(n \log n)$ time due to sorting, Delaunay triangulation, and binary search tree operations. Given the partitioning of the point set into subcells, each input to the MWISP algorithm can be constructed as the union of a constant number of subcells. Finally, the total size of the constant-diameter MWISP instances is $O(n)$, since each point from the original point sets appears in a constant number—a function of the fixed ε—of such instances.

To prove that the returned solution I^* is indeed a $(4 + \varepsilon)$-approximation, we use a probabilistic argument. Let i, j be picked uniformly at random from $0, \ldots, k - 1$ and let OPT denote the optimal solution. For every cell C, we have

$$w(I_{i,j}(C)) \geq \frac{w(OPT \cap C^-)}{4}.$$

Consequently, by summing over all grid cells,

$$w(I_{i,j}) = \sum_C w(I_{i,j}(C)) \geq \frac{1}{4} \sum_C w(OPT \cap C^-).$$

We now bound $E[w(I_{i,j})]$. Let $\rho(p)$ denote the probability that a given point p is contained in some contracted cell. Since $w(p)$ does not depend on the choice of i, j, we can write

$$4\, E[w(I_{i,j})] \;\geq\; E\left[\sum_C w\left(OPT \cap C^-\right)\right] \;=\; \sum_{p \in OPT} \rho(p)w(p).$$

Note that, for all $p \in P$, $\rho(p)$ corresponds to the ratio between the areas of C^- and C, namely

$$\rho(p) = \frac{\text{area}(C^-)}{\text{area}(C)} = \left(\frac{k-2}{k}\right)^2.$$

Therefore, by using inequality (1), we obtain

$$E[w(I_{i,j})] \geq \frac{1}{4}\left(\frac{k-2}{k}\right)^2 \sum_{p \in OPT} w(p) \geq \frac{1}{4}\left(\frac{4}{4+\varepsilon}\right)w(OPT) = \frac{1}{4+\varepsilon}\,w(OPT).$$

Since I^* has maximum weight among the independent sets $I_{i,j}$, it follows that $w(I^*)$ is at least as large as their average weight. Therefore, I^* satisfies

$$w(I^*) \geq E[w(I_{i,j})] \geq \frac{1}{4+\varepsilon}\,w(OPT),$$

closing the proof.   □

4   Minimum Dominating Set

In this section, we show how to obtain a linear-time $(4+\varepsilon)$-approximation to the MDSP (in fact, a generalization of it). We start by presenting a 4-approximation for point sets of constant diameter, and then we use the shifting strategy to obtain the desired $(4+\varepsilon)$-approximation. We say that a point p dominates a point q if $\|pq\| \leq 2$. Given two sets of points D and P', we say that D is a P'-dominating set if every point in P' is dominated by some point in D.

We now define a more general version of the MDSP, which we refer to as the *minimum partial dominating set problem (MPDSP)*. Such a generalization is necessary to properly apply the shifting strategy. In the MPDSP, we are given a set P of n points and also a subset $P' \subseteq P$. The goal is to find the smallest P'-dominating subset $D \subseteq P$.

In order to analyze our algorithm, we prove a geometric lemma that shows that the set-theoretic difference between a unit circle and two unit disks that are sufficiently close to it and form a sufficiently big angle consists of one or two "small" arcs. Given a point p, let $\bigcirc_p$ denote the unit disk centered at p, and $\partial\bigcirc_p$ denote its boundary circle.

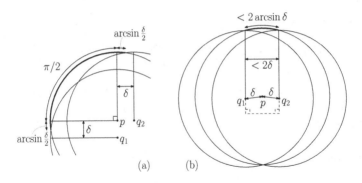

Fig. 2. Proof of Lemma 3

Lemma 3. *Given $\delta > 0$ and three points $p, q_1, q_2 \in \mathbb{R}^2$ with (i) $\|pq_1\| \leq \delta$, (ii) $\|pq_2\| \leq \delta$, and (iii) the smallest angle $\angle q_1 p q_2$ is greater than or equal to $\pi/2$, we have that:*

(1) the portion $T = (\partial \bigcirc_p) \setminus (\bigcirc_{q_1} \cup \bigcirc_{q_2})$ of the boundary $\partial \bigcirc_p$ consists of one or two circular arcs;

(2) if T consists of one circular arc, then the arc length is less than or equal to $\pi/2 + 2\arcsin(\delta/2)$; and

(3) if T consists of two circular arcs, then each arc length is less than $2\arcsin \delta$.

Proof. Statement (1) is clearly true. We start by proving statement (2). The arc length $\|T\|$ is maximized as the angle $\angle q_1 p q_2$ decreases while the distances $\|pq_1\|, \|pq_2\|$ are kept constant, therefore it suffices to consider the case when $\angle q_1 p q_2 = \pi/2$. The arc T centered at p can be decomposed into three arcs by rays in directions $q_1 p$ and $q_2 p$, as shown in Fig. 2(a). The central arc measures $\pi/2$, while each of the other two arcs measures $\arcsin(\delta/2)$, proving statement (2).

Next, we prove statement (3). Let T_1, T_2 denote the two arcs that form T with $\|T_1\| \geq \|T_2\|$. The arc length $\|T_1\|$ is maximized in the limit when $\|T_2\| = 0$, as shown in Fig. 2(b). The rays connecting q_1 and q_2 to the two extremes of T_1 are parallel, and therefore $\|T_1\| < 2\arcsin \delta$. □

We are now able to prove the following theorem, which presents our 4-approximation algorithm for point sets of constant diameter.

Theorem 4. *Given two sets of points P and P' as input, with $P' \subseteq P$, $|P| = n$, and $\operatorname{diam}(P) = O(1)$, the MPDSP can be 4-approximated in $O(n)$ time in the real-RAM.*

Proof. First, we determine a bounding box of constant size for P, as we did in the algorithm for the MWISP. Within this bounding box, we create a grid with cells of diameter $\gamma = 0.24$ (any positive γ satisfying

$$\sqrt{8 - 8\cos\left(\frac{\frac{\pi}{2} + 2\arcsin(\frac{\gamma}{2})}{2}\right)} + \gamma < 2$$

suffices). Note that the number of grid cells is constant, and therefore the points of P can be partitioned among the grid cells in $O(n)$ time (even without using the floor function or hashing). Then, we build the subset $Q \subseteq P$ as follows. For each non-empty grid cell, we add to Q the (at most four) extreme points inside the cell, i.e., those presenting minimum or maximum coordinate in either dimension. Ties are broken arbitrarily. Since there is a constant number of grid cells and we include in Q at most four points per cell, we have $|Q| = O(1)$. Afterwards, we determine the smallest P'-dominating subset $D^* \subseteq Q$. To do that, we examine the subsets of Q, from smallest to largest, verifying if all points of P' are dominated, until we find the dominating set D^*, which is returned as the approximate solution. Since Q has a constant number of points, this procedure takes $O(n)$ time.

Now we show that the returned solution D^* is indeed a 4-approximation. We argue that, given a P'-dominating set $D \subseteq P$, there is a P'-dominating set $D' \subseteq Q$ with $|D'| \leq 4\,|D|$. To build the set D' from D, we proceed as follows. For each point $p \in D$, if $p \in Q$, we add p to D'. Otherwise, since the set Q contains points of extreme coordinates in both x and y axes, in the cell of p, there are two points $q_1, q_2 \in Q$ such that (i) $\|pq_1\| \leq \gamma$, (ii) $\|pq_2\| \leq \gamma$, and (iii) the smallest angle $\angle q_1 p q_2$ is at least $\pi/2$. We add these two points q_1, q_2 to D'.

By Lemma 3, the portion $T = (\partial \bigcirc_p) \setminus (\bigcirc_{q_1} \cup \bigcirc_{q_2})$ of $\partial \bigcirc_p$ consists of one or two circular arcs. We first consider the case where T consists of one circular arc. Let R be the set of points from P' which are dominated by p, but not by q_1 or q_2. If R is empty, then no extra point needs to be added to D'. Otherwise, the line ℓ which contains p and bisects T separates R into two (possibly empty) sets R_1, R_2. If $R_1 \neq \emptyset$, let p_3 be an arbitrary point of R_1. Since Q contains a point in the same cell as p_3, there is a point q_3 with $\|p_3 q_3\| \leq \gamma$. We add the point q_3 to D'. Analogously, if $R_2 \neq \emptyset$, let p_4 be an arbitrary point of R_2 and let $q_4 \in Q$ be a point with $\|p_4 q_4\| \leq \gamma$. We add the point q_4 to D'.

We now show that the four points $q_1, q_2, q_3, q_4 \in Q$ dominate all points dominated by p. Consider a point v that is dominated by p but not by q_1 or q_2. The point v must be inside the circular crown sector depicted in Fig. 3(a) and described as follows. Because v is dominated by p, we have $\|pv\| \leq 2$. By Lemma 3, the arc length $\|T\| < 1.82$. Also, $\|pv\| > 1$, because otherwise the unit circles centered at p and v would intersect forming an arc of length at least $2\pi/3$, which is greater than $\|T\|$, in which case v is dominated by q_1 or q_2. Finally, since v is closer to p than it is to q_1 or q_2, it follows that v must be between the lines that connect p to the endpoints of T. This circular crown sector is bisected by the line ℓ. Using the law of cosines, we calculate the diameter of each circular crown sector as $d = \sqrt{8 - 8\cos(\|T\|/2)} < 1.76$. Therefore, for any point v inside the circular crown sector, the point q_3 (or q_4, analogously) that is within distance at most γ from a point inside the same sector dominates v, as $\|vq_3\| \leq d + \gamma < 2$.

Finally, if T consists of two circular arcs T_1, T_2 centered in p, then we start by adding those same points q_1, q_2 to D', as if T consisted of only one arc. Then, if necessary, we add new points q_3, q_4 to D' as follows. The points that are dominated by p but not by q_1 or q_2 must be within distance 1 of either T_1 or T_2.

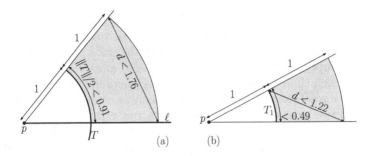

Fig. 3. Proof of Theorem 4

Let p_3, p_4 be arbitrary points that are within distance 1 of T_1 or T_2, respectively, but are not dominated by q_1 or q_2. If such points p_3, p_4 exist, then there are two points q_3, q_4 in Q that are within distance at most γ from respectively p_3, p_4. By Lemma 3, the largest arc among T_1, T_2 measures at most 0.49. The proof that all points dominated by p are dominated by q_1, q_2, q_3, or q_4 is analogous to the case where T consists of a single arc, using the circular crown sector illustrated in Fig. 3(b).

Since D^* is minimum among all subsets of Q that are P'-dominating sets, D^* is a 4-approximation for the MPDSP.   □

The following theorem uses the shifting strategy [8] to obtain a $(4 + \varepsilon)$-approximation for point sets of arbitrary diameter.

Theorem 5. *Given two sets of points P and P' as input, with $P' \subseteq P$ and $|P| = n$, the MPDSP can be $(4 + \varepsilon)$-approximated in $O(n)$ time on a real-RAM with constant-time hashing and the floor function. Without these operations, it can be done in $O(n \log n)$ time.*

Proof. Let k be the smallest integer such that

$$\left(\frac{k+2}{k}\right)^2 \leq 1 + \frac{\varepsilon}{4}.$$

We consider grids with square cells of side $2k$. We say a grid is *rooted at* a point (x, y) if there is a grid cell with corner at (x, y). Given a cell C, the square region C^+, called the *expansion* of C, is formed by C and all points within L_∞ distance at most 2 from C.

The algorithm proceeds as follows. For i, j from 0 to $k - 1$, we create a grid with cells of side $2k$ rooted at $(2i, 2j)$ and, for each cell C in the grid, we use Theorem 4 to 4-approximate the MPDSP with point sets $P \cap C^+, P' \cap C$, obtaining a solution $D_{i,j}(C)$. The dominating set $D_{i,j}$ is constructed as the union of the dominating sets $D_{i,j}(C)$ for all grid cells C. We return the smallest dominating set $D_{i,j}$ that is found, call it D^*. The remainder of the proof is similar to the proof of Theorem 2 and is omitted due to space limitations.   □

The MDSP is the special case of the MPDSP in which $P' = P$, and thus it can be $(4 + \varepsilon)$-approximated in linear time by the same algorithm.

5 Minimum Vertex Cover

In this section, we show how to obtain a linear-time approximation scheme to the MVCP. We start by presenting an approximation scheme for point sets of constant diameter, and then we use the shifting strategy to generalize the result to arbitrary diameter. Differently than in the previous two problems, the size of a minimum vertex cover for a point set of constant diameter is not upper bounded by a constant. Therefore, strictly speaking, a coreset for the problem does not exist. Nevertheless, it is possible to use coresets to approach the problem indirectly.

Given a graph $G = (V, E)$ with n vertices, it is well known that I is an independent set if and only if $V \setminus I$ is a vertex cover. While a maximum independent set corresponds to a minimum vertex cover, a constant approximation to the maximum independent set does not necessarily correspond to a constant approximation to the minimum vertex cover. However, in certain cases, an even stronger correspondence holds, as we show in the following proof.

Theorem 6. *Given a set P of n points as input, with $\mathrm{diam}(P) = O(1)$, the MVCP can be $(1 + \varepsilon)$-approximated in $O(n)$ time in the real-RAM, for constant $\varepsilon > 0$.*

Proof. Our algorithm considers two cases, depending on the value of n. If

$$n < \left(1 + \frac{3}{4\varepsilon}\right) \frac{(\mathrm{diam}(P) + 2)^2}{4},$$

then n is constant, and we can solve the MVCP optimally in constant time.

Otherwise, we use Theorem 1 to obtain a 4-approximation I to the maximum independent set. We now show that $V = P \setminus I$ is a $(1 + \varepsilon)$-approximation to the minimum vertex cover. Let I_{OPT}, V_{OPT} respectively be the maximum independent set and the minimum vertex cover. Note that $|V| = n - |I|$ and $|V_{OPT}| = n - |I_{OPT}|$. By a simple packing argument, dividing the area of a disk of diameter $\mathrm{diam}(P) + 2$ by the area of a unit disk,

$$|I_{OPT}| \leq \frac{(\mathrm{diam}(P) + 2)^2}{4},$$

and consequently

$$n \geq \left(1 + \frac{3}{4\varepsilon}\right) |I_{OPT}| = \left(1 + \frac{3}{4\varepsilon}\right) (n - |V_{OPT}|).$$

Manipulating the previous inequality, we obtain

$$n \leq \frac{4\varepsilon + 3}{3} |V_{OPT}|. \tag{2}$$

Since I is a 4-approximation to I_{OPT},

$$|V| = n - |I| \leq n - \frac{|I_{OPT}|}{4} = \frac{4n - |I_{OPT}|}{4} = \frac{3n + |V_{OPT}|}{4}. \tag{3}$$

Combining (2) and (3), we can write $|V| \leq (1 + \varepsilon)|V_{OPT}|$, as desired.   □

Table 1. Comparison of new and previous approximation algorithms

Previous/new results	MWISP	MDSP	MVCP
Previous approximation ratio in $o(n^4)$ time	5 [10]	4.889 [5]	$1 + \varepsilon$ [11]
Our approximation ratio in $O(n)$ time	$4 + \varepsilon$	$4 + \varepsilon$	$1 + \varepsilon$
Previous time for the same approximation	$O(n^4)$ [12]	$O(n^6 \log n)$ [9]	$O(n)$ [11]

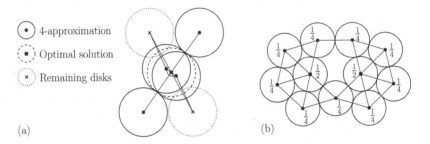

(a)   (b)

Fig. 4. (a) Example where the approximation ratio for the MDSP is exactly 4 (b) Coin graph used in the example where the approximation ratio for the MWISP is 3.25

Using the shifting strategy we obtain the following result. The proof is similar to that of Theorem 2 and is omitted due to space limitations.

Theorem 7. *Given a set P of n points in the plane as input, the MVCP can be $(1 + \varepsilon)$-approximated in $O(n)$ time on a real-RAM with constant-time hashing and the floor function, for constant $\varepsilon > 0$. Without these operations, it can be done in $O(n \log n)$ time.*

6   Conclusion

We introduced a method to obtain linear-time approximation algorithms for problems on unit-disk graphs and other geometric intersection graphs. The central idea of the method is a technique to obtain approximate solutions when the inputs are point sets of constant diameter. For the MWISP and the MDSP, the proposed algorithms provide improved approximation factors when compared not only to existing linear-time algorithms, but also to sub-quartic-time algorithms, as shown in Table 1.

While the approximation ratio for the MWISP and the MDSP is 4 (for constant diameter inputs), we only know that the analysis is tight for the MDSP. Figure 4(a) shows an MDSP instance where our algorithm does not achieve an approximation ratio better than 4, even if we reduce the grid size and search for extreme points in a larger number of directions. In contrast, for the MWISP, the best lower bound we are aware of is 3.25, as shown in the following example. Let P_1 be the weighted point set from Fig. 4(b), where all adjacent vertices are at distance exactly 2. Create another set P_2 by multiplying the coordinates of the points in P_1 by $1 + \varepsilon$, while multiplying their weights by $1 - \varepsilon$, for arbitrarily small $\varepsilon > 0$. The set P_2 forms an independent set of weight just smaller than 3.25, while the maximum independent set in P_1 has weight 1. Since each vertex in P_2 has a smaller weight and is arbitrarily close to a vertex of P_1, the vertices of P_2 will be disregarded by the algorithm for the input instance $P_1 \cup P_2$.

Several open problems remain. Can we obtain an approximation ratio better than 4 in (close to) linear time for the MWISP, or at least for its unweighted version? Can the linear-time approximation scheme for the MVCP be generalized for the weighted version? Are the point coordinates really necessary, or is it possible to devise similar graph-based algorithms? Also, can we use our method to obtain better linear-time approximations to related problems on unit disk graphs such as finding the minimum-weight dominating set or the minimum connected dominating set?

References

1. Agarwal, P.K., Har-Peled, S., Varadarajan, K.R.: Geometric approximation via coresets. In: Goodman, J.E., Pach, J., Welzl, E. (eds.) Combinatorial and Computational Geometry. Cambridge University Press, Cambridge (2005)
2. Appel, K., Haken, W.: Solution of the four color map problem. Sci. Am. **237**(4), 108–121 (1977)
3. Bentley, J., Stanat, D., Williams Jr., E.H.: The complexity of finding fixed-radius near neighbors. Inf. Process. Lett. **6**(6), 209–212 (1977)
4. Breu, H., Kirkpatrick, D.G.: Unit disk graph recognition is NP-hard. Comput. Geom. **9**(1–2), 3–24 (1998)
5. da Fonseca, G.D., de Figueiredo, C.M.H., Sá, V.G.P., Machado, R.C.S.: Efficient sub-5 approximations for minimum dominating sets in unit disk graphs. WAOA 2012, Theor. Comput. Sci. **540–541**, 70–81 (2014)
6. De, M., Das, G., Carmi, P., Nandy, S.: Approximation algorithms for a variant of discrete piercing set problem for unit disks. Int. J. Comput. Geom. Appl. **6**(23), 461–477 (2013)
7. Hochbaum, D.S., Maass, W.: Approximation schemes for covering and packing problems in image processing and VLSI. J. ACM **32**(1), 130–136 (1985)
8. Hunt III, H.B., Marathe, M.V., Radhakrishnan, V., Ravi, S., Rosenkrantz, D.J., Stearns, R.E.: NC-approximation schemes for NP- and PSPACE-hard problems for geometric graphs. J. Algorithms **26**, 238–274 (1998)
9. Jallu, R.K., Prasad, P.R., Das, G.K.: Minimum dominating set for a point set in $\mathbb{R}^2$. preprint, arXiv:1312.7243 (2014)
10. Marathe, M.V., Breu, H., Hunt III, H.B., Ravi, S.S., Rosenkrantz, D.J.: Simple heuristics for unit disk graphs. Networks **25**(2), 59–68 (1995)
11. Marx, D.: Efficient approximation schemes for geometric problems? In: Brodal, G.S., Leonardi, S. (eds.) ESA 2005. LNCS, vol. 3669, pp. 448–459. Springer, Heidelberg (2005)
12. Matsui, T.: Approximation algorithms for maximum independent set problems and fractional coloring problems on unit disk graphs. In: Akiyama, J., Kano, M., Urabe, M. (eds.) JCDCG 1998. LNCS, vol. 1763, pp. 194–200. Springer, Heidelberg (2000)
13. Nieberg, T., Hurink, J., Kern, W.: Approximation schemes for wireless networks. ACM Trans. Algorithms **4**(4), 49:1–49:17 (2008)

The Minimum Feasible Tileset Problem

Yann Disser, Stefan Kratsch, and Manuel Sorge[✉]

Technische Universität Berlin, Berlin, Germany
{yann.disser,stefan.kratsch,manuel.sorge}@tu-berlin.de

Abstract. We consider the MINIMUM FEASIBLE TILESET problem: Given a set of symbols and subsets of these symbols (scenarios), find a smallest possible number of pairs of symbols (tiles) such that each scenario can be formed by selecting at most one symbol from each tile. We show that this problem is NP-complete even if each scenario contains at most three symbols. Our main result is a 4/3-approximation algorithm for the general case. In addition, we show that the MINIMUM FEASIBLE TILESET problem is fixed-parameter tractable both when parameterized with the number of scenarios and with the number of symbols.

1 Introduction

Consider the general assignment problem where several devices (e.g., workers, robots, microchips, ...) each can be used in one of k functions/modes (e.g., employing different skills, tools, instruction sets, ...) at a time. Given a set of scenarios, the goal is to assign k different functions to each device, such that, for each scenario, all functions requested by the scenario are available simultaneously. In this paper, we initiate the study of this problem for $k = 2$ and the case that each function is requested at most once by each scenario. Formally, we study the following problem (we use "*tile*" instead of "*device*" to intuitively capture the fact that a device/tile has two modes/sides).

MINIMUM FEASIBLE TILESET
Input: A universe of symbols F, scenarios $\mathcal{S} \subseteq 2^F \setminus \{F\}$, and $\ell \in \mathbb{N}$.
Problem: Is there a tileset $\mathcal{T}$ of at most ℓ tiles $T \in \binom{F}{2}$ that is feasible for all scenarios in $\mathcal{S}$?

In the above, we refer to (multi-)sets of tiles as *tilesets*. A tileset $\mathcal{T}$ is *feasible* for scenario S, if we can produce all symbols in S by taking at most one symbol from each tile in $\mathcal{T}$. Formally, a tileset $\mathcal{T}$ is feasible for a scenario $S \subset F$ if there is a mapping $\phi \colon \mathcal{T} \to F$, such that $\phi(T) \in T$ for all $T \in \mathcal{T}$, and $S \subseteq \phi[\mathcal{T}] := \{\phi(T) \mid T \in \mathcal{T}\}$. By definition, no scenario contains all symbols of F. Note that such a scenario would require $|F|$ tiles, making the problem trivial. Similarly, we may assume that all symbols in F appear in at least one scenario, otherwise we

Y. Disser—Supported by the Alexander von Humboldt-Foundation.
S. Kratsch—Supported by the German Research Foundation (DFG), KR 4286/1.
M. Sorge—Supported by the German Research Foundation (DFG), NI 369/12.

© Springer International Publishing Switzerland 2015
E. Bampis and O. Svensson (Eds.): WAOA 2014, LNCS 8952, pp. 144–155, 2015.
DOI: 10.1007/978-3-319-18263-6_13

can simply remove each symbol that does not occur in any scenario. Finally, the requirement that tiles contain no less than two symbols can be met by arbitrarily assigning a second symbol to all tiles of cardinality one.

Apart from practical motivations MINIMUM FEASIBLE TILESET is appealing from a structural point of view. In this work we exhibit equivalent definitions for the problem which are interesting in their own right. At first glance, MINIMUM FEASIBLE TILESET is a covering problem since we must cover all scenarios using tiles that can each cover one of their two symbols in each scenario. It turns out that the problem can also be phrased as a packing/partitioning problem, but with an objective function different from the classical one in terms of number of packed objects or sets (see Sect. 3). In addition, having tiles be symbol sets of size two suggests a graph interpretation where we are asked to find a minimum set of edges such that for each scenario there is an orientation where each vertex has indegree at least one. We favor the tileset formulation, since it most naturally generalizes to the original assignment problem with tiles of larger sizes and scenarios which contain multiple copies of the same symbols. Also, the MINIMUM FEASIBLE TILESET interpretation appears suitable for studying the effect of parameters, such as the number of symbols/scenarios, on the complexity.

Results and Outline. We analyze the structure of the graph that has the tiles of a minimum cardinality tileset as its edges, and show that this graph is always (wlog.) a forest. In fact, only the component structure of this forest matters: We may replace trees by arbitrary trees spanning the same components without affecting the feasibility of the corresponding tileset (Sect. 2). This lets us view MINIMUM FEASIBLE TILESET as a partitioning problem, which in turn allows us to prove NP-completeness even when scenarios have size at most three (Sect. 3). As our main result, we complement the hardness with a 4/3-approximation algorithm (for scenarios of arbitrary sizes) inspired by the component structure of the optimum solution (Sect. 4). Finally, we show that the problem is fixed-parameter tractable with respect to the number of scenarios (Sect. 5) and the number of symbols (Sect. 6), respectively. Due to space constraints, we defer proofs for results marked by ⋆ to a full version of the paper.

Related Work. The problem most closely related to MINIMUM FEASIBLE TILE-SET is arguably SET PACKING, as 3-SET PACKING appears as a subproblem in our approximation algorithm and also as the source problem for our NP-hardness reduction. SET PACKING has been extensively studied for both approximability and parameterized complexity (see, e.g., [1,5,19] and [6,17] for some recent results). The main difference between the two problems is that SET PACKING is a maximization problem whereas MINIMUM FEASIBLE TILESET seeks to minimize the size of a feasible tileset—a measure that is only indirectly related to the number of sets (scenarios). In particular, SET PACKING becomes trivial for a bounded number of sets, whereas for MINIMUM FEASIBLE TILESET we get a nontrivial polynomial-time algorithm via integer linear programming.

As mentioned above, the MINIMUM FEASIBLE TILESET problem can equivalently be seen as designing an edge-minimal graph on the set of symbols such that, for each scenario, the edges (tiles) can be oriented in such a way that all symbols in the scenario have indegree at least one. The question whether a *given* graph admits an orientation with certain properties has been studied in various settings. For example, Biedl et al. [2] proposed an approximation algorithm for finding a balanced acyclic orientation. Another natural constraint on an orientation that has been studied is to prescribe degrees for each vertex [8,10,14].

More abstractly, we are looking for a graph on the set of symbols that fulfills a certain constraint for each scenario. The case where the subgraph induced by each scenario has to be connected is well-studied [3,4,9,13,15]. In particular, it is NP-hard to find the minimum number of edges needed [9] and to decide whether a planar solution [3,15] or a solution of treewidth at most three [13] exists.

2   Graph Structure of Tilesets

The tiles in a tileset T over a universe of symbols F can be viewed as the edges of the undirected (multi-) graph $G(T) := (F, T)$. In this section, we establish that there always exist optimal tilesets with a simple graph structure. This is made formal in the following lemma which will be useful in later sections.

Lemma 1 ($\star$). *Let F be a universe of symbols, S a family of scenarios over F, and T a tileset feasible for S. There is a tileset $T' \subseteq \binom{F}{2}$ feasible for S such that $|T'| \leq |T|$ and $G(T')$ is a forest.*

Note that each connected component of $G(T')$ has size at least two because each symbol occurs in at least one scenario and hence is incident with at least one edge. We show that only the partition of the symbols induced by the component structure of a tileset matters, but not the exact topology of each of the trees.

Theorem 1 ($\star$). *Let S be a family of scenarios and T be a tileset over symbols F. If $G(T)$ is a forest, then T is feasible for S if and only if no connected component C of $G(T)$ is fully contained in any scenario $S \in S$, i.e., $C \not\subseteq S$ for all scenarios $S \in S$ and all connected components C of $G(T)$.*

3   NP-Completeness of Minimum Feasible Tileset

In this section we establish the following completeness result.

Theorem 2. MINIMUM FEASIBLE TILESET *is* NP-*complete, even if each scenario has size at most three.*

Let us check that MINIMUM FEASIBLE TILESET is contained in NP: A feasible tileset can be encoded using polynomially many bits with respect to $|F|$. Verifying feasibility comes down to solving one bipartite matching problem for each scenario on an auxiliary graph that has an edge between each symbol in the scenario and every tile containing that symbol, which is possible in polynomial time.

It remains to prove NP-hardness. For this, we first give a reduction from the following partition problem, and later prove this problem to be NP-hard.

FINE CONSTRAINED PARTITION
Input: A universe U, constraints $\mathcal{V} \subseteq 2^U \setminus \{U\}$, and $p \in \mathbb{N}$.
Problem: Is there a partition $\mathcal{P}$ of U, $|\mathcal{P}| \geq p$, such that $P \not\subseteq V$ for all parts $P \in \mathcal{P}$ and all $V \in \mathcal{V}$?

Lemma 2. MINIMUM FEASIBLE TILESET *and* FINE CONSTRAINED PARTITION *are equivalent if we identify scenarios and constraints.*

Proof. We claim that an instance $(F, \mathcal{S}, \ell)$ of MINIMUM FEASIBLE TILESET admits a solution if and only if the instance $(F, \mathcal{S}, |F| - \ell)$ of FINE CONSTRAINED PARTITION admits a solution.

"⇒": By Lemma 1 there is a feasible tileset $\mathcal{T}'$ for $\mathcal{S}$ of cardinality at most ℓ such that $G(\mathcal{T}')$ is a forest. The connected components $C_1, \ldots, C_k$ of $G(\mathcal{T}')$ induce a partition that is a solution for the FINE CONSTRAINED PARTITION instance: By Theorem 1 we indeed have $C_i \not\subseteq S$ for all connected components C_i, $i \in [k]$, and scenarios $S \in \mathcal{S}$. Furthermore, since there are at most ℓ edges in $G(\mathcal{T}')$ and each connected component is a tree, we have $\ell \geq \sum_{i=1}^{k} |C_i| - 1 = |F| - k$. Hence, our partition has at least $k \geq |F| - \ell$ parts.

"⇐": Let $\mathcal{P} = \{P_1, \ldots, P_p\}$ be a solution for the FINE CONSTRAINED PARTITION instance. We construct a tileset $\mathcal{T}$ by setting $G(\mathcal{T})[P_i]$ to an arbitrary spanning tree for each $i \in [p]$. Since $P_i \not\subseteq S$ for each $S \in \mathcal{S}$ and each $i \in [p]$, by Theorem 1, $\mathcal{T}$ is feasible for $\mathcal{S}$. The number of tiles in $\mathcal{T}$ is $\sum_{i=1}^{p} |P_i| - 1 = |F| - p \leq |F| - (|F| - \ell) = \ell$, as required.   □

Note that the corresponding optimization problems are dual to each other in the sense that one is to minimize ℓ and the other to maximize $|F| - \ell$. We are now ready to give a reduction to FINE CONSTRAINED PARTITION from EXACT COVER BY 3-SETS, which is well known to be NP-hard [12], hence, completing the proof of Theorem 2.

EXACT COVER BY 3-SETS
Input: A universe X and a family $\mathcal{C}$ of three-element sets $C \in \binom{X}{3}$.
Problem: Is there an *exact cover* for X, i.e., a partition of X into a family $\mathcal{C}' \subseteq \mathcal{C}$ of disjoint sets?

Lemma 3. *There is a polynomial-time reduction from* EXACT COVER BY 3-SETS *to* FINE CONSTRAINED PARTITION *with constraints of size at most three.*

Proof. Let an instance $(X, \mathcal{C})$ of EXACT COVER BY 3-SETS be given. Without loss of generality, we may assume that $|X| = 3q$ for some integer q, as otherwise no exact cover exists. We construct an instance of FINE CONSTRAINED PARTITION with universe X asking for a partition of size at least q. First, we add constraints $\mathcal{V}_2 = \binom{X}{2}$ that exclude every two-element subset of X from all solution partitions. Since every solution partition needs to contain at least q parts and $|X| = 3q$, each such partition consists of sets of size exactly three. Next, we exclude partitions that contain sets outside of $\mathcal{C}$ by simply adding the constraints $\mathcal{V}_{\bar{\mathcal{C}}} = \binom{X}{3} \setminus \mathcal{C}$. This concludes the construction of the FINE CONSTRAINED PARTITION instance $(X, \mathcal{V}_2 \cup \mathcal{V}_{\bar{\mathcal{C}}}, q)$. Clearly, this takes polynomial time.

Now, if there is a partition $\mathcal{P}$ with at least q parts for the FINE CONSTRAINED PARTITION instance, by the above, we know that each of its parts is a set in $\mathcal{C}$. Hence $\mathcal{P}$ is an exact cover of X for family $\mathcal{C}$. Conversely, let $\mathcal{C}' \subseteq \mathcal{C}$ be an exact cover for X. Then $|\mathcal{C}'| \geq q$ and for all $C \in \mathcal{C}'$ and $V \in \mathcal{V}_2 \cup \mathcal{V}_{\bar{c}}$ we have $C \not\subseteq V$, because C has size three and is not from $\mathcal{V}_{\bar{c}}$. Hence also the FINE CONSTRAINED PARTITION instance has a solution. □

4 A 4/3-Approximation for Minimum Feasible Tileset

In this section, we propose an approximation algorithm for MINIMUM FEASIBLE TILESET with unbounded scenario size. Motivated by the structural insights of Sect. 2, we construct a tileset that induces a forest in the corresponding graph, with the property that none of its components are contained in a single scenario. Since a component of size k requires $k - 1$ tiles, we additionally aim for small components in order to keep the resulting tileset small.

We first take as many components of size two as possible among all disjoint sets of two symbols that are not both contained in the same scenario. This can easily be achieved by computing a maximum matching in the graph that has an edge for each candidate component. Similarly, among all remaining symbols, we try to form many (disjoint) components of size three, without creating components that are contained in a single scenario. For this, we employ a simple greedy strategy, that repeatedly takes any possible component until no possible candidates remain. (While there are better packing strategies available for sets of size three, we will see that improving the packing strategy alone does not improve our approximation ratio.) Finally, for each leftover symbol we add an individual tile (pairing that symbol in such a way as to prevent cycles).

We give a more formal listing in Algorithm A. We use $\bar{F}_i(F') = \{C \in \binom{F'}{i} \mid \forall S \in \mathcal{S} : C \not\subseteq S\}$ to denote the family of all sets of symbols in F' that are of size i and not fully contained in a single scenario. In the following, we identify connected components with their sets of vertices.

Algorithm A. 4/3-approximation for minimum feasible tilesets

Input: A set F of symbols and a set $\mathcal{S}$ of scenarios, where $\mathcal{S} \subseteq 2^F \setminus \{F\}$.
Output: A set of tiles $\mathcal{T}$.
$\mathcal{T}_2 \leftarrow$ maximum matching in graph $G(\bar{F}_2(F))$.
$\mathcal{P} \leftarrow$ greedy set packing of $\bar{F}_3(F \setminus \bigcup_{t \in \mathcal{T}_2} t)$.
$\mathcal{T}_3 \leftarrow \bigcup_{\{f_1, f_2, f_3\} \in \mathcal{P}} \{\{f_1, f_2\}, \{f_2, f_3\}\}$.
if $\mathcal{T}_2 \cup \mathcal{T}_3 \neq \emptyset$ **then** take $f_{\text{root}} \in \bigcup_{t \in \mathcal{T}_2 \cup \mathcal{T}_3} t$
else take $f_{\text{root}} \in F$.

$\mathcal{T}_1 \leftarrow \{\{f, f_{\text{root}}\} \mid f \in F \setminus \bigcup_{t \in \mathcal{T}_2 \cup \mathcal{T}_3} t \,, f \neq f_{\text{root}}\}$.
return $\mathcal{T} = \mathcal{T}_1 \cup \mathcal{T}_2 \cup \mathcal{T}_3$.

Theorem 3. *Algorithm A computes a 4/3-approximation for* MINIMUM FEASI-
BLE TILESET.

Proof. We first argue that the set of tiles $\mathcal{T} = \mathcal{T}_1 \cup \mathcal{T}_2 \cup \mathcal{T}_3$ computed by
Algorithm A is feasible for $\mathcal{S}$. First observe that $G(\mathcal{T})$ is a forest. This is true,
because $G(\mathcal{T}_2 \cup \mathcal{T}_3)$ consists of trees of sizes 2 and 3, $G(\mathcal{T}_1)$ is a star, and
$\mathcal{T}_1 \cap (\mathcal{T}_2 \cup \mathcal{T}_3)$ contains at most one node (f_{root}). Using Theorem 1 it only
remains to show that no connected component C of $G(\mathcal{T})$ is contained in any
scenario $S \in \mathcal{S}$, i.e. $C \cap S \subsetneq C$. By definition of Algorithm A this is true for all
connected components of the graph $G(\mathcal{T}_2 \cup \mathcal{T}_3)$. If $\mathcal{T}_2 \cup \mathcal{T}_3 \neq \emptyset$, then each component of $G(\mathcal{T})$ is a superset of a component of $G(\mathcal{T}_2 \cup \mathcal{T}_3)$, and is thus not contained
in any scenario. If $\mathcal{T}_2 \cup \mathcal{T}_3$ is empty, then $G(\mathcal{T}) = G(\mathcal{T}_1)$ consists of a single component that is not contained in any scenario, since, by definition, $F \notin \mathcal{S}$. Thus
$\mathcal{T}$ is feasible for $\mathcal{S}$.

We now bound the size of $\mathcal{T}$ with respect to a minimum cardinality tileset $\mathcal{T}^*$.
To do this we *distribute* virtual currency *(gold)* to the symbols in F, such that
the total gold distributed is 4/3 times the size of $\mathcal{T}^*$. We later use this gold to
pay one unit of gold to certain symbols that these can in turn use to *provide
for* (at most) one tile of $\mathcal{T}$ that involves this symbol. To complete the proof, we
establish that each tile of $\mathcal{T}$ is provided for by one of its two symbols.

Let $G^* := G(\mathcal{T}^*)$ be the graph induced by $\mathcal{T}^*$ and $\bar{F}_i^*$ be the set of con-
nected components of size $i \in \{2, \ldots, |F|\}$ in G^*. By Lemma 1, we may assume
that G^* is a forest. Furthermore, because each symbol appears in at least one
scenario, graph G^* does not contain components of size 1. Since the symbols in
a component of size $i > 1$ are part of exactly $i - 1$ tiles in $\mathcal{T}^*$, we may distribute
all available gold by giving $4/3 \cdot \frac{i-1}{i}$ gold to each symbol in a component of $\bar{F}_i^*$,
for all $i \in \{2, \ldots, |F|\}$. This gold is used to pay symbols in what follows. We
call a symbol $s \in F$ *sufficiently paid* if one of the following holds: (i) s is paid,
(ii) s appears in a tile $T \in \mathcal{T}_2$ and the other symbol of T is paid, or (iii) s appears
in a tile $T \in \mathcal{T}_3$ and the other two symbols in the same component of $G(\mathcal{T}_3)$ are
paid. Below, we show how to sufficiently pay all symbols. This completes the
proof, since then all tiles in $\mathcal{T}_1 \cup \mathcal{T}_2 \cup \mathcal{T}_3$ can be provided for (note that then each
tile in $\mathcal{T}_1$ contains its own paid symbol). We call a component of G^* *sufficiently
paid*, if all its symbols are sufficiently paid. Let $F_{\geq 4}^* := F \setminus \bigcup_{C \in \bar{F}_2^* \cup \bar{F}_3^*} C$ be the
set of all symbols not in components of size two or three in G^*. In paying the
symbols we will maintain the invariant that each element of $\bar{F}_2^* \cup \bar{F}_3^* \cup F_{\geq 4}^*$ is
either sufficiently paid, or it still holds its gold (all its symbols still hold their
gold, respectively).

We define a graph $H = (V, E)$ that has the components in $\bar{F}_2^* \cup \bar{F}_3^*$ as its
vertices, as well as the symbols that are not part of these components, i.e., $V =
\bar{F}_2^* \cup \bar{F}_3^* \cup F_{\geq 4}^*$. In this way, each vertex of H represents up to three symbols. For
each tile $T \in \mathcal{T}_2$ we introduce an edge connecting the vertices of H representing
the two symbols of T, possibly introducing self-loops. Since $\mathcal{T}_2$ is a matching,
and since the vertices in H represent at most three symbols each, all vertices
in H have degree at most 3. We partition the edges of H into paths, cycles, and
self-loops, and show for each how to use the gold remaining at its vertices to pay

all symbols in the components of $G^\star$ that are intersected by the path/cycle/self-loop. We will ensure that every symbol (except possibly f_{root}) on a tile in $\mathcal{T}_1$ is paid. Since each symbol on a tile of $\mathcal{T}_2$ appears only exactly on this and no other tile of $\mathcal{T}_2 \cup \mathcal{T}_3$, it is thus sufficient to pay only one of the two symbols on each tile of $\mathcal{T}_2$.

Let $\mathcal{P}$ be the set of all paths in H connecting (different) vertices of degree 1 or 3 with internal nodes of degree 2. Consider the paths in $\mathcal{P}$ one by one. We use the gold available along path $P \in \mathcal{P}$ of length k as follows. Let N_2, N_3 be the number of internal nodes of P that represent 2 and 3 symbols, respectively. Note that P has no inner nodes that represent a single symbol, since $\mathcal{T}_2$ is a matching, and hence $k = 1 + N_2 + N_3$. Also, P is the only path visiting these inner nodes and hence they all still hold their gold. Let $N_1^{\text{end}}, N_2^{\text{end}}, N_3^{\text{end}} \leq 2$ be the number of endpoints of P that still hold gold and represent 1, 2, and 3 symbols, respectively. Similarly, let N_0^{end} be the number of endpoints without gold. By our invariant, the symbols or components represented by the endpoints without gold left have already been sufficiently paid before. We make sure that all other nodes along P are sufficiently paid. We do this by, for all tiles that form the path P, paying one of the two corresponding symbols, and, in addition, paying *every* further symbol represented by nodes along P. Note that this preserves the invariant. The total cost is

$$C^- = k + N_2^{\text{end}} + 2N_3^{\text{end}} + N_3 - N_0^{\text{end}} = 1 + N_2^{\text{end}} + 2N_3^{\text{end}} + N_2 + 2N_3 - N_0^{\text{end}}. \quad (1)$$

Using that each endpoint of P that contributes to N_1^{end} represents a symbol that is part of a component in $G^\star$ of size $i \geq 4$, we get that the gold available at this symbol is at least $\frac{4}{3} \cdot \frac{i-1}{i} \geq 1$. Hence, the gold available to us is at least

$$C^+ = \frac{4}{3}\left(N_2^{\text{end}} + 2N_3^{\text{end}} + N_2 + 2N_3 + \frac{3}{4}N_1^{\text{end}}\right). \quad (2)$$

Since $N_0^{\text{end}} + N_1^{\text{end}} + N_2^{\text{end}} + N_3^{\text{end}} = 2$, we get

$$C^+ - C^- = 1 - \frac{2}{3}N_2^{\text{end}} - \frac{1}{3}N_3^{\text{end}} + \frac{1}{3}N_2 + \frac{2}{3}N_3.$$

Hence, we have $C^+ \geq C^-$, unless $N_2^{\text{end}} = 2$ and $N_0^{\text{end}} = N_1^{\text{end}} = N_3^{\text{end}} = N_2 = N_3 = 0$, i.e. P is of length one, connecting two tiles $p_1, p_2 \in \bar{F}_2^\star$ by an edge which corresponds to a tile $t \in \mathcal{T}_2$. To see that this case cannot occur, observe that, first, p_1 and p_2 are of degree 1 in H. Second, since $\mathcal{T}^\star$ is feasible, no component of $G^\star$ is contained in a single scenario (Theorem 1), and thus $p_1, p_2 \in \bar{F}_2^\star \subseteq \bar{F}_2(F)$. This is a contradiction to $\mathcal{T}_2$ being a maximum matching in graph $G(\bar{F}_2(F))$, as the matching can be augmented by removing t and adding p_1 and p_2.

Similarly to the above, we can consider all cycles in H with at most one node of degree 3 one by one. (Note that cycles with at least two nodes of degree 3 contain a path as before.) If a cycle of length k does not contain a node of degree 3, or the node of degree 3 is not yet sufficiently paid (and thus still holds its gold), the cost for the cycle and its available gold are

$$C^- = k + N_3 = N_2 + 2N_3 = \frac{3}{4}C^+ < C^+,$$

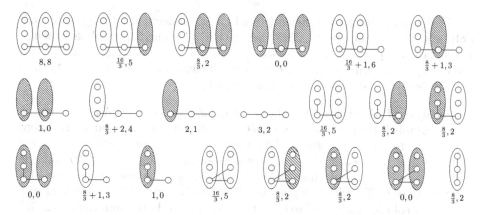

Fig. 1. Possible intersections of components of $G(\mathcal{T}_3)$ (arcs) and $G^\star$ (ellipses). Shaded components have been sufficiently paid previously. Configurations are labeled by the available gold C^+ and the required gold C^-. Symmetrical configurations are omitted.

where N_2, N_3 are the numbers of nodes of P that represent 2 and 3 symbols, respectively. If the node of degree 3 has no gold left, then it has already been sufficiently paid and $C^- = N_2 + 2N_3 - 3 < \frac{3}{4}C^+$. In either case, the available gold allows to sufficiently pay all nodes along the cycle. Finally, each self-loop in H connects two symbols in the same component C of size 2 or 3 in $G^\star$. If $|C| = 2$, the gold available among the two symbols is $C^+ = \frac{4}{3}$, while we require only $C^- = 1$ unit of gold. If $|C| = 3$, we have $C^+ = \frac{8}{3}$ and $C^- = 2$.

After processing all paths, cycles, and self-loops all nodes of H intersecting a tile of $\mathcal{T}_2$ are sufficiently paid. In particular, since $\mathcal{T}_2$ is a maximum matching, all components in $\bar{F}_2^\star$ are sufficiently paid. In the next step we ensure that all components of $\bar{F}_3^\star$ are sufficiently paid. By construction, every element of $\bar{F}_3^\star$, that is not sufficiently paid yet, intersects at least one tile of $\mathcal{T}_3$. We can thus consider the components of $G(\mathcal{T}_3)$ one by one and make sure to sufficiently pay each element of $\bar{F}_3^\star$ that intersects the considered component of $G(\mathcal{T}_3)$.

Consider a component of $G(\mathcal{T}_3)$ involving the three symbols f_1, f_2, f_3 (cf. Fig. 1 in the following). Let $\mathcal{C}_3 \subseteq \bar{F}_3^\star$ be the set of components of size 3 in $G^\star$ that involve at least one of these symbols and have not yet been sufficiently paid (i.e., still hold their gold). Further, let N_n be the number of symbols among $\{f_1, f_2, f_3\} \cap F_{\geq 4}^\star$ that are not yet sufficiently paid. Since all components in $\bar{F}_2^\star$ are sufficiently paid, the gold we have available is at least $C^+ \geq \frac{4}{3}(2|\mathcal{C}_3| + \frac{3}{4}N_n)$. We ensure that (at least) two symbols among f_1, f_2, f_3 are paid, as well as all other symbols appearing in $\mathcal{C}_3$. In this way, each component in $\mathcal{C}_3$ is sufficiently paid. Note that this preserves our invariant that each element of $\bar{F}_2^\star \cup \bar{F}_3^\star \cup F_{\geq 4}^\star$ is either sufficiently paid, or still holds its gold. The cost for paying the symbols f_1, f_2, f_3 is at most 2. Since in addition to f_1, f_2, f_3 there are $3|\mathcal{C}_3| + N_n - 3$ symbols needing pay in $\bigcup_{C \in \mathcal{C}_3} C \cup \{f_1, f_2, f_3\}$, and because $|\mathcal{C}_3| \leq 3$, the total cost is

$$C^- \leq 3|\mathcal{C}_3| + N_n - 1 \leq \frac{8}{3}|\mathcal{C}_3| + N_n \leq C^+.$$

At this point, we have sufficiently paid all components in $\bar{F}_2^\star \cup \bar{F}_3^\star$ using gold only from these components. This means that all remaining symbols that are not sufficiently paid yet have at least $\frac{4}{3} \cdot \frac{4-1}{4} = 1$ gold available, which we can use to pay these symbols themselves. Now all elements of $\bar{F}_2^\star \cup \bar{F}_3^\star \cup F_{\geq 4}^\star$ have been sufficiently paid and the proof is complete. $\square$

Our analysis of Algorithm A is tight in three different spots: (i) A path of length 1 in the graph H defined above that visits a component of size 2 and a component of size 3 of the optimum solution $\mathcal{T}$ may lead to 4 tiles in our solution compared to the 3 tiles required in the optimum solution, i.e., Eqs. (1) and (2) coincide if $N_2^{\mathrm{end}} = N_3^{\mathrm{end}} = 1$ and all other terms vanish. (ii) The first intersection of a component of $G(\mathcal{T}_3)$ with components of $G^\star$ illustrated in Fig. 1 may lead to 8 tiles in our solution compared to the 6 tiles required in the optimum solution. (iii) Each symbol of a component of size 4 in $G^\star$ might result in a single tile for this symbol only, in which case the optimum solution requires 3 tiles for the symbols of the component, while our solution requires 4 tiles. To improve Algorithm A we have to address each of these three bottlenecks. For (i), we either would have to alter the matching $\mathcal{T}_2$ to prevent the described situation, or combine the analysis to account for the loss in other places. The aspect (ii) can easily be prevented by employing a more sophisticated set packing algorithm (e.g., the $(4/3 + \varepsilon)$-approximation of Cygan [5]). Finally, to avoid (iii), we would need to pack sets of size 4 similarly to our packing of sets of size 3. In addition to requiring one more level of analysis, this would also complicate the other levels, as we would have to include sets of size 4 in our reasoning there.

5   Bounded Number of Scenarios

We prove that MINIMUM FEASIBLE TILESET can be solved in polynomial time when the number $|\mathcal{S}|$ of scenarios is bounded. More precisely, we provide an algorithm that solves any instance $(F, \mathcal{S}, k)$ in time $f(|\mathcal{S}|)|(F, \mathcal{S}, k)|^c$, i.e., in time $\mathcal{O}(|(F, \mathcal{S}, k)|^c)$ for bounded values of $|\mathcal{S}|$. In other words, MINIMUM FEASIBLE TILESET is fixed-parameter tractable with respect to the number of scenarios.

Our algorithm works by first translating the input instance $(F, \mathcal{S}, \ell)$ into an integer linear program (ILP) in such a way that the ILP is feasible (i.e., contains at least one integer point) if and only if $(F, \mathcal{S}, \ell)$ admits a feasible tileset with at most ℓ tiles. The ILP uses $\mathcal{O}(|\mathcal{S}|^{|\mathcal{S}|})$ variables. Lenstra [18] proved that deciding feasibility of any ILP is fixed-parameter tractable with respect to the number p of variables; the currently fastest algorithm has $\mathcal{O}^*(p^{\mathcal{O}(p)})$ running time and was obtained by Frank and Tardos [11], modifying an algorithm by Kannan [16]. Using this, we can prove the following result.

Theorem 4 ($\star$). MINIMUM FEASIBLE TILESET *on instances with at most k scenarios can be solved in time* $\mathcal{O}^*(k^{\mathcal{O}(k^{k+1})})$.

6   Bounded Number of Symbols

We analyze the influence of the number of symbols $|F|$ on the complexity of solving an instance $(F, \mathcal{S}, \ell)$ of MINIMUM FEASIBLE TILESET. It is easy to see that the problem becomes solvable in polynomial time when F is bounded: The instance is trivial if $\ell \geq |F|$ since, in that case, we can afford to dedicate a separate tile for each symbol. Otherwise, there are only $\mathcal{O}(|F|^{2\ell}) \subseteq \mathcal{O}(|F|^{2|F|})$ ways to fix ℓ tiles. As mentioned in Sect. 3, each candidate tileset can be verified by solving a bipartite matching problem for each scenario, on a graph that has an edge between each symbol in the scenario and every tile containing that symbol. This yields an overall runtime of $\mathcal{O}^*(|F|^{2|F|})$, and, hence, fixed-parameter tractability in $|F|$. Using structural insights of Sect. 2 we are able to improve on this naive running time.

Theorem 5 ($\star$). *Instances $(F, \mathcal{S}, \ell)$ of MINIMUM FEASIBLE TILESET can be solved in time $\mathcal{O}^*(3^{|F|})$.*

Note that, as every symbol occurs in a scenario, $\ell \geq |F|/2$. Hence, Theorem 5 gives a fixed-parameter algorithm also for parameter ℓ.

After this fixed-parameter tractability result, and taking into account the trivial bound of $2^{|F|}$ for the number of scenarios (giving a worst-case size of instances of $\mathcal{O}(2^{|F|}|F|)$), it is natural to ask whether polynomial-time preprocessing can simplify input instances to size polynomial in $|F|$. We show that this is impossible unless $\mathsf{NP} \subseteq \mathsf{coNP/poly}$ (and the polynomial hierarchy collapses). More generally, we prove that for the restricted case d-MINIMUM FEASIBLE TILESET, where scenarios have size at most d, no polynomial-time algorithm can achieve a size of $\mathcal{O}(k^{d-\varepsilon})$. Note that this restricted case has an essentially matching upper bound of $|\mathcal{S}| < (|F| + 1)^d = \mathcal{O}(|F|^d)$.[1] As a consequence there is no reduction to size polynomial in $|F|$ for the general MINIMUM FEASIBLE TILESET problem: Any size $\mathcal{O}(k^c)$ preprocessing for MINIMUM FEASIBLE TILESET could be used for d-MINIMUM FEASIBLE TILESET, for any $d > c$, and violate the lower bound.

Theorem 6. *Let $d \geq 3$ and ε be a positive real. There is no polynomial-time algorithm that reduces every instance of d-MINIMUM FEASIBLE TILESET to an equivalent instance (possibly of a different problem) of size $\mathcal{O}(|F|^{d-\varepsilon})$, unless $\mathsf{NP} \subseteq \mathsf{coNP/poly}$.*

To prove Theorem 6 we employ a similar result by Dell and Marx [6] for EXACT COVER BY d-SETS, which is defined as follows.[2]

> EXACT COVER BY d-SETS
> **Input:** A universe X and a family $\mathcal{C}$ of d-element sets $C \in \binom{X}{d}$.
> **Problem:** Is there an *exact d-set cover* for X, i.e., a partition of X into a family $\mathcal{C}' \subseteq \mathcal{C}$ of disjoint sets?

---

[1] A compression to $\mathcal{O}(|F|^d)$ size can be achieved by specifying one bit for each possible scenario in $\mathcal{S}$ and setting it to one if the scenario is present and zero otherwise.

[2] Dell and Marx called this problem PERFECT d-SET MATCHING.

Note that the original result by Dell and Marx [6] is given in terms of the size k of an exact d-set cover. Clearly, $k = \frac{|U|}{d}$ and, thus, we have $\mathcal{O}(k^{d-\varepsilon}) = \mathcal{O}(|U|^{d-\varepsilon})$ and may instead phrase the result in terms of $|U|$. Furthermore, their result builds on work by Dell and van Melkebeek [7] and, thus, extends to any polynomial time algorithms (rather than just kernels) whose output instances can be with respect to a different problem. We give the following paraphrased version of the result.

Theorem 7 (Dell and Marx [6]). *Let $d \geq 3$ and ε be a positive real. There is no polynomial-time algorithm that reduces every instance $(U, \mathcal{H})$ of* EXACT COVER BY d-SETS *to an equivalent instance of size $\mathcal{O}(|U|^{d-\varepsilon})$ (possibly with respect to a different problem), unless* NP $\subseteq$ coNP/poly.

The following lemma, together with Theorem 7, directly implies Theorem 6.

Lemma 4 ($\star$). *There is a polynomial-time reduction from* EXACT COVER BY d-SETS *to* MINIMUM FEASIBLE TILESET *such that instances $(X, \mathcal{C})$ are mapped to instances $(F, \mathcal{S}, \ell)$ with $F = X$ and scenario size at most d.*

We now consider a more general setting: In the GENERALIZED MINIMUM FEASIBLE TILESET problem we are also given a set of symbols and a set of scenarios, but here each scenario may be a *multi-set* of symbols (or, equivalently, each scenario is a function $S \colon F \to \mathbb{N}$ indicating the number of copies of each symbol f needed for S). We prove that GENERALIZED MINIMUM FEASIBLE TILESET can be solved in time $\mathcal{O}^*(|F|^{\mathcal{O}(|F|^2)})$. Note that for this problem the solution size ℓ may be much larger than $|F|$ and similarly the number of scenarios cannot in general be bounded in $|F|$.

Theorem 8 ($\star$). GENERALIZED MINIMUM FEASIBLE TILESET *can be solved in time $\mathcal{O}^*(|F|^{\mathcal{O}(|F|^2)})$, i.e., it is fixed-parameter tractable with respect to $|F|$.*

7   Conclusion

We initiated the study of the MINIMUM FEASIBLE TILESET problem and exposed an interesting combinatorial structure. We proved the problem to be NP-complete even in the restricted case with scenarios of size at most three. On the positive side, we showed that the MINIMUM FEASIBLE TILESET problem admits a 4/3-approximation algorithm and that it is fixed-parameter tractable with respect to the number of scenarios and number of symbols. The latter algorithm works also for the GENERALIZED MINIMUM FEASIBLE TILESET problem where each scenario can contain multiple copies of a symbol and we believe that it can be further generalized to work also for the original assignment problem where also tiles of larger (but constant) size are allowed. It would be interesting to see whether our other positive results transfer to this more general setting. We note that our approximation algorithm relies heavily on the structural observations from Sect. 2 which do not seem to generalize well. Our integer linear program for a fixed number of scenarios does not seem easily adaptable either.

References

1. Bansal, N., Caprara, A., Sviridenko, M.: A new approximation method for set covering problems, with applications to multidimensional bin packing. SIAM J. Comput. **39**(4), 1256–1278 (2009)
2. Biedl, T., Chan, T., Ganjali, Y., Hajiaghayi, M., Wood, D.: Balanced vertex-orderings of graphs. Discrete Appl. Math. **148**(1), 27–48 (2005)
3. Buchin, K., van Kreveld, M.J., Meijer, H., Speckmann, B., Verbeek, K.: On planar supports for hypergraphs. J. Graph Algorithms Appl. **15**(4), 533–549 (2011)
4. Chen, J., Komusiewicz, C., Niedermeier, R., Sorge, M., Suchý, O., Weller, M.: Effective and efficient data reduction for the subset interconnection design problem. In: Cai, L., Cheng, S.-W., Lam, T.-W. (eds.) ISAAC 2013. LNCS, vol. 8283, pp. 361–371. Springer, Heidelberg (2013)
5. Cygan, M.: Improved approximation for 3-dimensional matching via bounded path-width local search. In: Proceedings of the 54th Annual IEEE Symposium on Foundations of Computer Science (FOCS), pp. 509–518 (2013)
6. Dell, H., Marx, D.: Kernelization of packing problems. In: Proceedings of the 23rd Annual ACM-SIAM Symposium on Discrete Algorithms (SODA), pp. 68–81 (2012)
7. Dell, H., van Melkebeek, D.: Satisfiability allows no nontrivial sparsification unless the polynomial-time hierarchy collapses. In: Proceedings of the 42nd ACM Symposium on Theory of Computing (STOC), pp. 251–260 (2010)
8. Disser, Y., Matuschke, J.: Degree-constrained orientations of embedded graphs. In: Chao, K.-M., Hsu, T., Lee, D.-T. (eds.) ISAAC 2012. LNCS, vol. 7676, pp. 506–516. Springer, Heidelberg (2012)
9. Du, D.-Z., Miller, Z.: Matroids and subset interconnection design. SIAM J. Discrete Math. **1**(4), 416–424 (1988)
10. Frank, A., Gyárfás, A.: How to orient the edges of a graph. Colloquia mathematica societatis Janos Bolyai **18**, 353–364 (1976)
11. Frank, A., Tardos, É.: An application of simultaneous diophantine approximation in combinatorial optimization. Combinatorica **7**(1), 49–65 (1987)
12. Garey, M.R., Johnson, D.S.: Computers and Intractability. A Guide to the Theory of NP-Completeness. W.H.Freeman and Company, New York (1979)
13. Gottlob, G., Greco, G.: On the complexity of combinatorial auctions: structured item graphs and hypertree decomposition. In: Proceedings of the 8th ACM Conference on Electronic Commerce (EC), pp. 152–161 (2007)
14. Hakimi, S.: On the degrees of the vertices of a directed graph. J. Frankl. Inst. **279**(4), 290–308 (1965)
15. Johnson, D.S., Pollak, H.O.: Hypergraph planarity and the complexity of drawing Venn diagrams. J. Graph Theory **11**(3), 309–325 (1987)
16. Kannan, R.: Minkowski's convex body theorem and integer programming. Math. Oper. Res. **12**, 415–440 (1987)
17. Koutis, I.: Faster algebraic algorithms for path and packing problems. In: Aceto, L., Damgård, I., Goldberg, L.A., Halldórsson, M.M., Ingólfsdóttir, A., Walukiewicz, I. (eds.) ICALP 2008, Part I. LNCS, vol. 5125, pp. 575–586. Springer, Heidelberg (2008)
18. Lenstra, H.W.: Integer programming with a fixed number of variables. Math. Oper. Res. **8**, 538–548 (1983)
19. Sviridenko, M., Ward, J.: Large neighborhood local search for the maximum set packing problem. In: Fomin, F.V., Freivalds, R., Kwiatkowska, M., Peleg, D. (eds.) ICALP 2013, Part I. LNCS, vol. 7965, pp. 792–803. Springer, Heidelberg (2013)

Online Ad Assignment with an Ad Exchange

Wolfgang Dvořák[(✉)] and Monika Henzinger

Faculty of Computer Science, University of Vienna, Vienna, Austria
wolfgang.dvorak@univie.ac.at

Abstract. Ad exchanges are becoming an increasingly popular way to sell advertisement slots on the internet. An ad exchange is basically a spot market for ad impressions. A publisher who has already signed contracts reserving advertisement impressions on his pages can choose between assigning a new ad impression for a new page view to a contracted advertiser or to sell it at an ad exchange. This leads to an online revenue maximization problem for the publisher. Given a new impression to sell decide whether (a) to assign it to a contracted advertiser and if so to which one or (b) to sell it at the ad exchange and if so at which reserve price. We make no assumptions about the distribution of the advertiser valuations that participate in the ad exchange and show that there exists a simple primal-dual based online algorithm, whose lower bound for the revenue converges to $R_{ADX} + R_A(1 - 1/e)$, where R_{ADX} is the revenue that the optimum algorithm achieves from the ad exchange and R_A is the revenue that the optimum algorithm achieves from the contracted advertisers.

1 Introduction

The market for display ads on the internet is worth billions of dollars and continues to rise. Not surprisingly, there are multiple ways of selling display advertisements. Traditionally, publishers signed long-term contracts with their advertisers, fixing the number of *impressions*, i.e. assigned ad slots views, as well as their price. In the last few years, however, spot markets, so called *Ad Exchanges* [8], have been developed, with Amazon, Ebay, and Yahoo (to just name a few) all offering their own ad exchange. Thus, every time a user requests to download a page from a publisher, the publisher needs to decide (a′) which of the ad impressions on this page should be assigned to which contracted advertiser, and (b′) which should be sold at the ad exchange and at which *reserve price*[1].

Ad exchanges are interesting for publishers as (1) basically an unlimited number of ad impressions can be sold at ad exchanges, and (2) if the publishers have additional information about the user, they might sell an impression at a much higher price at the ad exchange than they could receive from their contracted advertisers. As ad impressions that did not receive a bid at or above

---

[1] The reserve price is the minimum required price at which an impression is sold at an ad auction. If no offer is at or above the reserve price, the impression is not sold.

© Springer International Publishing Switzerland 2015
E. Bampis and O. Svensson (Eds.): WAOA 2014, LNCS 8952, pp. 156–167, 2015.
DOI: 10.1007/978-3-319-18263-6_14

the reserve price at the ad exchange can still be assigned to contracted advertisers, a revenue-maximizing publisher can offer *every* ad impression first at an ad exchange at a "high enough" reserve price and then afterwards assign the still unsold impressions to contracted advertisers. The question for the advertiser becomes, thus, (a') what reserve price to choose, and (b') to which advertisers to assign the unsold impressions. We model this setting as an online problem and achieve the following two results: If the revenue achievable by the ad exchange for each ad impression is known, we give a constant competitive algorithm. Then we show how to convert this algorithm into a second algorithm that works in the setting where the revenue achievable from the ad exchange is not known. Assume that the auction executed at the ad exchange fulfills the following property P: If an ad impression is sold at the ad exchange, then the revenue achieved is independent of the reserve price chosen by the publisher. Thus, the reserve price influences only *whether* the ad impression is sold, *not* the price that is achieved. For example, a first price auction with reserve prices fulfills this condition. If the auction at the ad exchange fulfills this condition, then our second algorithm is constant competitive when compared with the optimum offline algorithm.

When modeling contracted advertisers we use the *model with free disposal* introduced in [4]: Each advertiser a comes with a number n_a and the revenue that an algorithm receives from a consists of the n_a *most valuable* ad impressions assigned to a. Additional impressions assigned to a do not generate any revenue.

More formally we define the following *Online Ad Assignment Problem with Free Disposal and an Ad Exchange*. There is a set of contracted advertisers A and an ad exchange α. Each advertiser a comes with a number n_a of ad impressions such that a pays only for the n_a *most valuable* ad impressions assigned to a, or for all assigned ad impressions if fewer than n_a are assigned to a. To simplify the notation we set $n_\alpha = \infty$. Now a finite sequence $\mathcal{S} = S_0, S_1, \ldots$ of sets S_l with $l = 0, 1, \ldots$, of ad impressions arrives in order. When S_l arrives, the weights $w_{i,a}$ for each $i \in S_l$ and $a \in A \cup \{\alpha\}$ are revealed and the online algorithm has to assign each $i \in S_l$ *before* further sets S_{l+1}, S_{l+2}, etc. arrive. Let $\mathcal{A} : I \to A \cup \{\alpha\}$ be an *assignment* of impressions to advertisers. An assignment is *valid* if no two impressions in the same set S_l are assigned to the same advertiser $a \in A$. Let $I_\mathcal{A}(a)$ be the set of n_a impressions with highest weight assigned to advertiser a by $\mathcal{A}$. Then the *revenue $R(\mathcal{A})$* of $\mathcal{A}$ is $\sum_{a \in A \cup \{\alpha\}} \sum_{i \in I_\mathcal{A}(a)} w_{i,a}$. The goal of the algorithm is to produce a valid assignment $\mathcal{A}$ with maximum revenue $R(\mathcal{A})$. The *competitive ratio* of an online algorithm is the minimum over all sequences $\mathcal{S}$ of the ratio of the revenue achieved by the online algorithm on $\mathcal{S}$ and the revenue achieved by the optimal offline algorithm on $\mathcal{S}$, where the latter algorithm is given all of $\mathcal{S}$ *before* it makes the first decision.

Feldman et al. [4] studied a special case of our problem, namely the setting *without an ad exchange* and where each set S_l has size one, i.e. where the impressions arrive consecutively. For that setting they gave a primal-dual based 0.5 competitive algorithm whose competitive ratio converges to $(1 - 1/e)$ ratio when *all* the n_a values go to infinity. More precisely let $n_A = \min_{a \in A} n_a$. Then their algorithm is $1 - (\frac{n_A}{n_A+1})^{n_A}$-competitive. They also showed that this ratio is tight when considering deterministic algorithms [4]. Let R_a for an advertiser

$a \in A \cup \{\alpha\}$ be the revenue that the optimal algorithm receives from a. We extend their results in several ways. (1) We consider a setting with one advertiser, called ad exchange, that has infinite capacity[2]. Moreover, we allow multiple ad slots on a page, with the condition that no two can be assigned to the same advertiser, i.e. for us $|S_l|$ can be larger than 1. (2) The revenue of our algorithm depends *directly* on the n_a value, not on n_A. More precisely, if no ad exchange exists, our algorithm receives a revenue of at least $\sum_a (1 - (\frac{n_a}{n_a+1})^{n_a}) R_a$. When an ad exchange is added, our algorithm achieves a revenue of at least $R_\alpha + \sum_a (1 - (\frac{n_a}{n_a+1})^{n_a}) R_a$. (3) We show how to modify our algorithm for the setting where $w_{i,\alpha}$ is unknown for all i. In this setting our algorithm computes a reserve price and sends *every* impression first to the ad exchange. The reserve price is set such that if the auction executed at the ad exchange fulfills property P then the above revenue bounds continue to hold, i.e. it achieves a revenue of at least $R_\alpha + \sum_a (1 - (\frac{n_a}{n_a+1})^{n_a}) R_a$.

Techniques. Our algorithm is a modification of the standard primal-dual algorithms in [4] but it is itself not a standard primal-dual algorithms as it does not construct a feasible primal and dual solution to a *single* LP. Instead in the analysis we use several primal and dual LPs, one for each advertiser a and use the dual solutions to upper bound R_a. However, the corresponding primal feasible solution is *not* directly related to the revenue the algorithm achieves from a. Instead, the solution constructed by the algorithm is a feasible solution for a primal program that is the combination of all individual LPs. This property is strong enough to give the claimed bounds. The crucial new ideas in our algorithms are (i) the observation that when deciding to whom an ad slot is assigned the publisher should be biased towards advertiser with large n_a and in particular towards the ad exchange and (ii) that based on the structure of the algorithm it can be easily modified to compute an reserve price for the auction in the ad exchange if the $w_{i,\alpha}$ values are unknown.

Further Related Work. We briefly sketch prior work on the question whether the publisher should assign an impression to a contracted advertiser or an ad exchange. In [2] a scenario is studied, where the $w_{i,a}$ follow a joint distribution and no disposal is allowed. Gosh et al. [5] assume that for each impression i the $w_{i,\alpha}$ values follow a known distribution and the contracted advertisers have a quality value depending on $w_{i,\alpha}$. They study the trade-off between the quality of the impressions assigned to the advertisers and revenue from the ad exchange. The work in [1], like our work, does not make Bayesian assumptions but studies online algorithms in the worst case setting. The main difference is that there the contracted advertisers also arrive online and that there is no free disposal.

Finally, Devanur et al. [3] extend [4] to the scenario with multiple ad slots on a page and constraints on ads being assigned together, but they neither consider ad-exchanges nor consider the different capacities n_a in the competitive ratio.

Structure of the Paper. In Sect. 2 we discuss why the algorithm from [4] is not satisfying in our setting and present a simple online algorithm for the 1-slot case, which

---

[2] It is straightforward to extend the algorithm and its analysis to multiple ad exchanges.

we improve in Sect. 3 to achieve a revenue of at least $R_\alpha + \sum_a (1 - (\frac{n_a}{n_a+1})^{n_a}) R_a$. In Sect. 4 we generalize this algorithm to the multi-slot setting. Finally, in Sect. 5 we show how to adapt it if the $w_{i,\alpha}$ values are unknown.

2   A Simple 1-Slot Online Algorithm

In Sects. 2 and 3 we consider algorithms for the 1-slot setting, i.e., where each S_l just contains a single impression i. Given an instance of such an online ad assignment problem we can build an equivalent instance where all capacities $n_a = 1$. Simply replace each advertiser a by n_a copies $a_1, \ldots a_{n_a}$ with the capacities 1 and for each impression i set $w_{i,a_p} = w_{i,a}$ for all $1 \le p \le n_a$. Thus in this section we assume $n_a = 1$ for each $a \in A$. Then we formulate the offline problem as an integer linear program (ILP), where the variable $x_{i,a}$ is set to 1 if i is assigned to advertiser a and to 0, otherwise.

$$\text{\textbf{Primal: } max} \sum_{i,a \in A \cup \{\alpha\}} w_{i,a}\, x_{i,a}$$

$$\sum_{a \in A \cup \{\alpha\}} x_{i,a} \le 1 \quad \forall i$$

$$\sum_i x_{i,a} \le 1 \quad \forall a \in A$$

The first type of constraints ensures that each impression is assigned to at most one advertiser, while the second type of constraints ensures that each $a \in A$ is assigned at most one impression. It has the following dual LP.

$$\text{\textbf{Dual: } min} \sum_i z_i + \sum_{a \in A} \beta_a$$

$$z_i + \beta_a \ge w_{i,a} \quad \forall i, \forall a \in A$$

$$z_i \ge w_{i,\alpha} \quad \forall i$$

For notational convenience we assume an additional variable β_α which remains 0 for the whole algorithm. We next consider a straight forward generalization of the online algorithm in [4], called Algorithm 1, to our setting. This algorithm constructs a feasible integral solution for the Primal LP, corresponding to an ad assignment, and a feasible solution for the dual LP that is used to bound the revenue of the optimal assignment.

Algorithm 1

---

1. Initialize $\beta_a = 0$, $\beta_\alpha = 0$
2. When impression i arrives
   (a) Compute $j = \operatorname*{argmax}_{a \in A \cup \{\alpha\}} \{w_{i,a} - \beta_a\}$.
   (b) if $j = \alpha$ then set $x_{i,\alpha} = 1$ and $z_i = w_{i,\alpha}$.
   (c) if $j \in A$ then set $x_{i,j} = 1$, $\forall\, i' \ne i$: $x_{i',j} = 0$, $z_i = w_{i,j} - \beta_j$ and $\beta_j = w_{i,j}$.

Algorithm 1 constructs feasible solutions for both the Primal and the Dual: when impression i is assigned to advertiser j then $x_{i,j}$ is set to 1, β_j is set to $w_{i,j}$, and z_i is set to $\max_{a \in A \cup \{\alpha\}} \{w_{i,a} - \beta_a\}$. Note that the loss in revenue of Algorithm 1 compared to the optimal assignment *exclusively* comes from the impression assigned to advertisers in A. However, the above algorithm does not guarantee that impressions are sent to ad exchange when the optimal algorithm does. Thus the optimal offline assignment might send many impressions to the ad exchange, while the online assignment of the above algorithm does not and thus might only be an $1/2$ approximation. Such a situation is given in Example 1.

Example 1. Consider $A = \{a\}$ with $n_a = 1$ and impressions $1 \leq i \leq n$ with $w_{i,\alpha} = 1 - \epsilon$ and $w_{i,a} = i$. Then the revenue $R(\mathcal{A})$ of Algorithm 1 after n impressions is n, while the optimal assignment achieves $n + (n-1)(1-\epsilon)$, where $(n-1)(1-\epsilon)$ is achieved by the ad exchange. For $\epsilon \to 0$ and $n \to \infty$ the ratio $R(\mathcal{A})/R(OPT)$ is $1/2$ although half of the revenue in the optimal assignment OPT comes from the ad exchange.

Thus the algorithm from [4] is only $1/2$-competitive, even when an ad exchange, i.e., an advertiser with infinite capacity, is added.

Given an ad assignment $\mathcal{A}$ let $R_\alpha(\mathcal{A})$ denote the revenue the assignment gets from impressions assigned to the ad exchange and let $R_A(\mathcal{A})$ denote the revenue the assignment gets from impressions assigned to contracted advertisers. Thus we have $R(\mathcal{A}) = R_\alpha(\mathcal{A}) + R_A(\mathcal{A})$. Additionally, we use OPT to denote the optimal assignment. We present next Algorithm 2, an online algorithm that receives as revenue at least $R_\alpha(OPT) + (1/2)R_a(OPT)$, which is already an improvement over Algorithm 1. It is based on the observation that *assigning an impression that should be sent to the ad exchange to an advertiser in A is worse than sending an impression that should go to an advertiser in A to the ad exchange.* Thus, the algorithm is biased towards the ad exchange. Specifically the algorithm assigns an impression to an advertiser $a \in A$ only if it gives at least double the revenue on a than on α.

Theorem 1. *Let $\mathcal{A}$ be the ad assignment computed by Algorithm 2 then $R(\mathcal{A}) \geq R_\alpha(OPT) + 1/2 \cdot R_A(OPT)$.*

Proof. Let I^A_{OPT}, resp. I^α_{OPT}, be the impressions assigned to A, resp. α, by the optimal (offline) assignment OPT. We give an LP P_A for the advertisers A and impressions I^A_{OPT} and its dual D_A such that any feasible solution for D_A gives an upper bound d_A for $R_A(OPT)$.

Algorithm 2

---

1. Initialize $\beta_a = 0$ for all $a \in A \cup \{\alpha\}$
2. When impression i arrives
   (a) Compute $j = \operatorname*{argmax}_{a \in A} \{w_{i,a} - \beta_a\}$.
   (b) if $\{w_{i,j} - \beta_j\} > 2 \cdot w_{i,\alpha}$ then assign i to j and set $\beta_j = w_{i,j}$.
   (c) if $\{w_{i,j} - \beta_j\} \leq 2 \cdot w_{i,\alpha}$ then assign i to α.

$$\textbf{Primal } P_A: \max \sum_{i \in I_{OPT}^A, a \in A} w_{i,a}\, x_{i,a} \qquad \textbf{Dual } D_A: \min \sum_{i \in I_{OPT}^A} z_i + \sum_{a \in A} \beta_a$$

$$\sum_{a \in A} x_{i,a} \leq 1 \quad \forall i \in I_{OPT}^A \qquad z_i + \beta_a \geq w_{i,a} \;\; \forall i \in I_{OPT}^A \;\; \forall a \in A$$

$$\sum_{i \in I_{OPT}^A} x_{i,a} \leq 1 \quad \forall a \in A$$

Note that the summation in P_A and the constraints in D_A are only over impressions in I_{OPT}^A. The objective value of the optimal solution of D_A, is an upper bound for the objective of P_A, and thus also for $R_A(OPT)$. However, there is no direct relationship between $R_A(\mathcal{A})$ and the objective of P_A for $\mathcal{A}$, as $\mathcal{A}$ might also assign impressions from I_{OPT}^α to A.

To upper bound $R_A(OPT)$ we construct a feasible solution for D_A. We do this in a iterative fashion, that is whenever Algorithm 2 assigns an impression $i \in I_{OPT}^A$ we update the feasible solution for D_A as follows: (i) For the β_a variables we use the values currently set by the Algorithm 2; (ii) For the variable z_i we set $z_i = w_{i,j} - \beta_j^o$, where β_a^o is the value of β_a before i is assigned. As $w_{i,j} - \beta_j^o = max_{a \in A}\{w_{i,a} - \beta_a\}$, all the constraints for i are satisfied. Hence, doing this for all $i \in I_{OPT}^A$ gives a feasible solution for D_A and its objective d_A fulfills $d_A \geq R_A(OPT)$.

Let $\Delta d_A(i)$ be the increase of the objective d_A when the algorithm assigns impression i, i.e., the change in d_A caused by the change in the β-values and the assignment of the z_i value. For notational convenience we also define $\Delta d_\alpha(i) = w_{i,\alpha}$ if $i \in I_{OPT}^\alpha$ and $\Delta d_\alpha(i) = 0$ otherwise. Furthermore, let $\Delta R(\mathcal{A}, i)$ be the increase in revenue of the algorithm when it assigns i. Note that $\sum_{i \in I} \Delta d_A(i) = d_A$, $\sum_{i \in I} \Delta d_\alpha(i) = R_\alpha(OPT)$ and $\sum_{i \in I} \Delta R(\mathcal{A}, i) = R(\mathcal{A})$.

We need to show that $R(\mathcal{A}) \geq R_\alpha(OPT) + 1/2 \cdot d_A$. For this it suffices to show that for each $i \in I$ it holds that

$$\Delta R(\mathcal{A}, i) \geq \Delta d_\alpha(i) + 1/2 \cdot \Delta d_A(i).$$

To prove this let β_a^n, resp. β_a^o, to denote the value of β_a after, resp. before i is assigned. We distinguish the cases (i) $i \in I_{OPT}^A$ and (ii) $i \in I_{OPT}^\alpha$ and use the fact that β_a is such that $\beta_a = 0$ if no impression was assigned to a and otherwise $\beta_a = w_{i',a}$, where i' is the impression currently assigned to a

(i) First consider the case $i \in I_{OPT}^A$, which implies $\Delta d_\alpha(i) = 0$. Thus, we have to show that $\Delta R(\mathcal{A}, i) \geq 1/2 \cdot \Delta d_A(i)$.

1. If Algorithm 2 assigns i to an $j \in A$ recall that we set $z_i = w_{i,j} - \beta_j^o$ and the algorithm sets $\beta_j^n = w_{i,j}$. Thus $\Delta d_A(i) = 2 \cdot (w_{i,j} - \beta_j^o)$ and $\Delta R(\mathcal{A}, i)$ is given by $w_{i,j}$ minus the value of the impression we have to drop (if any), given by β_a^0. As this values is stored in β_j^o we get $\Delta R(\mathcal{A}, i) = w_{i,j} - \beta_j^o$ and thus $\Delta R(\mathcal{A}, i) \geq 1/2 \cdot \Delta d_A(i)$.

2. If Algorithm 2 assigns i to α (although the OPT does not), we know from Step 2c that $\{w_{i,j} - \beta_j\} \leq 2w_{i,\alpha}$, where $j = \underset{a \in A}{\text{argmax}} \{w_{i,a} - \beta_a\}$. As we set

$z_i = w_{i,j} - \beta_j^o$ and the algorithm keeps all β_a unchanged we get $\Delta d_A(i) = w_{i,j} - \beta_j^o$ and as we assign i to α we have $\Delta R(\mathcal{A}, i) = w_{i,\alpha}$. Thus $\Delta R(\mathcal{A}, i) = w_{i,\alpha} \geq 1/2 \{w_{i,j} - \beta_j\} = 1/2\, \Delta d_A(i)$.

Thus, for $i \in I_{OPT}^A$ it holds that $\Delta R(\mathcal{A}, i) \geq 1/2 \cdot \Delta d_A(i) = \Delta d_\alpha(i) + 1/2 \cdot \Delta d_A(i)$.

(ii) Now consider $i \in I_{OPT}^\alpha$, which implies $\Delta d_\alpha(i) = w_{i,\alpha}$. Recall that no z-values are involved in this case. We show that $\Delta R(\mathcal{A}, i) \geq w_{i,\alpha} + 1/2 \cdot \Delta d_A(i)$.

1. If Algorithm 2 assigns i to the ad exchange then the β_a are not changed. Thus $\Delta d_A(i) = 0$ and $\Delta R(\mathcal{A}, i)$ is simply $w_{i,\alpha}$. Hence, $\Delta R(\mathcal{A}, i) \geq w_{i,\alpha} + 1/2 \cdot \Delta d_A(i)$.
2. If Algorithm 2 assigns i to an $a \in A$ we have $\{w_{i,a} - \beta_a^o\} > 2w_{i,\alpha}$ and the algorithm sets $\beta_a^n = w_{i,a}$. Thus $\Delta d_A(i) = w_{i,a} - \beta_a^o$. Furthermore, $\Delta R(\mathcal{A}, i)$ is given by $w_{i,a}$ minus the value of the impression we have to drop (if any), given by β_a^0. Thus $\Delta R(\mathcal{A}, i) = (w_{i,a} - \beta_a^o) = (w_{i,a} - \beta_a^o)/2 + (w_{i,a} - \beta_a^o)/2 \geq w_{i,\alpha} + 1/2 \cdot \Delta d_A(i)$.

Thus, for $i \in I_{OPT}^\alpha$ it holds that $\Delta R(\mathcal{A}, i) \geq w_{i,\alpha} + 1/2 \cdot \Delta d_A(i) = \Delta d_\alpha(i) + 1/2 \cdot \Delta d_A(i)$. Combined we obtain that

$$R(\mathcal{A}) = \sum_{i \in I} \Delta R(\mathcal{A}, i) \geq \sum_{i \in I} \left(\Delta d_\alpha(i) + \frac{\Delta d_A(i)}{2} \right) \geq R_\alpha(OPT) + \frac{R_A(OPT)}{2}.$$

$\square$

3   An Online 1-Slot Algorithm Exploiting High Capacities

In this section we generalize the result from Sect. 2 to the setting where each advertiser $a \in A$ has an individual limit n_a for the number of ad impressions he is willing to pay for and we present Algorithm 3 that achieves an improvement in revenue for advertisers a with large n_a.

In Algorithm 3 we consider variables β_a which, for $a \in A$, are always set s.t.

$$\beta_a = \frac{1}{n_a(e_{n_a} - 1)} \sum_{j=1}^{n_a} w_j \left(1 + \frac{1}{n_a} \right)^{j-1} \tag{1}$$

where the w_j's are the weights of the impressions assigned to a in non-increasing order and $e_{n_a} = (1 + 1/n_a)^{n_a}$. That is, β_a stores a weighted mean of the n_a most valuable impressions assigned to a. Again we keep $\beta_\alpha = 0$ in the whole algorithm. Next we consider how assigning a new impression to a affects β_a.

Lemma 1 ([4]). *Consider a new impression i being assigned to advertiser a. Let β_a^o, resp. β_a^n denote the value of β_a before, resp. after i was assigned and v the value of the impression dropped from β_a (0 if no impression is dropped), then*

$$\beta_a^n - \beta_a^o \leq \frac{\beta_a^o}{n_a} - \frac{v \cdot e_{n_a}}{n_a(e_{n_a} - 1)} + \frac{w_{i,a}}{n_a(e_{n_a} - 1)}.$$

Algorithm 3

1. Initialize $\beta_a = 0$ for all $a \in A \cup \{\alpha\}$
2. When impression i arrives
   (a) Compute $x = \underset{a \in A \cup \{\alpha\}}{\operatorname{argmax}} \{c_a \cdot (w_{i,a} - \beta_a)\}$
   (b) assign i to x and update β_x according to (1)

where weights c_a are defined as $c_a = \begin{cases} 1 - \frac{1}{e^{n_a}} & a \in A \\ 1 & a = \alpha \end{cases}$

Notice that in Algorithm 3 for each $a \in A$ we have that $1/2 \leq c_a < 1 - 1/e$. We use $R_a(\mathcal{A})$ for $a \in A \cup \{\alpha\}$ to denote the revenue the assignment $\mathcal{A}$ gets from advertiser a. Thus, $R(\mathcal{A}) = \sum_{a \in A \cup \{\alpha\}} R_a(\mathcal{A})$.

Theorem 2. *Let $\mathcal{A}$ be the assignment computed by Algorithm 3 then $R(\mathcal{A}) \geq \sum_{a \in A \cup \{\alpha\}} c_a \cdot R_a(OPT)$.*

Theorem 2 will be a direct consequence of Theorem 3.

Finally let us briefly discuss whether the constants c_a are chosen optimally. From a result in [6] on online algorithms for b-matchings it follows immediately that the constants c_a in Theorem 2 are optimal for deterministic algorithms. Moreover, in [7] it is shown that even randomized algorithms cannot achieve a better competitive ratio than $(1-1/e)^3$. So for large values of n_a even randomized algorithms cannot improve over Algorithm 3.

4   A Multi-slot Online Algorithm

In practice publishers often have several ad slots at a single page and want to avoid to show multiple ads from the same advertiser on the same page to avoid annoying their users. This can be modeled as follows: A sequence $\mathcal{S} = S_0, S_1, \ldots$ of *sets* of impressions arrive in an online manner. Each set S has be assigned (a) before any future sets have arrived, and (b) such that non two impressions in S are assigned to the same advertiser in A. Note that we allow multiple impressions from S to be assigned to the ad exchange as we expect the ad exchange to return different advertisers for them. Let the set of all impressions $I = \sum_{S \in \mathcal{S}} S$. With Algorithm 4 we present an online algorithm for this setting with the same competitive ratio as Algorithm 3. Note, however, that, unlike Algorithm 3, it is compared to the optimal offline solution that respects the above restriction. More formally, we call a function $\mathbf{a} : S \rightarrow A \cup \{\alpha\}$ assigning impressions S to advertisers *valid* if there are no $i, i' \in S$, $i \neq i'$, $a \in A$ such that $\mathbf{a}(i) = \mathbf{a}(i') = a$. Our Algorithm 4 generates a valid assignment and is compared to the revenue of the *valid* assignment generated by the optimal offline algorithm. Notice that the computation of argmax in Algorithm 4 is a weighted bipartite matching problem and thus can be computed efficiently.

---

[3] In [7] the authors study the Adwords problem but in [4] it is argued that the given example can be also be interpreted as Online Ad Assignment problem.

Algorithm 4

---

1. Initialize $\beta_a = 0$ for all $a \in A \cup \{\alpha\}$
2. When impressions $S = \{i_1, \ldots, i_l\}$ arrive

    (a) Compute $\mathbf{b} = \underset{\text{valid } \mathbf{a}}{\operatorname{argmax}} \left\{ \sum_{i \in S} c_{\mathbf{a}(i)} \cdot (w_{i,\mathbf{a}(i)} - \beta_{\mathbf{a}(i)}) \right\}$

    (b) assign each i to $\mathbf{b}(i)$ and, if $\mathbf{b}(i) \in A$, update $\beta_{\mathbf{b}(i)}$ according to (1).

where weights c_a are defined as $c_a = \begin{cases} 1 - \frac{1}{e^{n_a}} & a \in A \\ 1 & a = \alpha \end{cases}$

Recall that $R_a(OPT)$ for $a \in A \cup \{\alpha\}$ is the revenue that an optimal assignment generates from advertiser a. We show the following performance bound.

Theorem 3. *Let $\mathcal{A}$ be the assignment computed by Algorithm 4 and OPT the optimal multi-slot ad assignment, then $R(\mathcal{A}) \geq \sum_{a \in A \cup \{\alpha\}} c_a \cdot R_a(OPT)$.*

Proof. We proceed as follows: First we give a linear program P_a and its dual D_a for each $a \in A$ such that the final objective value of any feasible solution of D_a is an upper bound of $R_a(OPT)$. Note, however, that there is no direct relationship between the final objective values of the P_a's and the revenue of the algorithm. However, we are able to construct a feasible solution for each D_a with objective value d_a such that the revenue $R(\mathcal{A})$ of the algorithm is at least $\sum_{a \in A \cup \alpha} c_a \cdot d_a$. Together with the observation that $d_a \geq R_a(OPT)$ and a bound d_α on $R_\alpha(OPT)$ this proves the theorem.

Let I^a_{OPT} be the impressions assigned to $a \in A \cup \{\alpha\}$ by the optimal (offline) assignment OPT. We use the following LPs for each $a \in A$.

Primal P_a: $\max \sum_{i \in I^a_{OPT}} w_{i,a}\, x_{i,a}$ **Dual D_a:** $\min \sum_{i \in I^a_{OPT}} z_i + n_a \beta_a$

$$x_{i,a} \leq 1 \quad \forall i \in I^a_{OPT} \qquad z_i + \beta_a \geq w_{i,a} \quad \forall i \in I^a_{OPT}$$

$$\sum_{i \in I^a_{OPT}} x_{i,a} \leq n_a$$

Note that the summation in the primal and the constraints in the Dual are only over the impressions *in* I^a_{OPT}, i.e., the impressions assigned by OPT to a. The objective value of the optimal solution for D_a is an upper bound for the objective of P_a, and thus also for $R_a(OPT)$. This implies that any feasible solution of D_a, also the one we construct next, gives an upper bound for $R_a(OPT)$. As there might be impressions assigned to a by the algorithm that do *not* belong to I^a_{OPT}, the objective value of P_a is, however, not necessarily related to $R_a(\mathcal{A})$.

Next we give a feasible solution for D_a for all $a \in A$, using the β_a values as currently set by the algorithm. More specifically, let $\mathbf{a}$ be the assignment of the impressions in S by the optimal solution. For each $i \in I$, we set $z_i = w_{i,\mathbf{a}(i)} - \beta_{\mathbf{a}(i)}$

exactly when the algorithm assigns i. Note that this results in a feasible dual solution for *all* a as each i belongs to exactly one set $I_{OPT}^{\mathbf{a}(i)}$ and z_i is chosen exactly so as to make the solution of $D_{\mathbf{a}(i)}$ feasible, together with the current $\beta_{\mathbf{a}(i)}$ values. As $\beta_{\mathbf{a}(i)}$ only increases in the course of the algorithm the solution remains feasible at the end of the algorithm. Let d_a be the value of this feasible solution for D_a for some $a \in A$. By the above observation $d_a \geq R_a(OPT)$.

For all $a \in A$ let $\Delta d_a(S)$ be the increase of the objective value d_a when the algorithms assigns S, i.e., the change in d_a caused by the change in the β_a-values *and* the assignment of the z_i-values for all $i \in S$. Note that $\sum_{S \in \mathcal{S}} \Delta d_a(S) = d_a$ and, thus, $R_a(OPT) \leq \sum_S \Delta d_a(S)$. For convenience we also define $\Delta d_\alpha(S) = \sum_{i \in S \cap I_{OPT}^\alpha} w_{i,\alpha}$. Furthermore, let $\Delta R(\mathcal{A}, S)$ be the increase in revenue of the algorithm when it assigns S. Thus $R(\mathcal{A}) = \sum_{S \in \mathcal{S}} \Delta R(\mathcal{A}, S)$.

We are left with showing that $R(\mathcal{A}) \geq \sum_{a \in A \cup \alpha} c_a \cdot d_a$. To prove that $R(\mathcal{A}) = \sum_{S \in \mathcal{S}} \Delta R(\mathcal{A}, S) \geq \sum_{a \in A \cup \alpha} c_a \cdot d_a = \sum_{S \in \mathcal{S}} \left(\sum_{a \in A \cup \alpha} c_a \cdot \Delta d_a(S) \right)$ it suffices to show that for each $S \in \mathcal{S}$ it holds that

$$\Delta R(\mathcal{A}, S) \geq \sum_{a \in A \cup \alpha} c_a \cdot \Delta d_a(S).$$

We show this next. To simplify the notation let $\Delta d(S) = \sum_{a \in A \cup \alpha} c_a \cdot \Delta d_a(S)$.

First consider $\Delta R(\mathcal{A}, S)$: For $a \in A$ let v_a be the value of the n_a-th valuable impression assigned to a (the impression we would "drop" by assigning a new one), and let $v_\alpha = 0$. If i is assigned to α then the gain in revenue is $w_{i,\mathbf{b}(i)}$ which equals $w_{i,\mathbf{b}(i)} - v_{\mathbf{b}(i)}$. If i is assigned to $a \in A$ then the gain in revenue is the difference between the revenue of the new impression and the impression we have to drop, i.e., $w_{i,\mathbf{b}(i)} - v_{\mathbf{b}(i)}$. Thus for S altogether it holds

$$\Delta R(\mathcal{A}, S) = \sum_{i \in S} (w_{i,\mathbf{b}(i)} - v_{\mathbf{b}(i)})$$

Now consider $\Delta d(S)$: Recall that $\mathbf{a}$ is the assignment of the optimal solution for the impressions S and let $\mathbf{b}$ be the assignment from Algorithm 4. For all $a \in A$ let β_a^o, β_a^n denote the value of β_a right *before*, resp. right *after* this assignment. Recall that for $a = \alpha$, it holds that $\beta_a = 0$ throughout the algorithm. Now note that

$$\Delta d(S) = \sum_{i \in S} \left(c_{\mathbf{a}(i)} \cdot (w_{i,\mathbf{a}(i)} - \beta_{\mathbf{a}(i)}^o) + c_{\mathbf{b}(i)} \cdot n_{\mathbf{b}(i)} \cdot (\beta_{\mathbf{b}(i)}^n - \beta_{\mathbf{b}(i)}^o) \right),$$

where the first term comes from the new variables z_i which we set to $w_{i,\mathbf{a}(i)} - \beta_{\mathbf{a}(i)}^o$ (to make $D_{\mathbf{a}(i)}$ feasible), and the second term comes from the updates of β_a. By the choice of $\mathbf{b}$ in the algorithm we get

$$\Delta d(S) \leq \sum_{i \in S} \left(c_{\mathbf{b}(i)} \cdot (w_{i,\mathbf{b}(i)} - \beta_{\mathbf{b}(i)}^o) + c_{\mathbf{b}(i)} \cdot n_{\mathbf{b}(i)} \cdot (\beta_{\mathbf{b}(i)}^n - \beta_{\mathbf{b}(i)}^o) \right)$$

$$= \sum_{i \in S} c_{\mathbf{b}(i)} \cdot \left((w_{i,\mathbf{b}(i)} - \beta_{\mathbf{b}(i)}^o) + n_{\mathbf{b}(i)} \cdot (\beta_{\mathbf{b}(i)}^n - \beta_{\mathbf{b}(i)}^o) \right).$$

Next we bound the contribution of each $i \in S$ separately by analyzing two cases:

- If $\mathbf{b}(i) = \alpha$ then we know that $\beta^o_{\mathbf{b}(i)} = \beta^n_{\mathbf{b}(i)} = v_{\mathbf{b}(i)} = 0$ and $c_{\mathbf{b}(i)} = 1$. Thus

$$c_{\mathbf{b}(i)} \cdot \left((w_{i,\mathbf{b}(i)} - \beta^o_{\mathbf{b}(i)}) + c_{\mathbf{b}(i)} \cdot n_{\mathbf{b}(i)} \cdot (\beta^n_{\mathbf{b}(i)} - \beta^o_{\mathbf{b}(i)}) \right) = (w_{i,\mathbf{b}(i)} - v_{\mathbf{b}(i)}).$$

- If $\mathbf{b}(i) \in A$ then we can apply Lemma 1 to bound $(\beta^n_{\mathbf{b}(i)} - \beta^o_{\mathbf{b}(i)})$ as follows

$$c_{\mathbf{b}(i)} \cdot \left((w_{i,\mathbf{b}(i)} - \beta^o_{\mathbf{b}(i)}) + n_{\mathbf{b}(i)} \cdot (\beta^n_{\mathbf{b}(i)} - \beta^o_{\mathbf{b}(i)}) \right)$$

$$\leq c_{\mathbf{b}(i)} \cdot \left((w_{i,\mathbf{b}(i)} - \beta^o_{\mathbf{b}(i)}) + \beta^o_{\mathbf{b}(i)} - \frac{v_{\mathbf{b}(i)} \cdot e_{n_{\mathbf{b}(i)}}}{e_{n_{\mathbf{b}(i)}} - 1} + \frac{w_{i,\mathbf{b}(i)}}{e_{n_{\mathbf{b}(i)}} - 1} \right)$$

$$= c_{\mathbf{b}(i)} \cdot \left(\frac{w_{i,\mathbf{b}(i)} \cdot e_{n_{\mathbf{b}(i)}}}{e_{n_{\mathbf{b}(i)}} - 1} - \frac{v_{\mathbf{b}(i)} \cdot e_{n_{\mathbf{b}(i)}}}{e_{n_{\mathbf{b}(i)}} - 1} \right) = (w_{i,\mathbf{b}(i)} - v_{\mathbf{b}(i)})$$

In the last step we used that $c_a = 1 - 1/e_{n_a}$ for $a \in A$. By the above we obtain

$$\Delta d(S) \leq \sum_{i \in S} (w_{i,\mathbf{b}(i)} - v_{\mathbf{b}(i)}) = \Delta R(\mathcal{A}, S).$$

Now consider that the set of impression is given by a series $(S_j)_{0 \leq j \leq n}$ of pairwise disjoint sets of impressions that show up simultaneously. By using the fact that the gain in the revenue, resp. the gain in the upper bound for the sum, for the sets S_j sum up to the total revenue of $\mathcal{A}$, resp. an upper bound for OPT we get:

$$R(\mathcal{A}) = \sum_{j=0}^{n} \Delta R(\mathcal{A}, S_j) \geq \sum_{j=0}^{n} \Delta d(S_j) = \sum_{j=0}^{n} \sum_{a \in A \cup \{\alpha\}} c_a \Delta d_a(S_j) \geq \sum_{a \in A \cup \{\alpha\}} c_a R_a(OPT)$$

$\square$

5   An Algorithm for Computing Reserve Prices

In our model we assumed the publisher knows exactly how much revenue he can get from the ad exchange, i.e., the $w_{i,\alpha}$ values are given for all $i \in I$. The critical reader may interpose that this is not the fact in the real world or in the ad exchange model proposed in [8]. Instead whenever sending an impression to the ad exchange an auction is run. However, the publisher can set a reserve price and if all the bids are below the reserve price then he can still assign it to one of the contracted advertisers.

One nice property of Algorithms 2 and 3 is that they allow to compute the minimal price we have to extract from the ad exchange such that it is better to assign an impression to the ad exchange than to a contracted advertiser. This price is given by $\max_{a \in A} \{c_a \cdot (w_{i,a} - \beta_a)\}$. It follows that this price is also a natural choice for the reserve price. Assume the auction executed at the ad exchange fulfills the following *property (P): If an ad impression is sold at the ad exchange, then the revenue achieved is independent of the reserve price chosen by the publisher.* Thus, the reserve price influences only whether the ad impression is sold, not the price that is achieved. Then Theorem 3 applies, i.e., the revenue of the algorithm is at least $\sum_{a \in A \cup \{\alpha\}} c_a \cdot R_a(OPT)$, even though the algorithm is not given the $w_{i,\alpha}$ values and it is compared to an optimal algorithm that does. The reason is that the algorithm makes exactly the same decisions and receives exactly the same revenue as Algorithm 3 that is given the $w_{i,\alpha}$ values.

Theorem 4. *Let A be the assignment computed by the Algorithm described above, i.e., without knowledge of the $w_{i,\alpha}$ values. If the auction at the ad exchange fulfills property P, then $R(A) \geq \sum_{a \in A \cup \{\alpha\}} c_a \cdot R_a(OPT)$.*

Acknowledgments. The research leading to these results has received funding from the European Research Council under the European Union's Seventh Framework Programme (FP/2007-2013)/ERC Grant Agreement no. 340506 and from the Vienna Science and Technology Fund (WWTF) through project ICT10-002.

The authors are grateful to Claire Kenyon and Moses Charikar for useful discussions on formulating the model.

References

1. Alaei, S., Arcaute, E., Khuller, S., Ma, W., Malekian, A., Tomlin, J.: Online allocation of display advertisements subject to advanced sales contracts. In: ADKDD 2009, pp. 69–77. ACM, New York (2009)
2. Balseiro, S., Feldman, J., Mirrokni, V., Muthukrishnan, S.: Yield optimization of display advertising with Ad exchange. In: EC 2011, pp. 27–28. ACM (2011)
3. Devanur, N.R., Huang, Z., Korula, N., Mirrokni, V.S., Yan Q.: Whole-page optimization and submodular welfare maximization with online bidders. In: EC 2013, pp. 305–322. ACM (2013)
4. Feldman, J., Korula, N., Mirrokni, V., Muthukrishnan, S., Pál, M.: Online Ad assignment with free disposal. In: Leonardi, S. (ed.) WINE 2009. LNCS, vol. 5929, pp. 374–385. Springer, Heidelberg (2009)
5. Ghosh, A., McAfee, P., Papineni, K., Vassilvitskii, S.: Bidding for representative allocations for display advertising. In: Leonardi, S. (ed.) WINE 2009. LNCS, vol. 5929, pp. 208–219. Springer, Heidelberg (2009)
6. Kalyanasundaram, B., Pruhs, K.: An optimal deterministic algorithm for online b-matching. In: Chandru, V., Vinay, V. (eds.) Foundations of Software Technology and Theoretical Computer Science. LNCS, vol. 1180, pp. 193–199. Springer, Berlin (1996)
7. Mehta, A., Saberi, A., Vazirani, U.M., Vazirani, V.V.: Adwords and generalized online matching. J. ACM **54**(5), 22 (2007)
8. Muthukrishnan, S.: Ad exchanges: research issues. In: Leonardi, S. (ed.) WINE 2009. LNCS, vol. 5929, pp. 1–12. Springer, Heidelberg (2009)

Minimum Linear Arrangement
of Series-Parallel Graphs

Martina Eikel, Christian Scheideler, and Alexander Setzer[✉]

University of Paderborn, Paderborn, Germany
{martinah,scheideler,asetzer}@mail.upb.de

Abstract. We present a factor $14D^2$ approximation algorithm for the minimum linear arrangement problem on series-parallel graphs, where D is the maximum degree in the graph. Given a suitable decomposition of the graph, our algorithm runs in time $O(|E|)$ and is very easy to implement. Its divide-and-conquer approach allows for an effective parallelization. Note that a suitable decomposition can also be computed in time $O(|E| \log |E|)$ (or even $O(\log |E| \log^* |E|)$ on an EREW PRAM using $O(|E|)$ processors).

For the proof of the approximation ratio, we use a sophisticated charging method that uses techniques similar to amortized analysis in advanced data structures.

On general graphs, the minimum linear arrangement problem is known to be NP-hard. To the best of our knowledge, the minimum linear arrangement problem on series-parallel graphs has not been studied before.

1 Introduction

The minimum linear arrangement problem is a well-known graph embedding problem, in which an arbitrary graph is mapped onto the line topology, such that the sum of the distances of nodes that share an edge is minimized. We consider the class of series-parallel graphs, which arises naturally in the context of parallel programs: modelling the execution of a parallel program yields a series-parallel graph, where sources of parallel compositions represent fork points, and sinks of parallel compositions represent join points (for the definition of a parallel composition, see Subsect. 1.1). Note that in this context, series-parallel graphs typically have a very low node degree: Since spawning child processes is costly, one would usually not spawn too many of them at a time.

1.1 Problem Statement and Definitions

Throughout this work, we consider undirected graphs only. The following definition of the minimum linear arrangement problem is based on [22]:

This work was partially supported by the German Research Foundation (DFG) within the Collaborative Research Center "On-The-Fly Computing" (SFB 901) and by the EU within FET project MULTIPLEX under contract no. 317532.

© Springer International Publishing Switzerland 2015
E. Bampis and O. Svensson (Eds.): WAOA 2014, LNCS 8952, pp. 168–180, 2015.
DOI: 10.1007/978-3-319-18263-6_15

Definition 1 (Linear Arrangement). *Given a graph $G = (V, E)$, let $n = |V|$. A linear arrangement π of G is a one-to-one function*

$$\pi : V \to \{1, \dots, n\}.$$

For a node $v \in V$, $\pi(v)$ is also called the position of v in π.

Definition 2 (Cost of a Linear Arrangement). *Given a graph $G = (V, E)$ and a linear arrangement π of G, we denote the cost of π by*

$$COST_\pi(G) := \sum_{\{u,v\} \in E} |\pi(u) - \pi(v)|.$$

Definition 3 (Minimum Linear Arrangement Problem). *Given a graph $G = (V, E)$ (the* input graph*), the* minimum linear arrangement problem *(MINLA) is to find a linear arrangement π that minimizes $COST_\pi(G)$.*

Next we define the class of series-parallel graphs, (the following is based on [11]):

Two-terminal Graph (TTG). A *two-terminal graph* $G = (V, E)$ is a graph with node set V, edge set E, and two distinct nodes $s_G, t_G \in V$ that are called source and sink, respectively. s_G and t_G are also called the *terminals* of G.

Series Composition. The *series composition* SC of $k \geq 2$ TTGs $X_1, \dots, X_k$ is a TTG created from the disjoint union of $X_1, \dots, X_k$ with the following characteristics: The sink t_{X_i} of X_i is merged with the source $s_{X_{i+1}}$ of X_{i+1} for $1 \leq i < k$. The source s_{X_1} of X_1 becomes the source s_{SC} of SC and the sink t_{X_k} of X_k becomes the sink t_{SC} of SC.

Parallel Composition. The *parallel composition* PC of $k \geq 2$ two-terminal graphs $X_1, \dots, X_k$ is a TTG created from the disjoint union of $X_1, \dots, X_k$ with the following two characteristics: The sources $s_{X_1}, \dots, s_{X_k}$ are merged to create s_{PC} and the sinks $t_{X_1}, \dots, t_{X_k}$ are merged to create t_{PC}.

Two-terminal Series-Parallel Graph (TTSPG). A *two-terminal series-parallel graph* G with source s_G and sink t_G is a graph that may be constructed by a sequence of series and parallel compositions starting from a set of copies of a single-edge two-terminal graph $G' = (\{s, t\}, \{\{s, t\}\})$.

Series-Parallel Graphs. A graph G is a *series-parallel graph* if, for some two distinct nodes s_G and t_G in G, G can be regarded as a TTSPG with source s_G and sink t_G.

Note that the series and parallel compositions are commonly defined over two input graphs only. However, it is not hard to see that our definition of a series-parallel graph is equivalent.

An example of a series-parallel graph is shown in Fig. 1.

1.2   Related Work

The MINLA was first stated by Harper [18]. Garey, Johnson, and Stockmeyer were the first to prove its NP-hardness on general graphs [16]. Ambühl, Mastrolilli, and Svensso showed that the MINLA on general graphs does not have a polynomial-time approximation scheme unless NP-complete problems can be solved in randomized subexponential time [3]. To the best of our knowledge, the two best polynomial-time approximation algorithms for the MINLA on general graphs are due to Charikar, Hajiaghayi, Karloff, and Rao [6], and Feige and Lee [13]. Both algorithms yield an $O(\sqrt{\log n} \log \log n)$-approximation of the MINLA. The latter algorithm is a combination of techniques of earlier works by Rao and Richa [24], and Arora, Rao, and Vazirani [4]. For planar graphs (which include the series-parallel graphs), Rao and Richa [24] also present a $O(\log \log n)$-approximation algorithm. Note that even though, for high degree graphs, these algorithms achieve a better approximation factor than the one we present in this work, there are some key differences between these algorithms and ours: First of all, the algorithm we present is a very simple divide-and-conquer algorithm and its functioning can be understood easily. The aforementioned algorithms, however, are much more complex and involve solving a linear or semidefinite program. Furthermore, our algorithm achieves a runtime of only $O(|E|)$ (if the series-parallel graph is given in a suitable format - otherwise, a more complex preprocessing is required that takes time $O(|E| \log |E|)$, but this can be parallelized down to $O(\log |E| \log^* |E|)$) making it suitable in situations where a low runtime is more important than the approximation guarantee. Still, for low graph degrees (which are reasonable to assume in certain applications), our algorithm even improves the approximation factor of Rao and Richa.

For special classes of graphs, the NP-hardness has been shown for bipartite graphs [12], interval graphs, and permutation graphs [8]. On the other hand, polynomial-time optimal algorithms have been found for hypercubes [18], trees [7], d-dimensional c-ary cliques [21], meshes [14], and chord graphs [25]. Note that many people claim that the MINLA is optimally solvable on outerplanar graphs, referring to [15]. However, the problem solved in [15] is different from the MINLA as we show in [26]. Note that the question whether the MINLA is NP-hard on series-parallel graphs is unsettled.

Applications of the MINLA include the design of error-correcting codes [18], machine job scheduling (e.g., [2]), VLSI layout (e.g., [1,9]), and graph drawing (e.g., [27]). For an overview of heuristics for the MINLA see the survey paper by Petit [23].

The class of series-parallel graphs, first used by MacMahon [20], has been studied extensively. It turns out that many problems that are NP-complete on general graphs can be solved in linear time on series-parallel graphs. Among these are the decision version of the dominating set problem [19], the minimum vertex cover problem, the maximum outerplanar subgraph problem, and the maximum matching problem [28]. Furthermore, since the class of series-parallel graphs is a subclass of the class of planar graphs, any problem that is already in P for that class of graphs can be solved optimally in polynomial time for series-parallel graphs as well (such as the max-cut problem [17]).

Another problem regarding series-parallel graphs is to decide, given an input graph G, whether it is series-parallel and, if so, to output the operations that recursively constructed the series-parallel graph. The first step is referred to as *series-parallel graph recognition* while the second step is referred to as *constructing a decomposition tree*. A parallel linear-time algorithm for this problem on directed graphs was first presented by Valdes, Tarjan, and Lawler [29]. Later, Eppstein [11] developed a parallel algorithm for undirected graphs using a so-called *nested ear decomposition*. The concept of an S-decomposition used in our analysis is technically similar to that concept, though we use a different notation more suitable for our purposes. The algorithm we propose for approximating the MINLA on series-parallel graphs also relies on a decomposition tree. For instances in which it is not given, the algorithm by Bodlaender and De Fluiter [5] can be used, since it runs on undirected graphs and outputs so-called *SP-tree* , which can be easily transformed into a format suitable for our algorithm.

1.3   Our Contribution

We describe a simple approximation algorithm for the minimum linear arrangement problem on series-parallel graphs with an approximation ratio of $14D^2$, where D is the degree of the graph, and a running time of $O(|E|)$ if the series-parallel graph is given in a suitable format. If the series-parallel graph is not given in the required format, this format can be computed in time $O(|E| \log |E|)$ (which can even be further parallelized down to $O(\log |E| \log^* |E|)$ on an EREW PRAM using $O(|E|)$ processors). However, for certain applications it is reasonable to assume that the graph is given in the right format, e.g., when the series-parallel graph is used to model the execution of a parallel program, the desired representation can be constructed along with the model. The simplicity and the structure of the algorithm allow for an efficient distributed implementation.

Moreover, our proof of the approximation ratio introduces a sophisticated charging method following an approach that is known from the amortized analysis of advanced data structures. This technique may be applied in other analyses as well.

2   Preliminaries

The algorithm we present is defined recursively and is based on a decomposition of the series-parallel graph into components. Therefore, prior to describing the algorithm, we introduce several definitions needed to formalize this decomposition.

The following definition is similar to the one in [5].

Definition 4 (SP-tree, Minimal SP-tree). *An SP-tree T of a series-parallel graph G is a rooted tree with the following properties:*

1. *Each node in T corresponds to a two-terminal subgraph of G.*
2. *Each leaf is a so-called L-node labelled as $L(k)$ and corresponds to a path with k edges.*
3. *Each inner node is a so-called S-node or P-node, and the two-terminal subgraph G' associated with an S-node (P-node) is the graph obtained by a series (parallel) composition of the graphs associated with the children of G', where*

the order of the children defines the order in which the series composition is applied (the order does not matter for a parallel composition).

4. The root node corresponds to G.

An SP-tree T of a series-parallel graph G is called minimal *if the following two conditions hold:*

1. *All children of an S-node are either P-nodes or L-nodes, but at least one is a P-node.*
2. *All children of a P-node are either S-nodes or L-nodes.*

It is easy to see that for any fixed series-parallel graph G, there exists a minimal SP-tree for G.

We are now ready to introduce the following three important notions:

Definition 5 (Simple Node Sequence, Parallel Component, Series Component). *Let G be a series-parallel graph and T be a minimal SP-tree of G. The sub-graph of G associated with a leaf $L(k)$ of T for $k \in \mathbb{N}$ is called a* simple node sequence. *The sub-graph of G associated with a P-node is called a* parallel component *of G. The sub-graph of G associated with a S-node is called a* series component *of G. Furthermore, any simple node sequence is called a series component, too.*

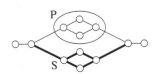

Fig. 1. Example of a simple series-parallel graph. P is a parallel component consisting of two series components (more precisely, two simple node sequences with two edges each). The thick edges belong to the subgraph induced by the series component S. It consists of two single-edge simple node sequences (on the left and right end) and a parallel component.

An illustration of the different types of components is given by Fig. 1. The definition of a minimal SP-tree implies the following: Each parallel component P is the result of a parallel composition of two or more series components. Furthermore, each series component S is the result of a series composition of two or more parallel components or simple node sequences, but not exclusively simple node sequences. This leads to the following definition:

Definition 6 (Child Component). *Let G be a series-parallel graph, let T be a minimal SP-tree, and let X and Y be two nodes in T such that Y is a child of X. Further, let C_i be the (series or parallel) component that is associated with Y and let C be the (parallel or series) component C that is associated with X. Then, C_i is called a* child component *of C, and we say: $C_i \in C$.*

For example, the two simple node sequences that induce the parallel component P in Fig. 1 are child components of P. One implication of this definition is that the terminals of a parallel component and its child components overlap.

For the rest of this work, we assume that for any fixed series-parallel graph G, the simple node sequences, series components and parallel components of G are uniquely defined by a fixed minimal SP-tree T. In the full version [10], we describe an efficient method to compute a minimal SP-tree according to our definition. It is basically an extension of an algorithm by Bodlaender and de Fluiter [5].

3   The Series-Parallel Graph Arrangement Algorithm

The Series-Parallel Graph Arrangement Algorithm (SPGAA) is defined recursively. In order to arrange the nodes of a series or parallel component C, the SPGAA first determines the order of its child components recursively, and then places the child components side by side in an order that depends on their size. For any component C, when the algorithm has just arranged the nodes of C, it holds that its source receives the leftmost position among all nodes of C and that its sink receives the position directly to the right of the source. However, later computations (in a higher recursion level) may re-arrange the terminals and pull them apart. More specific details are given in the corresponding subsections for the different types of components. Illustrations of all arrangements and all different cases can be found in the full version [10].

3.1   Arrangement of a Simple Node Sequence

For any simple node sequence L, we label the nodes of L from left to right by 1 to k. That is, the source receives label 1 and the sink receives label k. The arrangement of this sequence then is: $1, k, 2, k-1, 3, k-2, \ldots$. One can see that this arrangement fulfills the property that the source is on the leftmost position and that the sink is its right neighbor.

3.2   Arrangement of a Parallel Component

For any parallel component P with source u, sink v, and $m \geq 2$ child components $S_1, S_2, \ldots, S_m$ (note that any parallel component has at least two child components), the SPGAA recursively determines the arrangement of the child components. We denote the computed arrangement of S_i excluding the two terminal nodes (which would have been placed at the first two positions of the arrangement, see Subsect. 3.3) by S_i^-. W.l.o.g. let S_m be a biggest child component (w.r.t. the number of nodes in it). Then, the algorithm places u at the first position, v at the second position, the nodes of S_1^- to the right of that (in their order), and the nodes of S_i^- to the right of S_{i-1}^- for $i \in \{2, \ldots, m\}$.

3.3   Arrangement of a Series Component

For any series component S with source u, sink v, and $m \geq 2$ child components $P_1, P_2, \ldots, P_m$ (note that any series component has at least two child components, otherwise it would be a simple node sequence), the SPGAA first recursively determines the arrangement of the child components. Second, it puts u and v at the first two positions, in this order. The third step differs from the case of a parallel component: To keep the cost of the arrangement low while ensuring that a biggest child component P_a receives the rightmost position, the general order of the child components is: $P_1, P_2, \ldots, P_{a-1}, P_m, P_{m-1}, \ldots, P_{a+2}, P_{a+1}, P_a$. Here, the components from P_m to P_{a+1} are flipped (the order of their nodes is reversed). For $m = a$, the order is $P_1, P_2, \ldots, P_m$ and for $a = 1$, the components are ordered in reverse (i.e., $P_m, P_{m-1}, \ldots, P_1$) (where all components except for P_1 are flipped).

However, since each two neighboring child components P_i and P_{i+1} share a (terminal) node, it must be decided which of the two components may "keep" its node. The strategy here is as follows: Each component P_i (except for the first component, whose source has received the leftmost position already) keeps its source and lends its sink to P_{i+1} (of which it is a source), except for P_m (whose sink has been placed at the second position already). This may stretch existing edges, which we will keep track of in the analysis.

An illustration of the arrangement for the case $1 < a < m$ can be found in Fig. 2.

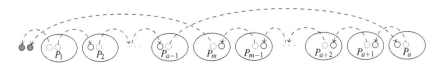

Fig. 2. Order in which the SPGAA arranges a series component consisting of m child components for $1 < a < m$ (where P_a is a biggest component). Dotted nodes indicate the position at which a node would be placed according to the previous recursion level. Dashed arrows indicate the change in position at the current recursion level.

4   Analysis

In this section, we prove the approximation ratio of $14 \cdot D^2$ for the Series-Parallel Graph Arrangement Algorithm described in Sect. 3. As a first step, we provide lower bounds on the *amortized cost* in an optimal arrangement for each kind of component. The amortized cost of a component is the sum of two values: First, the exclusive cost of this component (cost of the current component minus the individual cost of all child components). Second, some cost that has been accounted for in a lower recursion level. This cost is chosen such that the sum of all amortized costs does not contain this cost more than three times. We use these bounds to establish a lower bound on the total cost of an optimal solution. The details are described in Subsect. 4.2. As a second step, we state upper bounds on the exclusive costs generated at each recursion step of the SPGAA in order to determine an upper bound on the total cost in Subsect. 4.3. Last, we use both the lower bound as well as the upper bound to relate the cost of an optimal arrangement to that of an arrangement computed by the SPGAA. This is done in Subsect. 4.4. In addition to providing the approximation ratio of the SPGAA, we establish a polynomial runtime bound of our algorithm in Subsect. 4.5. Note that all the proofs in this section can be found in the full version [10].

4.1   Prerequisites

For the analysis, we need several notions, which we now introduce.

Definition 7 (Length of an Edge). *Given a graph $G = (V, E)$ and a linear arrangement π of G, let $u, v \in E$. The* **length of (u,v) in π**, *denoted by $length_\pi(u, v)$ is defined as:*

$$length_\pi(u, v) = |\pi(u) - \pi(v)|.$$

Definition 8. *Given a linear arrangement π of a series-parallel graph $G = (V, E)$ and a (series or parallel) component C in G, we define:*

Restricted Arrangement. *The* **arrangement π cted to C**, *denoted by $\pi(C)$ is obtained by removing all nodes from π that do not belong to C, as well as their incident edges, i.e., $\pi(C)$ maps the nodes from C to $\{1, \ldots, |C|\}$.*
Restricted Length of an Edge. *For any edge (u, v) that belongs to C, the* **length of (u,v) restricted to C**, *denoted by $length_{\pi(C)}(u,v)$, is the distance between u and v in $\pi(C)$.*
Restricted Cost of an Arrangement. *Let E_C be the set of all edges from G whose both endpoints are in C. The* **cost of C restricted to C**, *denoted by $R\text{-}COST_\pi(C)$, is defined as:*

$$R\text{-}COST_\pi(C) := \sum_{(u,v)\in E_C} length_{\pi(C)}(u, v).$$

Definition 9 (Exclusive Cost of a Series/Parallel Component). *Given a linear arrangement π of a series-parallel graph G and a (series or parallel) component C in G containing $m \geq 0$ child components $C_1, \ldots, C_m$, the* **exclusive cost of C in π**, *denoted by $E\text{-}COST_\pi(C)$, is defined as*

$$E\text{-}COST_\pi(C) := R\text{-}COST_\pi(C) - \sum_{i=1}^{m} R\text{-}COST_\pi(C_i).$$

Note that the exclusive cost of a simple node sequence S is equal to the restricted cost of S.

We can make the following observation regarding the relationship between the exclusive costs of the components and the total cost:

Observation 1. *Let G be a series-parallel graph and let π be a linear arrangement of G. Further, let $\mathcal{C}$ be the set of all (series or parallel) components in G. It holds:*

$$\sum_{C \in \mathcal{C}} E\text{-}COST_\pi(C) = COST_\pi(G).$$

In the analysis of the SPGAA, we need to find at least one path from s_P to t_P through P for each parallel component P such that any two such paths are edge-disjoint for two different parallel components. Therefore, we introduce the following notion of an *S-decomposition*, which yields these paths and is recursively defined as follows:

Definition 10 (A-path, S-path, S-decomposition). *Let P be an "innermost" parallel component in a series-parallel graph G (i.e., one whose child components are simple node sequences only) with source s, sink t, and k child components. Select an arbitrary simple path from s to t through P (i.e., select one of the simple node sequences). This path is called the* auxiliary path *or simply* A-path *of P. The remaining paths from s to t through P are called the* selected paths *or simply* S-paths *of P.*

Recursively, for an arbitrary parallel component P, with source s, sink t, and $m \geq 2$ child components $S_1, \ldots, S_m$, for each child component S_i, $1 \leq i \leq m$, select a simple path Q_i from s to t through S_i in the following way: If S_i is a simple node sequence, Q_i is the whole sequence. Otherwise, S_i is a series component, which consists of $k \geq 0$ simple node sequences and $l \geq 1$ parallel components (note that $k + l \geq 2$). Denote these child components by $P_1, \ldots P_{k+l}$ in the order in which they appear in S_i. Construct the path Q_i step by step: Start with P_1 and add P_1 completely to Q_i if P_1 is a simple node sequence. If, however, P_1 is a parallel component, select the A-path of P_1 and extend Q_i by it. Continue in the same manner up to P_{k+l}. After this, the whole path Q_i is constructed. Q_1 is called the A-path *of P and the remaining paths $Q_2, \ldots, Q_m$ are called the* S-paths *of P.*

The selection of S-paths (and A-paths accordingly) for all parallel components of G is called an S-decomposition *of G.*

An example of an S-decomposition can be found in the full version [10].

Intuitively, an auxiliary path of a parallel component P_j is a path through the whole component which is *reserved* to be used in higher recursion levels (to eventually become part of an S-path there). Any edges of an S-path are not used for any S-path or A-path in any higher recursion level.

The main contribution of the S-decomposition is that it gives a mapping from parallel components to paths through the respective components (the S-paths) such that all these paths are edge-disjoint. More formally:

Lemma 1. *For each series-parallel graph G, there exists an S-decomposition SD. Besides, in any S-decomposition, each edge belongs to at most one S-path in SD.*

Provided with the definition of an S-decomposition, we are ready to define the amortized cost as follows:

Definition 11 (Amortized Cost). *Let π_{OPT} be an (optimal) linear arrangement of a series-parallel graph G, let SD be an S-decomposition of G, and let S be a series component in G. Further, let E_S be the set that contains all edges of simple node sequences that are child components of S and all edges of S-paths of the child components of S that are parallel components. The* amortized cost of S, *denoted by $A\text{-}COST_{\pi_{OPT}}(S)$, is defined as:*

$$A\text{-}COST_{\pi_{OPT}}(S) := E\text{-}COST_{\pi_{OPT}}(S) + \sum_{\{x,y\}\in E_S} length_{\pi_{OPT}(S)}(x,y).$$

For any parallel component P in a series-parallel graph G and any optimal linear arrangement π_{OPT},

$$A\text{-}COST_{\pi_{OPT}}(P) := E\text{-}COST_{\pi_{OPT}}(P).$$

Note that the addend in the amortized cost for simple node sequences is zero (as the set E_S is empty in this case).

This definition will be helpful for the analysis of the minimum cost of an optimal arrangement. The amortized cost adds a certain value to the exclusive cost of a (series or parallel) component C, with the following property:

Lemma 2. *Let $G = (V, E)$ be a series-parallel graph, and π_{OPT} be an (optimal) linear arrangement for G. Further, let C be the set of all (series or parallel) components of G. It holds:*

$$\sum_{C\in\mathcal{C}} A\text{-}COST_{\pi_{OPT}}(C) \leq 3 \cdot \sum_{C\in\mathcal{C}} E\text{-}COST_{\pi_{OPT}}(C).$$

For the analysis of an optimal arrangement, we also need the following notation:

Definition 12 (Δ_C). *Given an (optimal) linear arrangement π_{OPT} of a series-parallel graph G, and a (series or parallel) component C in G, consider π_{OPT} restricted to C. We denote the smallest number of nodes to the left or to the right (depending on which number is smaller) of a terminal node of C in $\pi_{OPT}(C)$ by Δ_C.*

It is convenient to define:

Definition 13 (Cardinality of a Component). *For any series-parallel graph G and any (series or parallel) component C in G: $|C|$ is the number of nodes in C, $|C^{\ominus}|$ is the number of all nodes in C without the sink of C, and $|C^{-}|$ is the number of nodes in C without the two terminal nodes of C.*

4.2 A Lower Bound on the Total Cost of Optimal Solutions

In this subsection, we give lower bounds on the amortized costs of an optimal arrangement for simple node sequences, parallel components, and series components. In the end, we consolidate the results and state a general lower bound on the total cost of an optimal arrangement. For the proofs, we refer to the full version [10].

First of all, the following is a simple result about simple node sequences:

Lemma 3. *For any simple node sequence L in a series-parallel graph G in an optimal arrangement π_{OPT}, it holds:*

$$A\text{-}COST_{\pi_{OPT}}(L) \geq |L| - 1 + \Delta_L.$$

We now provide a lower bound on the amortized cost of series components:

Lemma 4. *For any series component S in a series-parallel graph G with $m \geq 2$ child components $P_1, \ldots, P_m$ in an optimal arrangement π_{OPT}, it holds:*

$$A\text{-}COST_{\pi_{OPT}}(S) \geq \frac{1}{2}\left(\sum_{i=1}^{m} |P_i^{\ominus}| - \max_i |P_i^{\ominus}|\right) + 1 + \sum_{i=1}^{m} \Delta_{P_i} - \Delta_S.$$

For the amortized cost of parallel components, we have the following result:

Lemma 5. *For any parallel component P in a series-parallel graph G with $m \geq 2$ child components $S_1, \ldots, S_m$, in an optimal arrangement π_{OPT}, it holds:*

$$A\text{-}COST_{\pi_{OPT}}(P) \geq \frac{1}{2}\left(\sum_{i=1}^{m} |S_i^{-}| - \max_i |S_i^{-}|\right) + \sum_{i=1}^{m} \Delta_{S_i} - \Delta_P.$$

These three lower bounds for the different types of components in any series-parallel graph can be combined into a single lower bound:

Corollary 1. *Let $G = (V, E)$ be an arbitrary series-parallel graph and π_{OPT} an optimal arrangement of G. Further, denote the total cost of π_{OPT} by $COST_{\pi_{OPT}}(G)$, the set of simple node sequences in G by L_G, the set of parallel components by P_G, the set of series components by S_G. Then, it holds:*

$$7 \cdot COST_{\pi_{OPT}}(G) \geq \sum_{L \in L_G} 2 \cdot (|L| - 1) + \sum_{P \in P_G}\left(\sum_{S_i \in P} |S_i^{-}| - \max_{S_i \in P}|S_i^{-}|\right)$$

$$+ \sum_{S \in S_G}\left(\sum_{P_i \in S} |P_i^{\ominus}| - \max_{P_i \in S}|P_i^{\ominus}|\right).$$

4.3 An Upper Bound on the Total Cost of SPGAA Arrangements

For the approximation ratio of the SPGAA, we also need to find an upper bound on the cost of arrangements computed by the SPGAA. One can show the following result:

Lemma 6. *Let $G = (V, E)$ be an arbitrary series-parallel graph and let π_{ALG} be an arrangement of G computed by the SPGAA. Furthermore, denote the total cost of π_{ALG} by $COST_{\pi_{ALG}}(G)$, the set of simple node sequences in G by L_G, the set of parallel components by P_G, the set of series components by S_G. Then, it holds:*

$$COST_{\pi_{ALG}}(G) \leq \sum_{L \in L_G} 2 \cdot (|L| - 1) + \sum_{P \in P_G} 2D^2 \cdot \left(\sum_{S_i \in P} |S_i^{-}| - \max_{S_i \in P}|S_i^{-}|\right)$$

$$+ \sum_{S \in S_G} 2D \cdot \left(\sum_{P_i \in S} |P_i^{\ominus}| - \max_{P_i \in S}|P_i^{\ominus}|\right).$$

4.4   The Approximation Ratio of $14 \cdot D^2$

Finally, based on the groundwork of the previous subsections, proving the main theorem of this chapter is straightforward.

Theorem 2. *For a series-parallel graph G, let π_{ALG} be the linear arrangement of G computed by the SPGAA, and let π_{OPT} be an optimal linear arrangement of G. It holds:*

$$COST_{\pi_{ALG}}(G) \leq 14 \cdot D^2 \cdot COST_{\pi_{OPT}}(G).$$

4.5   Runtime

Regarding the runtime of the SPGAA, one can show the following result:

Theorem 3. *On a series-parallel graph $G = (V, E)$, the SPGAA has a runtime of $O(|E|)$ if a minimal SP-tree of G is given as an input, and a runtime of $O(|E| \log |E|)$ otherwise.*

References

1. Adolphson, D., Hu, T.C.: Optimal linear ordering. SIAM J. Appl. Math. **25**(3), 403–423 (1973)
2. Adolphson, D.L.: Single machine job sequencing with precedence constraints. SIAM J. Comput. **6**(1), 40–54 (1977)
3. Ambühl, C., Mastrolilli, M., Svensson, O.: Inapproximability results for sparsest cut, optimal linear arrangement, and precedence constrained scheduling. In: Proceedings of FOCS, pp. 329–337. IEEE Computer Society (2007)
4. Arora, S., Rao, S., Vazirani, U.: Expander flows, geometric embeddings and graph partitioning. J. ACM **56**(2), 5 (2009)
5. Bodlaender, H.L., de Fluiter, B.: Parallel algorithms for series parallel graphs. In: Diaz, J., Serna, M. (eds.) ESA 1996. LNCS, vol. 1136, pp. 277–289. Springer, Heidelberg (1996)
6. Charikar, M., Hajiaghayi, M.T., Karloff, H., Rao, S.: L22 spreading metrics for vertex ordering problems. In: Proceedings of SODA, pp. 1018–1027. Society for Industrial and Applied Mathematics, Philadelphia (2006)
7. Chung, F.R.K.: Labelings of graphs. In: Beineke, L., Wilson, R. (eds.) Selected Topics in Graph Theory, vol. 3, pp. 151–168. Academic Press, New York (1988)
8. Cohen, J., Fomin, F.V., Heggernes, P., Kratsch, D., Kucherov, G.: Optimal linear arrangement of interval graphs. In: Královič, R., Urzyczyn, P. (eds.) MFCS 2006. LNCS, vol. 4162, pp. 267–279. Springer, Heidelberg (2006)
9. Díaz, J., Petit, J., Serna, M.: A survey of graph layout problems. ACM Comput. Surv. **34**(3), 313–356 (2002)
10. Eikel, M., Scheideler, C., Setzer, A.: Minimum linear arrangement of series-parallel graphs (full paper). ArXiv e-prints, October 2014
11. Eppstein, D.: Parallel recognition of series-parallel graphs. Inf. Comput. **98**(1), 41–55 (1992)
12. Even, S., Shiloach, Y.: NP-completeness of several arrangement problems. Department of Computer Science, Technion, Haifa, Israel, Technical Report 43 (1975)

13. Feige, U., Lee, J.R.: An improved approximation ratio for the minimum linear arrangement problem. Inf. Process. Lett. **101**(1), 26–29 (2007)
14. Fishburn, P., Tetali, P., Winkler, P.: Optimal linear arrangement of a rectangular grid. Discret. Math. **213**(1–3), 123–139 (2000)
15. Frederickson, G.N., Hambrusch, S.E.: Planar linear arrangements of outerplanar graphs. IEEE Trans. Circ. Syst. **35**(3), 323–333 (1988)
16. Garey, M., Johnson, D., Stockmeyer, L.: Some simplified NP-complete graph problems. Theoret. Comput. Sci. **1**(3), 237–267 (1976)
17. Hadlock, F.: Finding a maximum cut of a planar graph in polynomial time. SIAM J. Comput. **4**(3), 221–225 (1975)
18. Harper, L.H.: Optimal assignments of numbers to vertices. J. SIAM **12**(1), 131–135 (1964)
19. Kikuno, T., Yoshida, N., Kakuda, Y.: A linear algorithm for the domination number of a series-parallel graph. Discret. Appl. Math. **5**(3), 299–311 (1983)
20. Macmahon, P.A.: The combination of resistances. Electrician **28**, 601–602 (1892)
21. Nakano, K.: Linear layouts of generalized hypercubes. In: van Leeuwen, J. (ed.) WG 1993. LNCS, vol. 790, pp. 364–375. Springer, Heidelberg (1994)
22. Petit, J.: Experiments on the minimum linear arrangement problem. J. Exp. Algorithmics **8**, 2–3 (2003)
23. Petit, J.: Addenda to the survey of layout problems. Bull. EATCS **3**(105), 177–201 (2013)
24. Rao, S., Richa, A.W.: New approximation techniques for some ordering problems. In: Proceedings of SODA, pp. 211–218. Society for Industrial and Applied Mathematics, Philadelphia (1998)
25. Rostami, H., Habibi, J.: Minimum linear arrangement of chord graphs. Appl. Math. Comput. **203**(1), 358–367 (2008)
26. Setzer, A.: The planar minimum linear arrangement problem is different from the minimum linear arrangement problem. ArXiv e-prints, September 2014
27. Shahrokhi, F., Sýkora, O., Székely, L., Vrto, I.: On bipartite drawings and the linear arrangement problem. SIAM J. Comput. **30**(6), 1773–1789 (2001)
28. Takamizawa, K., Nishizeki, T., Saito, N.: Linear-time computability of combinatorial problems on series-parallel graphs. J. ACM **29**(3), 623–641 (1982)
29. Valdes, J., Tarjan, R.E., Lawler, E.L.: The recognition of series parallel digraphs. In: Proceedings of STOC, pp. 1–12. ACM, New York (1979)

Online Dual Edge Coloring of Paths and Trees

Lene M. Favrholdt$^{(\boxtimes)}$ and Jesper W. Mikkelsen

Department of Mathematics and Computer Science,
University of Southern Denmark, Odense, Denmark
{lenem,jesperwm}@imada.sdu.dk

Abstract. We study a dual version of online edge coloring, where the goal is to color as many edges as possible using only a given number, k, of available colors. All of our results are with regard to competitive analysis. For paths, we consider $k = 2$, and for trees, we consider any $k \geq 2$. We prove that a natural greedy algorithm called First-Fit is optimal among deterministic algorithms on paths as well as trees. This is the first time that an optimal algorithm for online dual edge coloring has been identified for a class of graphs. For paths, we give a randomized algorithm, which is optimal and better than the best possible deterministic algorithm. Again, it is the first time that this has been done for a class of graphs. For trees, we also show that even randomized algorithms cannot be much better than First-Fit.

1 Introduction

In the classical edge coloring problem, the edges of a graph must be colored using as *few colors* as possible, under the constraint that no two adjacent edges receive the same color. There is a dual version of the problem where a fixed number, k, of colors is given and the goal is to color as *many edges* as possible, using at most k colors. Sometimes the classical problem is called the *minimization* version and the dual problem is called the *maximization* version of the problem.

In this paper, we study the online version of the maximization problem. In the online version, the edges of the graph arrive one by one, each specified by its endpoints. Immediately upon receiving an edge, the algorithm must either color the edge with one of the k colors or reject the edge. The decision of which of the k colors to use or to reject the edge is irrevocable. We call this problem EDGE-k-COLORING. For any class, CLASS, of graphs, we let EDGE-k-COLORING(CLASS) denote the problem of EDGE-k-COLORING restricted to graphs of class CLASS. For instance, EDGE-2-COLORING(PATH) is the online problem of coloring as many edges as possible in a path using only two colors.

Quality Measure. We measure the quality of an online algorithm, A, for EDGE-k-COLORING using the standard notion of competitive ratio [11,15]. The competitive ratio compares the performance of A to that of an optimal offline algorithm,

L.M. Favrholdt and J.W. Mikkelsen—Supported in part by the Villum Foundation and the Danish Council for Independent Research, Natural Sciences.

© Springer International Publishing Switzerland 2015
E. Bampis and O. Svensson (Eds.): WAOA 2014, LNCS 8952, pp. 181–192, 2015.
DOI: 10.1007/978-3-319-18263-6_16

OPT. We denote by $A(\sigma)$ the number of edges colored by A when given a sequence, σ, of edges. Similarly, $\text{OPT}(\sigma)$ is the number of edges in σ colored by OPT. The algorithm A is said to be C-competitive if there exists a constant b such that $A(\sigma) \geq C \cdot \text{OPT}(\sigma) - b$ for any input sequence σ. The competitive ratio, $C_A(k)$, of A is the supremum over all C for which A is C-competitive. The competitive ratio of A for EDGE-k-COLORING(CLASS) is denoted by $C_A^{\text{CLASS}}(k)$.

Note that by this definition, $0 \leq C_A(k) \leq 1$. In particular, upper bounds on the competitive ratio are negative results and lower bounds are positive results.

If the inequality above holds even when $b = 0$, we say that A is *strictly C-competitive*. This gives rise to the notion of *strict competitive ratio*. The results in this paper are strongest possible in the sense that all positive results hold for the strict competitive ratio and all negative results hold for the competitive ratio.

For randomized algorithms, a similar definition of competitive ratio is used but $A(\sigma)$ is replaced by the expected value $E[A(\sigma)]$.

Notation and Terminology. We label the k colors $1, 2, \ldots, k$. For $1 \leq i \leq j \leq k$, define $C_{i,j} = \{i, i+1, \ldots, j\}$. At any fixed point in the processing of the input sequence, we denote by C_v the set of colors used at edges incident to the vertex v. A color $i \in C_{1,k}$ is said to be *available* at v if $i \notin C_v$. Two colorings of a graph are said to be *equivalent* if one can be obtained from the other by renaming the colors.

If v is a vertex in the input graph, we denote by $d(v)$ the number of edges incident to v. An *isolated edge* $e = (v, u)$ is an edge such that $d(v) = d(u) = 1$ at the time where e is revealed. For any m, we let $\langle e_1, e_2, \ldots, e_m \rangle$ denote a path with m edges and label the edges such that, for $2 \leq i \leq m - 1$, e_i is adjacent to e_{i-1} and e_{i+1}. A *star* with m edges is the complete bipartite graph $K_{1,m}$.

Algorithms. An algorithm is called *fair* if it never rejects an edge unless all of the k colors have already been used on adjacent edges. In [8], the following two fair and deterministic algorithms were studied:

First-Fit (FF) uses the lowest available color when coloring an edge. It can be viewed as the natural greedy strategy.

Next-Fit (NF) remembers the last used color c_{last}. When coloring an edge, it uses the first available color in the ordered sequence $\langle c_{\text{last}} + 1, \ldots, k, 1, \ldots, c_{\text{last}} \rangle$. For the very first edge, it uses the color 1.

For the EDGE-2-COLORING(PATH) problem, we introduce a new family of randomized algorithms: For $\frac{1}{2} \leq p \leq 1$, Rand_p is defined as follows. Whenever an isolated edge is revealed, Rand_p uses the color 1 with probability p and the color 2 with probability $1 - p$. All non-isolated edges are colored (with the only remaining color) if possible. Note that Rand_1 is identical to **First-Fit**.

Previous Results. In [8] it is shown that any fair algorithm for EDGE-k-COLORING has a competitive ratio of at least $2\sqrt{3} - 3 \approx 0.464$, and at most $\frac{1}{2}$ if it is deterministic. The lower bound is tight in the sense that **Next-Fit** has a competitive ratio of exactly $2\sqrt{3} - 3$. It remains an open problem if any algorithm has a competitive ratio better than $2\sqrt{3} - 3$. The authors of [8] also show that no algorithm (even when allowing randomization) has a competitive ratio better than $\frac{4}{7} \approx 0.57$.

The problem EDGE-k-COLORING(k-COLORABLE) is also studied in [8]. When the input graph is k-colorable, any fair algorithm is shown to have a competitive ratio of at least $\frac{1}{2}$. Again, the lower bound is tight because Next-Fit has a competitive ratio of $\frac{1}{2}$. The competitive ratio of First-Fit is shown to be $\frac{k}{2k-1}$. An upper bound of $\frac{2}{3}$ is given for deterministic algorithms in this case.

We remark that all of the negative results mentioned above hold even if the input graph is bipartite. Thus, contrary to offline edge coloring, the online EDGE-k-COLORING problem does not appear to be significantly easier when restricted to bipartite graphs.

It is well known that for $k = 1$ (i.e., for the matching problem), the greedy algorithm is an optimal deterministic algorithm with a competitive ratio of $\frac{1}{2}$.

The relative worst order ratio [3,4] of both the maximization and minimization version of online edge coloring is studied in [6]. For the maximization version, it is shown that First-Fit and Next-Fit are not (strictly) comparable. This is true even when the input is restricted to bipartite graphs. For the minimization version, First-Fit is proven better than Next-Fit.

The minimization version of online edge coloring is studied in [1]. If an online algorithm never introduces a new color unless forced to do so, it will never use more than $2\Delta - 1$ different colors on graphs of maximum degree Δ. It is shown in [1] that no (randomized) online algorithm can do better than this, even if the input graph is restricted to being a forest. On any graph, an optimal offline algorithm uses at most $\Delta + 1$ colors, and on trees, Δ colors suffice. Hence, any algorithm that introduces a new color only when necessary, has a competitive ratio of 2, and this is optimal.

The problem of online vertex coloring has received much attention, especially in the minimization version (see [12] for a survey). Edge coloring a path of m edges is equivalent to vertex coloring a path of m vertices. Thus, our results for EDGE-2-COLORING(PATH) are also valid for online dual vertex coloring of paths with 2 colors available.

A study of approximation algorithms for the offline version of EDGE-k-COLORING for multigraphs was initiated in [9]. This line of work has been continued in [5,10,13,14] for both simple graphs and multigraphs.

Our Contribution. For EDGE-2-COLORING(PATH), we give a $\frac{4}{5}$-competitive randomized algorithm and prove that this is optimal. We also show that no deterministic algorithm can be better than $\frac{2}{3}$-competitive and observe that this upper bound is tight, since First-Fit is $\frac{2}{3}$-competitive. This is the first example of a class of graphs for which a randomized algorithm for EDGE-k-COLORING is proven optimal and better than any deterministic algorithm.

For EDGE-k-COLORING(TREE) where $k \geq 2$, we prove that First-Fit is $\frac{k-1}{k}$-competitive and that no deterministic or fair algorithm can be better than this. Thus, an algorithm would have to be both randomized and unfair to achieve a better competitive ratio than `First-Fit`. However, we show that even such algorithms cannot be better than $\frac{k}{k+1}$-competitive. We also show that any fair algorithm is $\frac{2\sqrt{k}-2}{2\sqrt{k}-1}$-competitive and that the competitive ratio of `Next-Fit` is no better than this if k is a square number. This implies that the competitive ratio of any fair algorithm goes to 1 as k goes to infinity.

PATH and TREE are the first examples of graph classes for which an optimal deterministic algorithm for EDGE-k-COLORING has been identified.

Due to space restrictions, some details of the proofs have been omitted. These can be found in the full version of the paper [7].

2   A Charging Technique for Proving Positive Results

We will now describe a simple charging technique for proving lower bounds on the competitive ratio. The technique was first used for deterministic algorithms in [8]. For some C, $0 \leq C \leq 1$, our goal is to prove that a given (possibly randomized) algorithm A is C-competitive. Assume that the edges of a graph $G = (V, E)$ have been given in some order, σ, and let $E_{\text{OPT}} \subseteq E$ be the set of edges colored in some optimal solution.

The *initial value* $v_i(e)$ of an edge, $e \in E$, is $v_i(e) = \Pr[e$ is colored by A$]$. For deterministic algorithms, $v_i(e) \in \{0, 1\}$ for all $e \in E$. Note that by linearity of expectation, we have $E[\mathsf{A}(\sigma)] = \sum_{e \in E} v_i(e)$.

The *surplus* $v_+(e)$ of an edge, $e \in E$, (with respect to C) is

$$v_+(e) = \begin{cases} v_i(e) - C, & \text{if } e \in E_{\text{OPT}} \\ v_i(e), & \text{if } e \notin E_{\text{OPT}} \end{cases}$$

We let $E_+ \subseteq E$ and $E_- \subseteq E$ denote the sets of edges with positive and negative surplus, respectively. Clearly, $E_- \subseteq E_{\text{OPT}}$. For deterministic algorithms, E_- is exactly those edges in E_{OPT} that are not colored by the algorithm, and E_+ is the set of edges colored by the algorithm (assuming $C < 1$). The total positive surplus $\sum_{e \in E_+} v_+(e)$ will be redistributed among the edges in E_- according to some strategy. This strategy is what needs to be defined when applying the technique.

The *final value* $v_f(e)$ of an edge $e \in E_{\text{OPT}}$ is the total value of e after the redistribution of surplus. Since only surplus value is redistributed, $v_f(e) \geq C$ for all $e \in E_{\text{OPT}} \setminus E_-$. Thus, if it can be proven that $v_f(e) \geq C$ for all $e \in E_-$, then

$$E[\mathsf{A}(\sigma)] = \sum_{e \in E} v_i(e) = \sum_{e \in E} v_f(e) \geq \sum_{e \in E_{\text{OPT}}} v_f(e) \geq C \cdot \mathsf{OPT}(\sigma).$$

Thus, it follows that A is (strictly) C-competitive.

3   Coloring of Paths

In this section, we study the EDGE-k-COLORING problem when the input graph is a path. Clearly, this is only interesting if $k \leq 2$. In this paper, we consider solely the case where $k = 2$, but we remark that one can use the same techniques to obtain tight bounds on the competitive ratio when $k = 1$. Also, the results for PATH can be extended to graphs of maximum degree 2.

For EDGE-2-COLORING(PATH), our main result is a randomized algorithm with a competitive ratio of $\frac{4}{5}$ and a proof that this is optimal. Before considering randomized algorithms, we give tight lower and upper bounds on the competitive ratio of deterministic algorithms. The proofs of Propositions 1 and 2 are straightforward and have been omitted (see also [8]).

Proposition 1. *For* EDGE-2-COLORING(PATH), *Next-Fit is a worst possible fair algorithm with* $C_{NF}^{\mathrm{PATH}}(2) = \frac{1}{2}$.

Proposition 2. *For* EDGE-2-COLORING(PATH), *First-Fit is an optimal deterministic algorithm with* $C_{FF}^{\mathrm{PATH}}(2) = \frac{2}{3}$.

Knowing that no deterministic algorithm can be better than $\frac{2}{3}$-competitive, a natural question to ask is how good a randomized algorithm can be. To this end, we analyze the family of fair, randomized algorithms, Rand_p, defined in the introduction.

Theorem 1. *Let* $\frac{1}{2} \leq p \leq 1$. *Then,*

$$C_{Rand_p}^{\mathrm{PATH}}(2) = \min\left\{p^2 - p + 1, \frac{2}{3}(-p^2 + p + 1)\right\}.$$

Proof. For the upper bound, consider the following two adversary strategies for revealing the edges of a path $\langle e_1, \ldots, e_m \rangle$:

(i) The adversary first reveals all edges e_i with $i \equiv 1 \pmod 3$, followed by all edges e_i with $i \equiv 0 \pmod 3$. Finally, all the remaining edges are revealed.
(ii) The adversary first reveals all the odd numbered edges and thereafter all the even numbered edges.

One can show that (i) gives an upper bound of $\frac{2}{3}(-p^2 + p + 1)$ on the competitive ratio and that (ii) gives an upper bound of $p^2 - p + 1$ on the competitive ratio.

For the lower bound, fix $\frac{1}{2} \leq p \leq 1$. Let P be a path. Consider an edge e at the time of its arrival. If two edges adjacent to e have already been revealed, we say that e is a *critical edge*. Denote by E_{crit} the critical edges of P.

We will apply the charging technique described in Sect. 2. Note that all non-critical edges have an initial value of 1 and, hence, a surplus of $1 - C$. Thus, $E_- \subseteq E_{\mathrm{crit}}$.

For a non-critical edge e, it is easy to show inductively that the following holds: The probability of e being colored with the color 1 is p or $1 - p$. In the former case, we say that e is *odd* and in the latter case, we say that e is *even*. Note that an even edge must be adjacent to at least one odd edge.

Let e_{crit} be a critical edge. Denote by e_l and e_r the two edges adjacent to e_{crit}. The edge e_{crit} will be rejected if and only if e_l and e_r are colored with different colors. Also, the random variable denoting the color received by e_l is independent of the random variable denoting the color received by e_r. There are two possible cases:

Case 1: One of e_l and e_r is odd and the other is even. Without loss of generality, assume that e_l is odd and that e_r is even. By the discussion above, the probability of e_{crit} being colored is $p(1-p)+(1-p)p$. Since e_r is even, it must be adjacent to at least one non-critical edge e'_r. We transfer a value of $\frac{1}{2}(1-C)$ from each of e_l and e'_r to e_{crit} and a value of $1-C$ from e_r to e_{crit}. Transferring the entire surplus of $1-C$ from e_r to e_{crit} is possible, since e'_r is non-critical and therefore e_{crit} is the only critical edge adjacent to e_r. Thus, the final value of e_{crit} is $2p(1-p)+2(1-C)$. It follows that if C is at most $\frac{2}{3}(-p^2+p+1)$, then the final value of e_{crit} is at least C.

Case 2: e_l and e_r are both odd or both even. By transferring half of the surplus from e_l and e_r to e_{crit}, one can show that if C is at most p^2-p+1, then the final value of e_{crit} is at least C.

We conclude that if C is bounded from above by both p^2-p+1 and $\frac{2}{3}(-p^2+p+1)$ then, using the strategy described above, all edges in the path end up with a final value of at least C.   □

Theorem 1 shows that, for $p=\varphi/\sqrt{5}$, Rand_p has a competitive ratio of $\frac{4}{5}$ (where $\varphi=(1+\sqrt{5})/2$ is the golden ratio). We will now show that $\frac{4}{5}$ is the best possible competitive ratio of *any* algorithm.

Theorem 2. *If R is an algorithm for* EDGE-2-COLORING(PATH), *then*

$$C_R^{\text{PATH}}(2) \leq \frac{4}{5}.$$

Proof. We will use Yao's minimax principle [2,16]. To this end, we describe a randomized adversary. Let D be a deterministic algorithm and let $M \in \mathbb{N}$ be a large even number. The adversary will reveal the edges of a path, P, as follows: First, it reveals $M+1$ isolated edges, $\{e_1,\ldots,e_{M+1}\}$. Afterwards, the adversary picks uniformly at random a set of indices $S \subseteq \{2,\ldots,M+1\}$ such that $|S|=\frac{M}{2}$. For each index $i \in S$, the adversary reveals a single edge, e, connecting e_i and e_{i-1} (so that $\langle e_{i-1},e,e_i\rangle$ becomes a subpath of P). On the other hand, for each index $i \in \overline{S}$, the adversary reveals two edges, e and e', connecting e_i and e_{i-1} (so that $\langle e_{i-1},e,e',e_i\rangle$ becomes a subpath of P).

Suppose that for some index i, both e_i and e_{i-1} are colored by D (it can be shown that D does not gain anything by rejecting isolated edges). If $i \in S$, then e_i and e_{i-1} must be colored the same in order to avoid rejecting the edge connecting them. If $i \in \overline{S}$, then e_i and e_{i-1} must be colored differently in order to avoid rejecting one of the edges connecting them. Since $\Pr(i \in S)=\Pr(i \in \overline{S})=\frac{1}{2}$, this observation implies that the expected number of edges rejected by D is at least $\frac{M}{2}$. The total number of edges in P is $(M+1)+\frac{M}{2}+M=\frac{5}{2}M+1$. Thus, $E[\text{D}(P)] \leq 2M+1 < \frac{4}{5}\text{OPT}(P)+1$.   □

Theorems 1 and 2 together give the following corollary.

Corollary 1. *For $p=\frac{\varphi}{\sqrt{5}}$, Rand_p is optimal for* EDGE-2-COLORING(PATH) *with*

$$C_{\text{Rand}_p}^{\text{PATH}}(2) = \frac{4}{5}.$$

4   Coloring of Trees

We will now consider the EDGE-k-COLORING problem when the input graph is a tree. Our main result is a proof that First-Fit is an optimal deterministic algorithm. We also show that, for any fixed $k \geq 4$, First-Fit has a better competitive ratio than Next-Fit for EDGE-k-COLORING(TREE). First, we give a general upper bound for algorithms that are deterministic or fair.

Theorem 3. *If A is a deterministic or fair algorithm and $k \geq 2$, then*

$$C_A^{\text{TREE}}(k) \leq \frac{k-1}{k}.$$

Proof. We only describe the adversary strategy. The edges of a tree are revealed in N steps, for some large $N \in \mathbb{N}$. The set of edges revealed in the ith step constitute a star, S_i, with $k + 1$ edges and center vertex c_i. For $2 \leq i \leq N$, if at least one edge in S_{i-1} is colored, the adversary chooses $c_i = x$ for some colored edge (c_{i-1}, x) in S_{i-1}. Otherwise, it chooses $c_i = x$ for an arbitrary edge (c_{i-1}, x) in S_{i-1}. Note that the adversary is clearly able to identify a colored edge in S_{i-1}, if one exists: If A is deterministic, this is trivially true, and if A is fair, the first $k - 1$ edges of S_{i-1} will be colored. □

Using the charging technique of Sect. 2, we will show that Theorem 3 is tight by proving a matching lower bound for First-Fit. To this end, we introduce some terminology related to deterministic algorithms.

Let A be a deterministic algorithm for EDGE-k-COLORING, let $G = (V, E)$ be a graph, and suppose that A has been given the edges of G in some order. Recall that, since A is deterministic, E_+ denotes the set of edges colored by A, and E_- denotes the set of edges colored by OPT only. We partition E_+ into the set, E_+^d, of edges colored by both A and OPT (*double colored* edges) and the set, E_+^s, of edges colored by A only (*single colored* edges). Thus, $E_{\text{OPT}} = E_- \cup E_+^d$. For $x \in V$, let $E_+(x)$ be the edges in E_+ incident to x and let $d_+(x) = |E_+(x)|$. Define $E_-(x), E_+^d(x), E_+^s(x), d_-(x), d_+^d(x)$ and $d_+^s(x)$ similarly.

Theorem 4. *For $k \geq 2$, First-Fit is an optimal deterministic algorithm for* EDGE-k-COLORING(TREE) *with*

$$C_{\text{FF}}^{\text{TREE}}(k) = \frac{k-1}{k}.$$

Proof. Fix a tree $T = (V, E)$ and assume that the edges of E have been revealed to First-Fit in some order. For the analysis, we will view T as a rooted tree by choosing an arbitrary vertex to be the root. When writing $e = (x, y) \in E$, we imply that x is the parent vertex of y.

Following Sect. 2, we set $C = \frac{k-1}{k}$. An edge in E_+^d then has a surplus of $1 - C = \frac{1}{k}$ and an edge in E_+^s has a surplus of 1. On the other hand, an edge in E_- has an initial value of zero.

We will define a strategy to distribute the total positive surplus obtained by First-Fit among the edges in E_- such that each edge gets a final value of at least C. For ease of presentation, the strategy will be described in a stepwise manner:

Step 1: Consider in turn all edges $e = (v, u) \in E_+$. Let c be the color assigned to e by First-Fit and let $e' = (w, v)$ be the parent edge of e (if it exists).

(a) If $e' \in E_+^d$ and e' has been colored with a color $c' > c$, then e transfers a value of $\frac{1}{k}$ to w.

(b) Any surplus remaining at e is transferred to v.

For each vertex v, let $m(v)$ denote the value transferred to v in this step.

Step 2: Consider in turn all vertices $v \in V$.

(a) If the vertex v has a parent edge $e' \in E_-$, then v transfers a value of $\min \{m(v), \frac{k-1}{k}\}$ to e'.

(b) Any value remaining at v is distributed equally among the child edges of v belonging to E_-.

The following simple but useful properties of the strategy defined above will be used to prove the theorem. Each of the four facts gives a lower bound on the value transferred from an edge $e = (v, u)$ to its parent vertex, v. Let $e' = (w, v)$ be the parent edge of e (if it exists).

Fact 1: Assume that $e \in E_+^s$. If $e' \notin E_+^d$ or e' does not exist, then e contributes a value of 1 to $m(v)$. If $e' \in E_+^d$, then e contributes a value of at least $\frac{k-1}{k}$ to $m(v)$.

Fact 2: Assume that e is colored with the color c. If $e' \notin E_+^d$ or e' does not exist, then $m(v) \geq \frac{c}{k}$.

Fact 3: Assume that $e \in E_+^d$. If $e' \notin E_+^d$, then e contributes a value of $\frac{1}{k}$ to $m(v)$.

In order to state the next fact, we need to introduce some new terminology. For $v \in V$, let $\widehat{c}_v = \max \{\overline{C}_v \cup \{0\}\}$. That is, $\widehat{c}_v$ is the largest color available at v (and $\widehat{c}_v = 0$ if no colors are available). If an edge e incident to v is colored with a color $c > \widehat{c}_v$, then e is said to be a *high-colored* edge (with respect to v).

Fact 4: Assume that $e \in E_+^d$. If e is high-colored with respect to v, then the colored child edges of e contribute a total value of at least $\frac{k-d_+(v)}{k}$ to $m(v)$.

We will combine these facts to show that any edge $e = (x, y) \in E_-$ gets a final value of at least $\frac{k-1}{k}$. If $C_x = C_{1,k}$, then $\widehat{c}_x = 0$. Otherwise, $\widehat{c}_x \in C_y$, since First-Fit is fair. Hence, Fact 2 implies that $m(y) \geq \frac{\widehat{c}_x}{k}$. Thus, e receives a value of at least $\min\{\frac{k-1}{k}, \frac{\widehat{c}_x}{k}\}$ from y. In particular, we will assume that $\widehat{c}_x < k - 1$, since otherwise we are done. We will now turn to proving that e receives a value of at least $\frac{k-\widehat{c}_x-1}{k}$ from x. This will finish the proof, since it means that e gets a final value of at least $\frac{\widehat{c}_x}{k} + \frac{k-\widehat{c}_x-1}{k} = \frac{k-1}{k}$.

Let $e' = (z, x)$ be the parent edge of x (if it exists). The rest of the proof is split into three cases depending on which of the sets E_+^d, E_-, and E_+^s (if any) that contains e'.

Case 1: $e' \in E_+^d$. There must be $k - \widehat{c}_x$ high-colored edges incident to x. Thus, x has at least $k - \widehat{c}_x - 1$ high-colored child edges, and at least $k - \widehat{c}_x - 1 - d_+^s(x)$ of them belong to E_+^d. By Fact 4, x receives a value of at least $\frac{k-d_+(x)}{k}$ from the child edges of each of these at least $k - \widehat{c}_x - 1 - d_+^s(x)$ edges. Moreover, by

Fact 1, each of the $d_+^s(x)$ child edges of e' belonging to E_+^s contributes a value of $\frac{k-1}{k}$ to $m(x)$. Thus,

$$m(x) \geq (k - \widehat{c}_x - 1 - d_+^s(x))\frac{k - d_+(x)}{k} + d_+^s(x)\frac{k - 1}{k} \geq d_-(x)\frac{k - \widehat{c}_x - 1}{k},$$

where the last inequality follows by calculations using that $d_+(x) \geq k - \widehat{c}_x$, $d_+(x) - d_+^s(x) = d_+^d(x)$, and $k - d_+^d(x) \geq d_-(x)$. Thus, since no value is transferred from x to e' in Step 2(a), each child edge of x belonging to E_- receives a value of at least $\frac{k - \widehat{c}_x - 1}{k}$ from x in Step 2(b).

The two remaining cases ($e' \in E_+^s$ or e' does not exist) and ($e' \in E_-$) are treated similarly to Case 1. □

By Theorems 3 and 4, an algorithm for EDGE-k-COLORING(TREE) can only be better than **First-Fit**, if it is both randomized and unfair. However, the next result shows that even such algorithms cannot do much better than **First-Fit**.

Theorem 5. *If R is an algorithm for* EDGE-k-COLORING *and* $k \geq 2$, *then*

$$C_R^{\mathrm{TREE}}(k) \leq \frac{k}{k + 1}.$$

Proof. We give only the adversary strategy. The adversary first reveals the edges of a path $P = \langle e_1, \ldots, e_m \rangle$, for some large $m \in \mathbb{N}$. Let $v_1, \ldots, v_{m+1}$ be the vertices in the path such that $e_i = (v_i, v_{i+1})$, for $1 \leq i \leq m$. If $E[\mathrm{R}(P)] \leq \frac{k}{k+1}m$, the adversary reveals no more edges. If $E[\mathrm{R}(P)] > \frac{k}{k+1}m$, then for each i, $1 \leq i \leq m + 1$, the adversary reveals k edges constituting a star, S_i, with center vertex v_i. □

We now show that, for any fixed $k \geq 4$, **First-Fit** is better than **Next-Fit**, but the competitive ratio of any fair algorithm tends to 1 as k tends to infinity. We will use the notation introduced just before Theorem 4.

Theorem 6. *If F is a fair algorithm, then for any* $k \geq 2$,

$$C_F^{\mathrm{TREE}}(k) \geq \frac{2\sqrt{k} - 2}{2\sqrt{k} - 1}.$$

Proof. If F is randomized, the proof holds for any coloring the algorithm may produce, and hence, for the expected number of edges colored.

Let $T = (E, V)$ be a tree and assume that the edges of T have been revealed to F in some order. We will view T as a rooted tree by choosing an arbitrary vertex to be the root. As in the proof of Theorem 4, we let $e = (x, y)$ imply that x is the parent of y.

We will apply the charging technique from Sect. 2 to show that F is C-competitive, where $C = \frac{2\sqrt{k}-2}{2\sqrt{k}-1}$. To this end, we use the following redistribution strategy:

Step 1: Each edge $(v, u) \in E_+$ transfers its entire surplus to its parent vertex, v.

For each vertex v, let $m(v)$ denote the value transferred to v in this step.

Step 2: Consider in turn all vertices $v \in V$.
(a) If the vertex v has a parent edge $e' \in E_-$, then v transfers a value of $\min\{m(v), C\}$ to e'.
(b) Any value remaining at v is distributed equally among the child edges of v belonging to E_-.

For each edge e, let $m_v(e)$ denote the value transferred from v to e in this step.

This finishes the description of the strategy.

Fix an edge $e = (x, y) \in E_-$. We need to show that $m_x(e) + m_y(e) \geq C$. Since $m_y(e) = \min\{C, d_+(y) - C d_+^{\mathrm{d}}(y)\}$, this will always be the case unless $d_+(y) - C d_+^{\mathrm{d}}(y) < C$. One can show that if $d_+(y) - C d_+^{\mathrm{d}}(y) < C$, then $d_+(y) = d_+^{\mathrm{d}}(y)$. It follows that we only need to consider the case where $d_+(y) = d_+^{\mathrm{d}}(y)$, meaning that all of the edges incident to y which have been colored by F have also been colored by OPT. In particular, this implies that $m_y(e) = (1 - C)d_+(y)$. Consider now the value $m_x(e)$ transferred to e from x.

Case 1: The parent edge of x belongs to E_-. The edge e receives a value of $m_x(e) \geq \frac{d_+(x) - C d_+^{\mathrm{d}}(x) - C}{d_-(x) - 1}$ from the colored child edges of x. Since $d_+^{\mathrm{d}}(x) + d_-(x) \leq k$ (OPT can color at most k edges incident to x), it follows that $d_-(x) \leq k - d_+^{\mathrm{d}}(x)$. Thus, $m_x(e) \geq \frac{d_+(x) - C d_+^{\mathrm{d}}(x) - C}{k - d_+^{\mathrm{d}}(x) - 1}$. We claim that

$$m_x(e) + m_y(e) \geq \frac{d_+(x) - C d_+^{\mathrm{d}}(x) - C}{k - d_+^{\mathrm{d}}(x) - 1} + (1 - C)d_+(y) \geq C$$

Using $d_+^{\mathrm{d}}(x) \leq d_+(x)$ and $d_+(y) \geq k - d_+(x)$, it follows that in order to prove the claim, it suffices to show that $d_+(x) + (1 - C)(k - d_+(x))(k - d_+(x) - 1) \geq Ck$. The left hand side of this inequality is a quadratic polynomial in $d_+(x)$. Allowing $d_+(x)$ to be any real number, one can show that this quadratic polynomial attains its minimum value of Ck when $d_+(x) = k - \sqrt{k}$. In particular, the inequality is certainly true for all integer values of $d_+(x)$ and hence, $m_x(e) + m_y(e) \geq C$.

The three remaining cases, the parent edge of e belongs to E_+^{s}, belongs to E_+^{d}, or does not exist, are treated similarly to Case 1. □

We will show that the lower bound of Theorem 6 is essentially tight by providing a matching upper bound on the competitive ratio of Next-Fit when k is a square number.

Theorem 7. *If $k = n^2$ for some integer $n \geq 2$, then Next-Fit is a worst possible fair algorithm with*

$$C_{\mathrm{NF}}^{\mathrm{TREE}}(k) = \frac{2\sqrt{k} - 2}{2\sqrt{k} - 1}.$$

Proof. The lower bound follows from Theorem 6. For the upper bound, we define a tree $T = (V, E)$ and a subset $E' \in E$. We specify a coloring, $\mathscr{C}$, of E' with the property that each edge in $E \setminus E'$ is adjacent to edges of all k colors.

The tree T consists of N *bunches* of stars, for some large N. Each bunch contains a *large* star with $k - \sqrt{k}$ edges colored with $C_{1,k-\sqrt{k}}$ and $\sqrt{k} - 1$ *small* stars, each with $\sqrt{k}$ edges colored with $C_{k-\sqrt{k}+1,k}$. The center vertex of the large star in bunch i, $1 \leq i \leq N$, is called v_i. This finishes the description of E' and its coloring. For each bunch of stars, $E \setminus E'$ contains an edge between v_i and the center vertex of each of the small stars in the bunch. The ith bunch is connected to the $(i + 1)$th bunch by an edge from v_{i+1} to the center vertex of one of the small stars in the ith bunch. Note that, after assigning the coloring $\mathscr{C}$ to E', none of the edges in $E \setminus E'$ can be colored. This finishes the description of T.

The adversary will use k disjoint copies, $T_1, \ldots, T_k$, of T. The edges in the resulting graph can be given in an order such that Next-Fit colors each tree with a coloring equivalent to $\mathscr{C}$. Finally, the k disjoint trees are connected, using $k - 1$ edges between vertices that have degree one in the trees. Since $k \geq 4$, we must have $\sqrt{k} + 2 \leq k$ and so the maximum degree of the graph is k. Thus, since the graph has no cycles, OPT colors all edges of the graph. □

We will briefly consider the case where k is not a square number. Any fair algorithm for EDGE-1-COLORING(TREE) is just the greedy matching algorithm. It is observed in several papers that this algorithm is $\frac{1}{2}$-competitive (for all input graphs) and that no deterministic algorithm can do better, even when the input graph is a tree. If $k \geq 2$, but not a square number, then the lower bound from Theorem 6 can be slightly improved by using the fact that $d_+(x)$ must be an integer. In particular, one can show that all fair algorithms for EDGE-k-COLORING(TREE) are at least $\frac{1}{2}$-competitive for $k = 2$ and $\frac{2}{3}$-competitive for $k = 3$. Since these bounds match the upper bound from Theorem 3, we conclude that all fair algorithms have the same competitive ratio when $k \leq 3$.

If $k \geq 4$ (but not necessarily a square number), one can obtain the following upper bound by rounding $\sqrt{k}$ appropriately in the proof of Theorem 7: $C_{\text{NF}}^{\text{TREE}}(k) \leq \frac{\frac{k}{\lceil\sqrt{k}\rceil} + \lceil\sqrt{k}\rceil - 2}{\frac{k}{\lceil\sqrt{k}\rceil} + \lceil\sqrt{k}\rceil - 1}$. In particular, for any fixed $k \geq 4$, the competitive ratio of First-Fit is better than the competitive ratio of Next-Fit for EDGE-k-COLORING(TREE).

5 Open Problems

Finding optimal online algorithms for EDGE-k-COLORING in general and on other classes of graphs is an interesting open problem. We believe that the techniques used in the proofs of Theorems 4 and 6 can be generalized to, e.g., graphs of bounded degeneracy. In particular, graphs of bounded degeneracy can be oriented so that each vertex has bounded outdegree and the resulting digraph is acyclic. This makes it possible to use strategies for redistributing the surplus similar to the ones we have used for trees.

References

1. Bar-Noy, A., Motwani, R., Naor, J.: The greedy algorithm is optimal for on-line edge coloring. Inf. Process. Lett. **44**(5), 251–253 (1992)
2. Borodin, A., El-Yaniv, R.: Online Computation and Competitive Analysis. Cambridge University Press, Cambridge (1998)
3. Boyar, J., Favrholdt, L.M.: The relative worst order ratio for online algorithms. ACM Trans. Algorithms **3**(2), 22 (2007)
4. Boyar, J., Favrholdt, L.M., Larsen, K.S.: The relative worst-order ratio applied to paging. J. Comput. Syst. Sci. **73**, 818–843 (2007)
5. Chen, Z.-Z., Konno, S., Matsushita, Y.: Approximating maximum edge 2-coloring in simple graphs. Discret. Appl. Math. **158**(17), 1894–1901 (2010)
6. Ehmsen, M.R., Favrholdt, L.M., Kohrt, J.S., Mihai, R.: Comparing first-fit and next-fit for online edge coloring. Theor. Comput. Sci. **411**(16–18), 1734–1741 (2010)
7. Favrholdt, L.M., Mikkelsen, J.W.: Online dual edge coloring of paths and trees. ArXiv e-prints, 1405.3817 [cs.DS] (2014)
8. Favrholdt, L.M., Nielsen, M.N.: On-line edge-coloring with a fixed number of colors. Algorithmica **35**(2), 176–191 (2003)
9. Feige, U., Ofek, E., Wieder, U.: Approximating maximum edge coloring in multi-graphs. In: Jansen, K., Leonardi, S., Vazirani, V.V. (eds.) APPROX 2002. LNCS, vol. 2462, pp. 108–121. Springer, Heidelberg (2002)
10. Kamiński, M., Kowalik, L.: Approximating the maximum 3- and 4-edge-colorable subgraph. In: Kaplan, H. (ed.) SWAT 2010. LNCS, vol. 6139, pp. 395–407. Springer, Heidelberg (2010)
11. Karlin, A.R., Manasse, M.S., Rudolph, L., Sleator, D.D.: Competitive snoopy caching. Algorithmica **3**, 77–119 (1988)
12. Kierstead, H.A.: Coloring graphs on-line. In: Fiat, A., Woeginger, G.J. (eds.) Online Algorithms 1996. LNCS, vol. 1442, pp. 281–305. Springer, Heidelberg (1998)
13. Kosowski, A.: Approximating the maximum 2- and 3-edge-colorable subgraph problems. Discret. Appl. Math. **157**(17), 3593–3600 (2009)
14. Rizzi, R.: Approximating the maximum 3-edge-colorable subgraph problem. Discret. Math. **309**(12), 4166–4170 (2009)
15. Sleator, D.D., Tarjan, R.E.: Amortized efficiency of list update and paging rules. Commun. ACM **28**(2), 202–208 (1985)
16. Yao, A.C-C.: Probabilistic computations: Toward a unified measure of complexity (extended abstract). In: FOCS, pp. 222–227 (1977)

Online Packet Scheduling
Under Adversarial Jamming

Tomasz Jurdzinski[1]([✉]), Dariusz R. Kowalski[2], and Krzysztof Lorys[1]

[1] Institute of Computer Science, University of Wrocław, Wrocław, Poland
tju@cs.uni.wroc.pl
[2] Department of Computer Science, University of Liverpool, Liverpool, UK

Abstract. We consider the problem of scheduling packets of different lengths via a directed communication link prone to jamming errors. Dynamic packet arrivals and errors are modelled by an adversary. We focus on estimating competitive throughput of online scheduling algorithms. We design an online algorithm for scheduling packets of arbitrary lengths, achieving optimal competitive throughput in $(1/3, 1/2]$ (the exact value depends on packet lengths). Another algorithm we design makes use of additional resources in order to achieve competitive throughput 1, that is, it achieves at least as high throughput as the best schedule without such resources, for any arrival and jamming patterns. More precisely, we show that if the algorithm can run with double speed, i.e., with twice higher frequency, then its competitive throughput is 1. This demonstrates that throughput of the best online *fault-tolerant* scheduling algorithms scales well with resource augmentation. Finally, we generalize the first of our algorithms to the case of any $f \geq 1$ channels and obtain competitive throughput $1/2$ in this setting in case packets lengths are pairwise divisible (i.e., any larger is divisible by any smaller).

Keywords: Packet scheduling · Adversarial jamming · Online algorithms · Competitive throughput · Resource augmentation

1 Introduction

Motivation. Achieving high-level reliability in packet scheduling has recently become more and more important due to substantial increase of the scale of networks and higher fault-tolerant demands of many incoming applications. In the era of Internet of Things and nano-devices, it will no longer be possible to attend devices physically, and therefore the designed protocols must be stable and robust no matter of failure pattern. Imagine the problem of thousands of malfunctioning nano-capsules with overflown buffers that need to be somehow removed from the human body, or the consequences of lack of communication between AVs with humans onboard or medical devices incorporated into patients bodies, even if such case might happen with probability less than 1 %.

This work was supported by the Polish National Science Centre grant DEC-2012/06/M/ST6/00459.

© Springer International Publishing Switzerland 2015
E. Bampis and O. Svensson (Eds.): WAOA 2014, LNCS 8952, pp. 193–206, 2015.
DOI: 10.1007/978-3-319-18263-6_17

Our Approach. This paper studies a fundamental problem of online packet scheduling via *unreliable* link (also called a channel), when transmitted packets may be interrupted by jamming errors. This problem was recently introduced by Fernandez Anta et al. [4] and analyzed for two different packet lengths; in our work, an *arbitrary number of packet lengths* is considered. Packets arrive dynamically to one end of the link, called a sender, and need to be transmitted in full, i.e., without any in-between jamming error, to the other end (called a receiver). Jamming errors are immediately discovered by the sender. We analyze all possible scenarios, including worst case ones, which we model as a conceptually adversary who controls both packet arrivals and channel jamming. The adversary is *unrestricted*, in the sense that she may generate *any arrival and error pattern* in time. The main objective of the online scheduling protocol is to achieve as high throughput as possible under current scenario, where the throughput is the rate of the total length of successfully sent packets in time. (Other measures, such as queue sizes and packet latency, are not considered in this work — it is known that they both require higher speedup augmentation in order to achieve competitiveness, cf., [5,6]). Because of the online setting, we consider the *competitive throughput* measure, which is roughly a ratio between the throughput achieved by the online algorithm and the one reached by any other deterministic algorithm (even equipped with the knowledge of adversarial arrivals and errors).

Previous Work. The framework considered in this work was recently introduced in [4] in the context of *two* packet lengths. The authors showed that general offline version of this problem, in which the scheduling algorithm knows a priori when errors will occur, is NP-hard, cf., [4]. They also considered algorithms and upper limitations for relative throughput in case of *two* packet lengths. In particular, they proved that relative throughput of *any* online scheduling protocol cannot be bigger than $\overline{\rho}/(\rho + \overline{\rho})$, where ρ is the ratio between the largest and the smallest packet lengths and $\overline{\rho} = \lfloor \rho \rfloor$. (Note that the upper bound becomes $1/2$ if the bigger packet length is a multiplicity of the smaller packet length.) This upper bound can be achieved by a protocol scheduling a specific preamble of shorter packets followed by the Longest_First rule after every error, but cannot be reached by simpler protocols such as Longest_First itself or Shortest_First. Therefore, it remained open whether there is an online scheduling protocol reaching the relative throughput of (roughly) $1/2$ for *arbitrary* number of packet lengths; we answer this question in affirmative in this work, using alternative techniques. Moreover, as also shown in [4], randomization does not help, which motivates the study of deterministic algorithms. Recently, Fernandez Anta et al. [6] analyzed four popular scheduling algorithms in the same framework: FIFO, LIFO, Longest_First and Shortest_First. Among others, they proved that, for any packet lengths, the first three algorithms have relative throughput 0 while the last one has $\frac{1}{1+\rho} < \frac{1}{2}$, even for pairwise divisible packet lengths. Moreover, none of the four algorithms reaches the relative throughput 1 for any speedup smaller than ρ. All results in [4,6] hold also for competitive throughput (for sufficiently large additive constant).

Our Contribution. We design a deterministic online scheduling algorithm achieving optimal competitive throughput for an arbitrary number k of packet lengths $\ell_{\min} = \ell_1 < \ell_2 < \ldots < \ell_k = \ell_{\max}$ (Sect. 3). We first show a simpler version of the algorithm, for the case when packet lengths are pairwise divisible, i.e., any larger is divisible by any smaller (we call it *pairwise divisibility property*). We then extend the protocol so that it does not need to rely on such limitation about divisibility, and achieves the competitive throughput $\min_{1 \leq j < i \leq k} \left\{ \frac{\lfloor \rho_{i,j} \rfloor}{\lfloor \rho_{i,j} \rfloor + \rho_{i,j}} \right\}$, where $\rho_{i,j} = \ell_i / \ell_j$ is the ratio between the i-th and the j-th packet length. Note that this general formula for competitive throughput is in the range $(\frac{1}{3}, \frac{1}{2}]$, which is independent of ρ (in contrast to the four popular algorithms analyzed in [6]), and it reaches $\frac{1}{2}$ if and only if the pairwise divisibility property holds.

A natural question arrises whether a better throughput could be achieved under some additional resources provided to the scheduler, for example, speedup (e.g., using higher frequency). Unfortunately, the designed protocols do not achieve competitive throughput 1 even if speedup 2 is applied (it can be easily checked that the competitive throughput is at most 2/3 in such case, while the four popular algorithms analyzed in [6] require speedup at least ρ), which implies that they are not well-scalable (i.e., linearly) with resource augmentation.[1] Therefore we design another deterministic online protocol to optimize competitive throughput for speedup 2, provided pairwise divisibility property holds (Sect. 4). It is a generalization of the preamble protocols, proposed in [4] and [5] in the case of only two packet lengths.

Finally, we show how to generalize our algorithm achieving throughput $\frac{1}{2}$ for a single channel (without speedup) to the setting with any $f \geq 1$ channels. This extension is not straightforward, since different channels may have different error frequencies generated by the adversary, and therefore the central scheduler has to adapt separately to each channel capacity while keeping general progress with respect to the joint set of pending packets. To the best of our knowledge, this is the first work studying throughput of fault-tolerant dynamic scheduling on *many channels* against unrestricted adversary.

Due to limited space, many details are deferred to the full version of the paper [8].

Related Work. Packet scheduling [9] is one of the most fundamental problems in computer networks. A realistic approach involves *online* scheduling [7,11], and therefore a *competitive analysis* [1,13] is often used to evaluate the performance of proposed solutions. Online scheduling was considered in a number of models; for more information the reader is referred to [10] and [11].

There are relatively few approaches assuming both online packet arrivals and errors. Apart from the already mentioned work [4], the authors in [5] studied *buffer sizes* of online scheduling protocols on error-prone channel. Unlike the relative/competitive throughput measure, in order to be positively competitive

---

[1] Note that the considered speedup 2 is chosen because we claim linear scalability of competitive throughput with the increase of speedup, that is, starting from level 1/2 with no speedup we expect the competitive throughput to reach value 1 for speedup 2.

with the best scheduling algorithms with respect to the buffer sizes, additional resources need to be given to the online protocol, i.e., speedup (higher frequency). This form of resource augmentation appeared to be efficient: for some speedup smaller than 2 there is a deterministic online scheduling algorithm having roughly the same queue sizes as any other scheduling algorithm running without speedup. That work motivated us to consider resource augmentation technique, in the form of using some speedup (higher frequency), to reach at least the same throughput as the best scheduler without speedup for *any* execution.

Wireless packet scheduling was also considered in models with physical constraints included, such as radio networks or SINR. Anantharamu et al. [2] considered packet scheduling on a multiple access channel with signal interference, under a *restricted* adversarial patterns of packet arrivals and channel jamming. Richa et al. [12] who analyzed competitive throughput of randomized scheduling protocols on multiple access channels with signal interference against *adaptive, but still restricted, adversarial jamming*.

Andrews and Zhang [3] studied buffer stability (i.e., bounded buffers property) of online packet scheduling on a wireless channel, where both the channel conditions and the data arrivals are controlled by an adversary. They also assumed *bounded adversary*, as otherwise stability could not be reached in their model.

2   Model

We consider a uni-directional point-to-point link in which one end point, called a *sender*, transmits packets to the other end point, called a *receiver*. The sender is equipped with unlimited buffer (or a queue), in which the arriving packets are queued. Packets may be of different lengths, and may arrive at any time; we assume that time is continuous, and scheduling algorithm have access to packets as soon as they arrive. There are $k \geq 2$ different packet lengths, denoted by $\ell_{\min} = \ell_1 < \ell_2 < \ldots < \ell_k = \ell_{\max}$. For simplicity, we will use the names "ℓ_i-packets" and "packets ℓ_i" for packets of length ℓ_i, for any $1 \leq i \leq k$. In some parts of the paper we assume that ℓ_i/ℓ_j is an integer for any $1 \leq j < i \leq k$ (so called *pairwise divisibility property*). We denote $\rho = \ell_{\max}/\ell_{\min}$. We assume that all packets are transmitted at the same bit rate, hence the transmission time is proportional to the packet's length. The link is prone to jamming errors, that is, transmitted packets might be corrupted at any time point.

Arrival Models. We consider adversarial packet arrivals: the packets' arrival time and length are governed by an adversary. We define an adversarial arrival pattern as a collection of packet arrivals (i.e., packet id, length and arrival time) caused by the adversary.

Link Jamming Errors. We consider adversarial model of jamming errors, in which the adversary decides at which time to cause a jamming error on the link. The error at time t implies that a packet being transmitted at time t is broken, and the information about it is immediately delivered to the sender so that it breaks the current transmission and could schedule another packet (or re-schedule the one that was just broken). A corrupted packet transmission is unsuccessful, in the sense that it is not received by the receiver and it needs to be retransmitted in full (not necessarily right after the error — scheduling algorithm may decide to postpone it and transmit another packet instead). We assume that scheduling algorithms do not voluntarily stop transmitting packets before the end of the transmission, unless they get feedback about jamming error. An adversarial error pattern is defined as a collection of error events on the link caused by the adversary.

Adversarial models are typically used to argue about the algorithm's behavior in any possible scenario, in particular, in the worst-case ones.

Efficiency Metric: *Competitive Throughput*. We would like to measure throughput of the communication link(s) in terms of competitive analysis of online algorithms. Let $\mathcal{A}$ be an arrival pattern and $\mathcal{E}$ an error pattern. For a given deterministic algorithm ALG, let $L_{ALG}(\mathcal{A}, \mathcal{E}, t)$ be the total length of all the successfully transmitted (i.e., non-corrupted) packets by time t under arrival pattern $\mathcal{A}$ and error pattern $\mathcal{E}$. Let OFF be any offline algorithm that knows the exact arrival and error patterns.

For arrival pattern $\mathcal{A}$, adversarial error pattern $\mathcal{E}$ and time t, we define the *competitive throughput* $T_{ALG}(\mathcal{A}, \mathcal{E}, OFF, t)$ *of a deterministic algorithm ALG by time t with respect to OFF* as:

$$T_{ALG}(\mathcal{A}, \mathcal{E}, OFF, t) = \frac{L_{ALG}(\mathcal{A}, \mathcal{E}, t)}{L_{OFF}(\mathcal{A}, \mathcal{E}, t)} \ .$$

For completeness, $T_{ALG}(\mathcal{A}, \mathcal{E}, OFF, t)$ equals 1 if $L_{ALG}(\mathcal{A}, \mathcal{E}, t) = L_{OPT}(\mathcal{A}, \mathcal{E}, t) = 0$.

We define the *competitive throughput* of *ALG* in the adversarial arrival model to be the biggest value T_{ALG} satisfying the following equation for each offline algorithm OFF, each time t and some constant a (depending only on the model parameters, but not on t):

$$L_{ALG}(\mathcal{A}, \mathcal{E}, t) \geq T_{ALG} \cdot L_{OFF}(\mathcal{A}, \mathcal{E}, t) - a. \qquad (1)$$

Resource Augmentation: *Speedup*. In the second part of the paper, in Sect. 4, we consider resource augmentation technique. This technique was recently applied to fault-tolerant scheduling in [5] in the context of buffer stability metric. In particular, we compare the throughput of a given online algorithm under the assumption that this algorithm is run with a certain speedup $s > 1$, with the throughput of the best scheduling algorithm run without any speedup. From technical perspective, computing of the *competitive throughput under speedup* $s > 1$ follows the same definitions as given above, with the only difference that the value of $L_{ALG}(\mathcal{A}, \mathcal{E}, t)$ is calculated under assumption that *ALG* transmits packets s times faster (i.e., a packet of length ℓ can be transmitted in time period of length ℓ/s). In this work we focus on speedup $s = 2$.

Notation. We use the notations $[n, m] = \{p \in \mathbb{N} \mid n \leq p \leq m\}$ and $[n] = [1, n]$.

Algorithm 1. Greedy

1: **loop**
2:     **while** $\sum_{i=1}^{k} \ell_i n_i < \ell_k$ **do** Stay *idle*   ▷ n_i : number of awaiting ℓ_i-packets
3:     Transmit-group(k)
---

3   Packet Scheduling for k Packet Lengths

In this section we present an online algorithm, which is optimal for any number of packet lengths $k \geq 2$. First, for the ease of presentation, we present the algorithm Greedy under assumption that $\ell_i / \ell_{i-1} \in \mathbb{N}$ for $1 < i < k$. Later, in Sect. 3.1, we show how to remove this assumption by modifying the algorithm Greedy; the resulted algorithm is called MGreedy.

The main idea behind our algorithms is to keep transmitting as many short packets as possible (shortest-first strategy), subject to some balancing constraints. Observe that it is difficult for any offline algorithm OFF to get substantial advantage over any online algorithm ALG when ALG sends small packets from its queue. Thus, preference for small packets ensures that ALG can be as efficient as any OFF, as long as ALG has short packets in its queue. However, if OFF transmits large packets during transmission of small packets by ALG, it can afterwards transmit small packets when ALG does not have any of them in its queue. Simultaneously, when OFF is transmitting small packets, the adversary can generate errors preventing ALG from a successful transmission of large packets. Despite this disadvantage of a greedy approach, we show that an appropriate implementation of this strategy, using some balancing constraints, provides an optimal solution with respect to competitive throughput, and thus against any optimal way of scheduling under occurring arrival and failure patterns.

Our specific modification of the greedy shortest-first strategy is based on sending packets in groups, which altogether balance the length of the next larger packet. We explain it first for two types of packet lengths: ℓ_{min} and ℓ_{max}. If there are at least $\rho = \ell_{max}/\ell_{min}$ small packets (ℓ_{min}) in the queue, the algorithm builds a *group*, which consists of ρ of them, and keeps sending them until all of them are transmitted successfully. If there are less than ρ small packets in the queue at the moment when a transmission of a group is finished, a large packet (ℓ_{max}) is transmitted. However, whenever there are at least ρ small packets, the group of small packets is formed, independently of the fact whether a transmission of a large packet(s) is successful or not. This idea is then recursively applied for the case when there are $k > 2$ types of packets.

A pseudo-code of our greedy algorithm is presented as Algorithm 1, with its recursive subroutine given as Algorithm 2. In the pseudo-codes, n_i denotes the number of ℓ_i-packets which are currently (at the moment) waiting in the queue for transmission.

Performance Analysis of Algorithm Greedy. For the sake of analysis of algorithm Greedy, we introduce some new notations. First, let us assume that an arrival pattern and an injection pattern are chosen arbitrarily and are fixed,

---

Algorithm 2. Transmit-group(j)

1: **loop**
2:   **if** $\sum_{i=1}^{j-1} \ell_i n_i \geq \ell_j$ **then**   ▷ n_i : number of awaiting ℓ_i-packets
3:     **for** $a = 1$ **to** ℓ_j / ℓ_{j-1} **do** Transmit-group($j-1$)
       return
4:   Transmit ℓ_j; If the transmission is successful: **return**

---

so we could omit them from formulas in the further analysis. For an algorithm A, let $q_A(i,t)$ denote the sum of lengths of ℓ_i-packets in the queue of A at the moment t. That is, $q_A(i,t) = n_i \cdot \ell_i$ for a fixed time t. Moreover, let $q_A(< i,t) = \sum_{j<i} q_A(j,t)$ and we define $q_A(\leq i,t)$ analogously. Let $L_A(i,t)$ denote the length of packets ℓ_i successfully transmitted by time t. For a time period $\tau = [t_1, t_2]$, let $L_A(i,\tau) = L_A(i,t_2) - L_A(i,t_1)$. That is, $L_A(i,\tau)$ denotes the total length of ℓ_i-packets successfully transmitted in the interval τ (i.e., such that their transmissions are finished in τ). The notions $L_A(< i,t)$, $L_A(\leq i,t)$, $L_A(< i,\tau)$, and $L_A(\leq i,\tau)$ for time t and time interval τ are defined analogously to $q_A(< i,t)$, $q_A(\leq i,t)$, $q_A(< i,\tau)$ and $q_A(\leq i,\tau)$. We also use the above introduced notations without the first argument, i.e., $q_A(t)$, $q_A(\tau)$, $L_A(t)$, and $L_A(\tau)$, which are shorthands for $q_A(\leq k,t)$, $q_A(\leq k,\tau)$, $L_A(\leq k,t)$ and $L_A(\leq k,\tau)$, respectively.

An algorithm A is *busy* at time t if it is transmitting a packet at t, it has just finished a successful transmission at t, or its transmission is jammed by an error at t. Otherwise A is *idle* at t.

Our goal is to compare progress in sending packets of our algorithm Greedy and any algorithm OFF. We say that an algorithm A is *m-busy* (with respect to OFF) in a time period $\tau = [t_1, t_2]$ if the following conditions are satisfied:

1. A is busy at each time $t \in \tau$;
2. A does not (try to) transmit packets ℓ_i for $i > m$ during τ;
3. $q_A(i,t_1) \geq q_{OFF}(i,t_1)$ for each $i \in [m]$. (That is, at time t_1 A has no less packets of length ℓ_i in its queue than OFF, for each $i \leq m$.)

Now, we state technical results regarding periods in which Greedy is m-busy for some $m \in [k]$. These lemmas eventually lead to the proof of the fact that competitive throughput of Greedy is $1/2$ (provided $\ell_i / \ell_{i-1} \in \mathbb{N}$ for $i \in [2,k]$), which is optimal. First, we make an observation that, if Greedy does not use packets longer than ℓ_m for $m \in [k]$, then the total length of packets transmitted by Greedy is at least as large as the total length of packets of length at least ℓ_m transmitted by OFF (up to an additive constant).

Lemma 1. *Assume that Greedy is m-busy in a time period τ, $m \leq k$. Then, $L_{Greedy}(\tau) \geq L_{OFF}(\geq m,\tau) - \ell_k$.*

Next, we formulate a relationship between the length of packets transmitted in time period τ by Greedy and OFF up to the moment when Greedy is transmitting the longest packet used by itself during τ.

Lemma 2. *Assume that Greedy is m-busy in a time period $\tau = [t_1, t_2]$, $m \leq k$. Let $t \in \tau$ be any time at which Greedy starts transmitting a packet of length ℓ_m. Then,*

$$2L_{Greedy}([t_1, t]) \geq L_{OFF}([t_1, t]) + q_{OFF}(< m, t) - \ell_m - \ell_k.$$

Using previous lemmas, one can prove by induction a relationship between $L_{Greedy}(\tau)$ and $L_{OFF}(\tau)$ for periods τ which are m-busy for Greedy, where $m \in [k]$.

Lemma 3. *Assume that Greedy is m-busy in a time period τ, for $m \leq k$. Then,*

$$2L_{Greedy}(\tau) \geq L_{OFF}(\tau) - g_m$$

where g_m satisfies the relationships $g_1 = \ell_k$ and $g_{i+1} = g_i + 2\ell_{i+1} + 2\ell_k$ for $i \in [1, k-1]$.

Theorem 1. *The competitive throughput of Greedy is equal to $1/2$, provided $l_i/l_{i-1} \in \mathbb{N}$ for each $i \in [2, k]$.*

Proof. Let us fix $\mathcal{A}$ and $\mathcal{E}$. For any time t, let t' be the largest value among $t'' < t$ such that Greedy is idle at t''. Then,

- $L_{Greedy}(t') < L_{OFF}(t') + \ell_k$, since Greedy is idle only in the case that total length of packets in its queue is smaller than ℓ_k;
- Greedy is k-busy in $\tau = [t', t]$ and therefore $L_{Greedy}(\tau) \geq \frac{1}{2} L_{OFF}(\tau) - \frac{1}{2} g_k$ by Lemma 3.

By combining the above observations, we get $L_{Greedy}(t) \geq \frac{1}{2} L_{OFF}(t) - \ell_k - \frac{1}{2} g_k$, which gives the claimed result for the additive constant $a = \frac{1}{2} g_k + \ell_k$. □

Corollary 1. *The algorithm Greedy achieves optimal competitive throughput for packets' lengths $\ell_1 < \ldots < \ell_k$ such that $\ell_i/\ell_{i-1} \in \mathbb{N}$ for each $i \in [2, k]$.*

Proof. It is shown in [4] that competitive throughput of any online algorithm for two types of packets is at most $\frac{\lceil \ell_2/\ell_1 \rceil}{\lceil \ell_2/\ell_1 \rceil + \ell_2/\ell_1}$, which is equal to $\frac{1}{2}$ when $\ell_2/\ell_1 \in \mathbb{N}$. As an adversary can decide to schedule merely two types of packets among available k types, Theorem 1 implies optimality of competitive throughput of Greedy. □

3.1 Arbitrary Lengths of Packets

In this section we discuss an application of the ideas behind the algorithm Greedy to the general case, i.e., when the condition $\ell_i/\ell_{i-1} \in \mathbb{N}$ is not satisfied. Let $\rho_{i,j} = \ell_i/\ell_j$. A natural generalization of Greedy is that, instead of $\rho_{i,i-1}$ groups of packets of length ℓ_{i-1} on the i-th level of recursion, we choose $\lfloor \rho_{i,i-1} \rfloor$ groups of packets of total length (as close as possible to but not larger than) ℓ_{i-1} in order to "cover" ℓ_i. Then, we can apply the ideas of "covering" packets transmitted by OFF using groups of packets transmitted by Greedy. If $k = 2$, this approach gives an algorithm with competitive throughput $\frac{\lfloor \rho_{2,1} \rfloor}{\lfloor \rho_{2,1} \rfloor + \rho_{2,1}}$, which is optimal due to [4]. The lower bound from [4] naturally generalizes to the following result, since the adversary may inject only two types of packets.

Theorem 2. *The competitive throughput of any online scheduling algorithm is at most*

$$\min_{1 \leq j < i \leq k} \left\{ \frac{\lfloor \rho_{i,j} \rfloor}{\lfloor \rho_{i,j} \rfloor + \rho_{i,j}} \right\}.$$

However, for $k > 2$, the additional advantage of OFF over Greedy, following from rounding on various levels of recursion, could accumulate. In order to limit this effect, instead of transmitting $\lfloor \ell_i / \ell_{i-1} \rfloor$ groups of packets on the level $i - 1$, we keep sending groups on the level $i - 1$ as long as the sum of lengths of packets from the transmitted groups is not larger than $\ell_i - \ell_{i-1}$. This gives the following technical result. (For simplifying the arguments in the remaining part of the analysis, let us denote $\rho_{i,i-1} = \ell_i / \ell_{i-1}$ by simply ρ_i, for $i \in [2, k]$.)

Lemma 4. *Consider such a modification of Greedy that TransmitGroup(j) keeps calling TransmitGroup($j - 1$), for $j > 1$, as long as the total length of transmitted packets in the current execution of TransmitGroup(j) is at most $\ell_j - \ell_{j-1}$. The competitive throughput of this algorithm is at least* $\min_{i \in [2,k]} \left\{ \frac{\rho_i - 1}{2\rho_i - 1} \right\}$.

Proof. In Lemmas 1, 2 and 3, we repeatedly use the argument that, if Greedy does not use packets of length ℓ_i for $i > m$, then each such packet transmitted by OFF corresponds to a group of (shorter) packets transmitted by Greedy of total length ℓ_i. This observation can be preserved for the modified Greedy algorithm with a relaxation that a packet ℓ_i transmitted by OFF corresponds to a group of packets transmitted by Greedy of length at least $\ell_i - \ell_{i-1} = (1 - \rho_i)\ell_i$. This relaxation translates inequalities from Lemmas 1, 2 and 3 to:

$$\frac{\rho_m}{\rho_m - 1} \cdot L_{Greedy}(\tau) \geq L_{OFF}(\geq m, \tau) - \ell_k$$
$$\left(1 + \frac{\rho_m}{\rho_m - 1}\right) \cdot L_{Greedy}([t_1, t]) \geq L_{OFF}([t_1, t]) + q_{OFF}(< m, t) - \ell_m - \ell_k$$
$$\left(1 + \frac{\rho_m}{\rho_m - 1}\right) \cdot L_{Greedy}(\tau) \geq L_{OFF}(\tau) - g_m$$

If we apply the above inequalities instead of those from Lemmas 1, 2 and 3 in the proof of Theorem 1, we obtain the result claimed here. □

However, as a group of packets transmitted by Greedy "covering" the packet ℓ_i transmitted by OFF may contain packets of various lengths, the competitive throughput of the solution from Lemma 4 is difficult to compare with the upper bound from Theorem 2. In order to tackle this issue, we introduce yet another modification to the algorithm.

The main goal of this modification is to ensure that Greedy is transmitting packets of the same length for long periods of time and it changes to other length only if it is necessary. An execution of the algorithm is split into *stages*. In a stage, packets of total length (close to) $ck\ell_k$ are transmitted, where $c \in \mathbb{N}$ is a fixed large constant. At the beginning of a stage, the set C of candidates is determined as $C = \{i \mid n_i \ell_i \geq ck\ell_k\}$. Then, the *interesting* length $\ell_{i^\star}$ is set for parameter $i^\star = \min(C)$, and the algorithm starts transmitting packets $l_{i^\star}$. After each transmission, successful or not, the interesting length $i^\star$ is updated to

---

Algorithm 3. MGreedy $\hfill$ c : constant parameter

1: $C \leftarrow \{i \,|\, n_i \ell_i \geq ck\ell_k\}$ $\hfill$ $\triangleright$ n_i : number of awaiting ℓ_i-packets
2: **loop**
3: $\quad$ **while** $\{i \,|\, \ell_i n_i \geq ck\ell_k\} = \emptyset$ **do** Stay *idle*
4: $\quad\quad$ $C \leftarrow \{i \,|\, \ell_i n_i \geq ck\ell_k\}$
5: $\quad\quad$ $i^* \leftarrow \min(C)$
6: $\quad\quad$ **for** $a = 1$ **to** ck **do** $\ell' \leftarrow$ TransmitGroup$(j-1)$
7: $\quad\quad\quad$ $\ell \leftarrow \ell + \ell'$

Algorithm 4. TransmitGroup(j) $\hfill$ i^* : global variable of MGreedy and
TransmitGroup

1: $\ell \leftarrow 0$
2: **while** $\ell \leq \ell_j - \ell_{i^*}$ **do**
3: $\quad$ **if** $j > i^*$ **then** $\ell' \leftarrow$ TransmitGroup$(j-1)$
4: $\quad\quad$ $\ell \leftarrow \ell + \ell'$
5: $\quad$ **else**
6: $\quad\quad$ Transmit ℓ_j
7: $\quad\quad$ $C \leftarrow C \cup \{i \,|\, \ell_i n_i \geq ck\ell_k\}$ $\hfill$ $\triangleright$ n_i : number of awaiting ℓ_i-packets
8: $\quad\quad$ $i^* \leftarrow \min(C)$
9: $\quad\quad$ If a transmission of ℓ_j successful: $\ell \leftarrow \ell_j$
10: **return** ℓ

$i^* \leftarrow \min(\{i^*\} \cup \{i \,|\, \ell_i n_i \geq ck\ell_k\})$. (Note that the set of candidates $\{i \,|\, \ell_i n_i \geq ck\ell_k\}$ may change over time, as the adversary injects packets.)

Using the notion of the interesting length, we work in line with the original algorithm Greedy, with the following restrictions:

- no packet is transmitted as long as the interesting length is not determined (i.e., the set of candidates is empty);
- only a packet of length l_{i^*} can be transmitted.

As the total length of packets staying in the queue whose lengths are not interesting is at most $k \cdot ck\ell_k$, they do not have impact on the multiplicative constant defining the competitive throughput. Thus, for the analysis of the competitive throughput, we can assume that there are no packets of lengths which are *not* interesting at each time t. That is, there are no packets of lengths ℓ_i such that $i \notin C$. Then, the new algorithm MGreedy works exactly as the original algorithm Greedy. The pseudo-code of algorithm MGreedy and the modified sub-routine TransmitGroup(j), which now returns also some value ℓ, are given as Algorithms 3 and 4, respectively.

Performance Analysis of Algorithm MGreedy. We say that an execution of TransmitGroup(k) is *uniform* if the algorithm transmits packets of a fixed length ℓ_i during that executions of TransmitGroup(k) as well as during the executions of TransmitGroup(k) directly preceding it. A new key property of algorithm MGreedy compared with Greedy is that most of its executions of sub-routine TransmitGroup(k) are uniform.

Proposition 1. *At least $ck - 2k$ calls of TransmitGroup in a stage of MGreedy are uniform.*

Given the above observation, we can evaluate the competitive throughput of MGreedy.

Lemma 5. *The competitive throughput of the algorithm MGreedy is at least*

$$\min_{1 \leq j < i \leq k} \left\{ \frac{\lfloor \rho_{i,j} \rfloor}{\lfloor \rho_{i,j} \rfloor + \rho_{i,j}} \right\} \cdot c',$$

where c' is a constant depending on the parameter c such that c' can be arbitrarily close to 1 for large enough c.

As we can choose arbitrarily large c, Lemma 5 implies that the competitive throughput of MGreedy might be arbitrarily close to the upper bound from Theorem 2. In the following theorem, we state that one can guarantee the optimal competitive throughput. This result is obtained by a modification of MGreedy such that it gradually increases the constant c during its execution.

Theorem 3. *The optimal competitive throughput of an online algorithm is equal to*

$$\min_{1 \leq j < i \leq k} \left\{ \frac{\lfloor \rho_{i,j} \rfloor}{\lfloor \rho_{i,j} \rfloor + \rho_{i,j}} \right\}.$$

4   An Algorithm for a Scenario with Speedup

Now we return to the packets whose lengths fulfil divisibility property, i.e. $\ell_i/\ell_{i-1} \in \mathbb{N}$ for $1 < i < k$, and address the problem of increasing throughput by enabling algorithm to work with greater speed. We design an algorithm Prudent (Algorithm 5) which, working with speedup $s = 2$, achieves competitive throughput 1. This algorithm works in *phases*, where a phase is a time period between two consecutive errors. Behaviour of the algorithm in a phase is described as Algorithm 5. During each phase it tries to send packets of maximal length which do not exceed the total length of packets sent so far in that phase. It can be treated as a greedy strategy restricted by a "safety policy" that does not allow to send long packets unless the cost of their unsuccessful transmissions can be amortized by an advantage over an adversary gained during the earlier transmissions since the time of the last error.

One can easily show that such a strategy guarantees that an algorithm sends at least as many ℓ_k-packets as any OFF, provided the algorithm works with speedup $s = 2$. Then, one can show inductively that the following inequality holds for $i \in [1, k-1]$ at any time t: $L_{Prudent}(\geq k-i, t) \geq L_{OFF}(\geq k-i, t) - \frac{5}{2}i\ell_k$. As a simple consequence of this fact, we get the following theorem.

Theorem 4. *The competitive throughput of Algorithm Prudent working with speed-up 2 is equal to 1, provided $\ell_i/\ell_{i-1} \in \mathbb{N}$ for each $i \in [2, k]$.*

Algorithm 5. Prudent

1: **loop**
2:     **while** $\{i \mid \ell_i n_i \geq \ell_k\} = \emptyset$ **do** Stay *idle*   ▷ n_i : number of awaiting ℓ_i-packets
3:     let i be the smallest number such that: $n_i \ell_i \geq \ell_k$;
4:     **if** $i < k$ **then**
5:         transmit ℓ_{i+1}/ℓ_i packets ℓ_i;
6:         $L_{sent} \leftarrow \ell_{i+1}$
7:         **while** $L_{sent} < \ell_k$ **do**
8:             $j \leftarrow$ maximal number such that $n_j \ell_j \geq \ell_k - L_{sent}$ and $\ell_j \leq L_{sent}$
9:             transmit ℓ_{j+1}/ℓ_j packets ℓ_j
10:             $L_{sent} \leftarrow L_{sent} + \ell_{j+1}$
11:     **loop**
12:         transmit longest unsent packet

Algorithm 6. GreedyMC on channel $\psi \leq f$

1: **loop**
2:     **while** $\sum_{i=1}^{k} \ell_i n_i < f\ell_k$ **do** Stay *idle*   ▷ n_i : number of awaiting ℓ_i-packets
3:     Transmit-group-MC(k)

Algorithm 7. Transmit-group-MC(j) on channel $\psi \leq f$

1: **loop**
2:     **if** $\sum_{i=1}^{j-1} \ell_i n_i \geq f\ell_j$ **then**   ▷ n_i : number of awaiting ℓ_i-packets
3:         **for** $a = 1$ to ℓ_j/ℓ_{j-1} **do** Transmit-group-MC($j-1$)
return
4:     Transmit ℓ_j; If the transmission is successful: **return**

5   Packet Scheduling on $f > 1$ Channels

In this section we show how to adjust algorithm Greedy from Sect. 3 to the setting where $f > 1$ independent channels are available. All channels are controlled by the adversary, and the only difference between the model with a single channel and the one with $f > 1$ channels is that the scheduling algorithm needs to choose a channel for each scheduled packet. The analysis of the modified algorithm, called the GreedyMC, will be done under the same assumption as for the original Greedy algorithm, that is, $\ell_i/\ell_{i-1} \in \mathbb{N}$ for $1 < i \leq k$.

The modification and its analysis are not obvious, since the adversary may for example cause more errors on one channel and consequently the scheduler should try to schedule packets of shorter lengths on it, comparing to the other channels. Therefore, the following two modifications need to be done. First, the scheduler runs the algorithm GreedyMC, and its sub-routine Transmit-group-MC, separately for each channel; that is, as soon as channel $\overset{\circ}{\psi} \leq f$ gets free, the algorithm tries to find a suitable packet to schedule on ψ from all available packets in the queue. Second, these individually run copies of the algorithm take into account availability of packets for other channels, by generalising the formula triggering the next level of recursion from $\sum_{i=1}^{j-1} \ell_i n_i \geq \ell_j$ to $\sum_{i=1}^{j-1} \ell_i n_i \geq f\ell_j$, i.e., by multiplying the right side of the equation by the number of channels f. We also assume that the scheduled packet is removed from the queue, to avoid redundancy; it gets re-stored in case an error is reported on the channel during its transmission. A pseudo-code of GreedyMC is presented as Algorithm 6, with its recursive subroutine given as Algorithm 7.

6   Conclusions

We presented novel efficient and reliable algorithms for online scheduling of packets of different lengths. The first two protocols assure maximum possible throughput and they guarantee to be no more than twice (or three times, in case of no divisibility) worse than the throughput of any other scheduling algorithm. The result can be generalized to the setting of any $f > 1$ channels. Another algorithm guarantees at least as high throughput as the optimal one, when run with additional speed-up of 2, which suggests linear scalability with resource augmentation.

We believe that this study together with recent work [4,6] demonstrates algorithmic potential of the model, and will result in further fault-tolerant study of more complex scheduling problems, including deadlines, priorities, dependencies, and other features.

References

1. Ajtai, M., Aspnes, J., Dwork, C., Waarts, O.: A theory of competitive analysis for distributed algorithms. In: Proceedings of the FOCS, pp. 401–411 (1994)
2. Anantharamu, L., Chlebus, B.S., Kowalski, D.R., Rokicki, M.A.: Online parallel scheduling of non-uniform tasks: trading failures for energy. In: Proceedings of the INFOCOM, pp. 146–150 (2010)
3. Andrews, M., Zhang, L.: Scheduling over a time-varying user-dependent channel with applications to high-speed wireless data. J. ACM 52(5), 809–834 (2005)
4. Anta, A.F., Georgiou, C., Kowalski, D.R., Widmer, J., Zavou, E.: Measuring the impact of adversarial errors on packet scheduling strategies. In: Moscibroda, T., Rescigno, A.A. (eds.) SIROCCO 2013. LNCS, vol. 8179, pp. 261–273. Springer, Heidelberg (2013)
5. Anta, A.F., Georgiou, C., Kowalski, D.R., Zavou, E.: Online parallel scheduling of non-uniform tasks: trading failures for energy. In: Gasieniec, L., Wolter, F. (eds.) FCT 2013. LNCS, vol. 8070, pp. 145–158. Springer, Heidelberg (2013)
6. Anta, A.F., Georgiou, C., Kowalski, D.R., Zavou, E.: Asymptotic Competitive Analysis of Task Scheduling Algorithms on Fault-prone Machines. Manuscript (2014)
7. Awerbuch, B., Kutten, S., Peleg, D.: Competitive distributed job scheduling. In: Proceedings of the STOC, pp. 571–580 (1992)
8. Jurdzinski, T., Kowalski, D.R., Lorys, K.: Online packet scheduling under adversarial jamming. CoRR (2013)
9. Meiners, C.R., Torng, E.: Mixed criteria packet scheduling. In: Kao, M.-Y., Li, X.-Y. (eds.) AAIM 2007. LNCS, vol. 4508, pp. 120–133. Springer, Heidelberg (2007)

10. Pinedo, M.L.: Scheduling: Theory, Algorithms, and Systems. Springer, Berlin (2012)
11. Pruhs, K., Sgall, J., Torng, E.: Online Scheduling, pp. 115–124. CRC Press, Boca Raton (2003)
12. Richa, A., Scheideler, C., Schmid, S., Zhang, J.: Competitive throughput in multi-hop wireless networks despite adaptive jamming. In: Distributed Computing, pp. 1–13 (2012)
13. Sleator, D.D., Tarjan, R.E.: Amortized efficiency of list update and paging rules. Commun. ACM **28**(2), 202–208 (1985)

Generalized Hypergraph Matching via Iterated Packing and Local Ratio

Ojas Parekh[1](✉) and David Pritchard[2]

[1] Sandia National Laboratories, Albuquerque, NM 87185, USA
odparek@sandia.gov
[2] University of Southern California, Los Angeles, CA 90089, USA
dpritcha@usc.edu

Abstract. In k-hypergraph matching, we are given a collection of sets of size at most k, each with an associated weight, and we seek a maximum-weight subcollection whose sets are pairwise disjoint. More generally, in k-hypergraph b-matching, instead of disjointness we require that every element appears in at most b sets of the subcollection. Our main result is a linear-programming based $(k - 1 + \frac{1}{k})$-approximation algorithm for k-hypergraph b-matching. This settles the integrality gap when k is one more than a prime power, since it matches a previously-known lower bound. When the hypergraph is bipartite, we are able to improve the approximation ratio to $k - 1$, which is also best possible relative to the natural LP. These results are obtained using a more careful application of the *iterated packing* method.

Using the bipartite algorithmic integrality gap upper bound, we show that for the family of combinatorial auctions in which anyone can win at most t items, there is a truthful-in-expectation polynomial-time auction that t-approximately maximizes social welfare. We also show that our results directly imply new approximations for a generalization of the recently introduced bounded-color matching problem. We also consider the generalization of b-matching to *demand matching*, where edges have nonuniform demand values. The best known approximation algorithm for this problem has ratio $2k$ on k-hypergraphs. We give a new algorithm, based on local ratio, that obtains the same approximation ratio in a much simpler way.

1 Introduction

In a matching problem we want to find the maximum weight subcollection of pairwise disjoint sets within a given collection. Often these problems are studied with respect to the maximum set size k (i.e. on "k-hypergraphs"); matching is polynomial-time solvable for $k = 2$, while it is APX-hard for $k = 3$, even in special cases like *3-dimensional matching* [18].

Sandia National Laboratories is a multi-program laboratory managed and operated by Sandia Corporation, a wholly owned subsidiary of Lockheed Martin Corporation, for the U.S. Department of Energy's National Nuclear Security Administration under contract DE-AC04-94AL85000.

© Springer International Publishing Switzerland 2015
E. Bampis and O. Svensson (Eds.): WAOA 2014, LNCS 8952, pp. 207–223, 2015.
DOI: 10.1007/978-3-319-18263-6_18

The b-matching problem generalizes matching: the input specifies a limit b_v for every vertex, and we can select at most b_v sets containing each v; ordinary matching results when b is the all-1 vector. A b-matching instance can allow each set e to be selected multiple times up to some upper *capacity* limit c_e. *Simple* b-matching is the case where all capacities are unit. The *uncapacitated* case is where $c = \vec{\infty}$, i.e. there are no capacity limits.

One of our results considers the generalization of b-matching to *demand matching*, a notion originally introduced for graphs in [26]. For this problem each edge is given a demand value d_e, and we now constrain that for every vertex v, the sum of the d-values of the incident edges should be at most b_v. When d is the all-1 vector we recover the b-matching problem.

Hypergraphic matching problems are often studied via linear programming relaxations. In this paper we use only the naive LP relaxations. The worst-case ratio between the LP optimum and the optimal integral solution is called the *integrality gap*. An *LP-relative α-approximation algorithm* is one that produces (in polynomial time) an integral solution of value at least $1/\alpha$ times the LP's optimal value — this both upper bounds the integrality gap by α and gives an α-approximation algorithm. Many classical approximation algorithms are LP-relative; so the notion is not novel, rather, this terminology helps us be concise.

1.1   Results

Our main result is the following theorem.

Theorem 1. *There is an LP-relative $(k - 1 + \frac{1}{k})$-approximation algorithm for k-hypergraph b-matching, for any capacities.*

In [23] one of the authors announced, without a proof, a weaker result than the above theorem, namely an upper bound of $k-1+\frac{1}{k}$ on the integrality gap. Here we give an algorithm to find an integral solution matching this bound in polynomial time, requiring a significant extension of the techniques presented in [23].

For the special case $b = 1$, Füredi, Kahn and Seymour [15] proved an upper bound of $k - 1 + \frac{1}{k}$ in 1993, while Chan & Lau [10] recently gave the first polynomial-time algorithm matching this bound. Their technique does not directly extend to the k-hypergraph b-matching case. The technique that we use to prove Theorem 1 is *iterated packing*, the same technique from [23]. Part of the contribution of the present paper is to simplify and extend some of the approaches from [10,23]. Our main technical innovation is, using iterated packing, to explicitly specify particular additional solutions as ineligible for packing: not only solutions that would be ineligible for the *original* problem, rather we additionally prohibit solutions exceeding the ceiling of the current fractional solution.

Theorem 1 is tight for infinitely many k: when $k - 1$ is a prime power, as observed in [15], the projective plane $\mathsf{PG}(2, k-1)$ of order $k-1$ yields a matching lower bound of $k-1+\frac{1}{k}$ on the integrality gap. It is an interesting open question to settle the integrality gap for any other values of k.

We are able to determine the exact integrality gap for another interesting class of hypergraphs. Call a hypergraph *bipartite* ([1]; cf. [24]) if, for some distinguished subset U of vertices, every hyperedge contains exactly one vertex from U.

Theorem 2. *There is an LP-relative $(k-1)$-approximation algorithm for bipartite k-hypergraph b-matching, for any capacities.*

Chan and Lau [10] proved Theorem 2 in the special case that $b = 1$ and the instance is k-dimensional[1]. Proving Theorem 2 is similar to Theorem 1 plus extending an observation of [10] from k-dimensional hypergraphs to bipartite ones. Like Theorem 1, a matching integrality gap lower bound is known [14, p. 157] when $k - 1$ is a prime power: the hypergraphic dual of the affine geometry $\mathsf{AG}(2, k-1)$, i.e. a truncated projective plane, has integrality gap $k-1$.

We obtain the following interesting corollaries from the bipartite case. In the bounded-color k-hypergraph b-matching problem we are given an instance of the k-hypergraph b-matching problem along with a partition of the edge set into l color classes, $E = E_1 \cup \cdots \cup E_l$, and a positive integer w_i for $1 \le i \le l$. We seek a feasible k-hypergraph b-matching of maximum weight such that at most w_i edges from class E_i are selected for each i.

Corollary 1. *There is an LP-relative k-approximation for bounded-color k-hypergraph b-matching.*

Corollary 2. *For combinatorial auctions where each bidder can win at most $(k-1)$ items, there is a randomized polynomial-time mechanism that, in expectation, is both truthful and $(k-1)$-approximately maximizes social welfare.*

We are not aware of any prior results for this extremely natural class of combinatorial auctions, cf. [22, Chap. 12].

The proof of Corollary 2 uses the mechanism of Lavi and Swamy [21], where the distinguished vertices in the bipartite hypergraph correspond to the bidders. For this application, it is crucial that Theorem 2 gives an *LP-relative* approximation in polynomial time.

Finally, we give a new short proof of the following known theorem:

Theorem 3 [23]. *There is an LP-relative $2k$-approximation for k-hypergraph demand matching.*

Our simpler proof is based on the local ratio method, rather than the iterated packing used in [23]. We rely on a connection in [5, p. 12] between local ratio and iterated packing.

1.2   Related Work

As Tutte observed [28], both in edge-weighted graphs and in the cardinality case, uncapacitated graphic b-matching can be reduced to matching by replacing each

---

[1] A hypergraph is k-dimensional if for some k-partition of the ground set, every edge intersects every part exactly once.

vertex by b_v clones. Each edge uv is likewise cloned $b_u b_v$ times. This reduction has two problems: (1) the clones cause an exponential increase in the instance size (from $\lg \|b\|_1$ to $\|b\|_1$); and (2) it does not work in the capacitated case, since we need to prevent too many clones of the same edge from being selected. Cloning applies to hypergraphs, too, but has the same two problems. Algorithmically, we can often avoid (1) by not dealing with the clones explicity. For graphs we can fix problem (2): an edge-trisecting reduction [28] (see also [25, p. 562]) extends cloning to work on capacitated instances. But for hypergraphs, there is no known workaround for problem (2).

As a strawman, let us mention that one can reduce capacitated b-matching in k-hypergraphs to uncapacitated b-matching in $(k+1)$-hypergraphs, by inserting new vertices in each hyperedge and by moving each edge's capacity to the b value of its new vertex. One can even then apply cloning. But this is not that useful for us: e.g., we cannot use the previously-known $b = 1$ case of version Theorem 1 to even prove the nonconstructive version of Theorem 1 for general b, since this reduction increases the hyperedge size from k to $k + 1$.

Algorithmically, the *simple* (capacity $c = 1$) case of b-matching is the hardest. The proof is standard, by fixing the integer part of an optimal fractional solution.

Observation 1. *Given an (LP-relative) α-approximation to simple b-matching in k-hypergraphs, we can obtain the same quality of approximation for general capacities.*

Hypergraph Matching. Matching problems in k-uniform hypergraphs are well-studied algorithmically. For any fixed $\varepsilon > 0$ the best known approximation ratios are $\frac{k}{2} + \varepsilon$ for the unweighted version by Hurkens and Schrijver [17] and $\frac{k+1}{2} + \varepsilon$ for the weighted version by Berman [8]. In the case $k = 3$, the algorithmic results of [10] give an ε-improved approximation ratio of 2 for 3-dimensional matching. On the other hand, Hazan, Safra and Schwartz [16] showed that the problem is hard to approximate within a factor of $\Omega(\frac{k}{\log k})$ unless $\mathsf{P} = \mathsf{NP}$, even in the k-dimensional case.

Hypergraph b-Matching. For b-matching in k-hypergraphs, Krysta [20] gave a greedy $k+1$-approximation for the simple case, and Young & Koufogiannakis [19] gave a k-approximation for the uncapacitated version. Both of these approximation algorithms give LP-relative guarantees. An improvement in some cases was recently obtained by the *k-exchange system* framework of Feldman et al. [13]. The b-matchings form a *k-exchange system* (this is explicit only for $k = 2$ in [13]). In this way one can obtain a local search-based $(\frac{k+1}{2} + \varepsilon)$-approximation algorithm for weighted k-hypergraph b-matching. However, its running time is exponential in k and it does not give any LP-relative guarantee.

It may be tempting to think that the b-matching problem in hypergraphs is a simple extension of 1-matching in hypergraphs because the theory and algorithms for b-matching in graphs closely relate to those for 1-matchings. As evidenced by the results above, this does not appear to be the case. An approximation algorithm that runs in time polynomial in k with guarantee better than k

for k-hypergraph b-matching had been an open problem that we resolve with this work. Our methods are LP-based, whereas local search seems to give the best known results; however, the bounding techinques used in local search for hypergraph 1-matching do not seem to readily extend to the hypergraph b-matching case. For example, Arkin and Hassin [2] give a local search $(k - 1 + \varepsilon)$-algorithm for weighted k-hypergraph 1-matching; however, as a warmup they present a trivial bound of k — even this trivial bound does not easily extend to the k-hypergraph b-matching case.

Other Work. Pseudo-greedy methods similar to iterated packing have been successfully applied to several packing and coloring problems, including multicommodity flows on trees [11], independent sets in t-interval graphs [5], and weighted edge coloring of bipartite graphs [12]. Iterated packing is a means of obtaining an approximate convex decomposition; Carr and Vempala [9] have shown a strong connection between the latter and approximation ratios of LP-based approximation algorithms.

As mentioned earlier, a $2k$-approximation for k-hypergraph demand matching is known [23]; a better ratio of 3 is possible when $k = 2$ [23]. These nearly match (exactly match, when $k = 2$) the best known lower bound of $2k - 1$ [3] on the integrality gap of the natural LP relaxation (this construction does not require that $k - 1$ is a prime power). Bansal et al. [3] devised a deterministic $8k$-approximation and a randomized $(ek + o(k))$-approximation for the more general problem of approximating k-column-sparse packing integer programs.

Stamoulis very recently introduced the bounded-color matching problem (defined above in the more general hypergraph context) and devised a 2-approximation [27]. This result is also based on iterated packing. Stamoulis observes that the bounded-color matching problem is a special case of 3-hypergraph b-matching. In fact it is suggested in this paper that a polynomial-time $(k-1+\frac{1}{k})$-approximation for k-hypergraph b-matching may be possible. Our work was developed independently, and we observe that our results generalize Stamoulis's results, since the special hypergraph b-matching instances obtained by the reduction he suggests are bipartite, and we are able to leverage Theorem 2 to give a k-approximation for the more general bounded-color k-hypergraph b-matching problem, which we introduce here.

We will exploit the interplay between LP-relative approximation algorithms and convex decompositions — an equivalence between the two was shown by Carr & Vempala [9]. The Lavi-Swamy [21] mechanism combines techniques from [9] with the VCG mechanism.

We give an overview of iterated packing in the next section. There, we also introduce a structure theorem from [10] and its specialization to bipartite instances, versions of which will be used throughout the paper. Next, to further introduce the iterated packing methodology, we give an iterated packing proof of the same result, although it does not run in polynomial time. This is extended to b-matching in Sect. 4, which contains our main technical innovations. First an existential proof is given (Algorithm 2) and then finally Algorithm 3 proves

Theorems 1 and 2 constructively. Then in Sect. 7 we present the proof of Theorem 3, which is based on the local ratio method.

2   Iterated Packing Overview

The notion of an approximate convex decomposition is essential to iterated packing, as the latter iteratively builds such a decomposition for a given fractional solution. Here we present a slightly different notion of an approximate convex decomposition than usually considered.

Definition 1. *For $\alpha \geq 1$, define α-convex multipliers to be any collection of nonnegative reals whose sum is α. Likewise, we say that x is an α-convex combination of the points $\{x^i\}_i$ if there are α-convex multipliers $\{\lambda_i\}_i$ so that $x = \sum_i \lambda_i x^i$.*

The utility of α-convex combinations is that they provide a convenient way to talk about integrality gaps without rescaling as was done in [9] or [23].

Proposition 1 [9]. *If every feasible LP solution for a packing program can be written as an α-convex combination of integral feasible solutions, then its integrality gap is at most α.*

Proof. We need to show that for any nonnegative weight function w, if x^* is the fractional solution that maximizes $w(x^*)$, then there is an integral solution of weight at least $w(x^*)/\alpha$. A random solution from the α-convex combination representation of x^*, drawing x^i with probability λ_i/α, has expected weight $\sum_i \frac{\lambda_i}{\alpha} w(x^i) = w(x^*)/\alpha$. So one of the x^i has at least this weight.

(In fact [9] also proves an algorithmic converse, used also by the Lavi-Swamy framework [21] underlying Corollary 2.)

We will use Proposition 1 as follows: we develop a polynomial-time algorithm to write fractional hypergraph b-matchings as ρ-convex combinations of feasible integral b-matchings. Then by Proposition 1, we get the LP-relative ρ-approximation algorithm claimed in Theorems 1 and 2.

In [23] the idea of iterated packing was introduced. Each iteration, called a *packing step*, updates the current α-convex combination to a new one, increasing some terms of the combination on one coordinate.

Definition 2 (Packing Step). *Let us be given an α-convex combination $x = \sum_i \lambda_i x^i$ where the x^i are feasible integral solutions, an edge e to pack, and a target value $t \in \mathbb{R}_+$. We may think of a packing step as packing the edge e into some of the solutions x^i such that each resulting solution is still feasible and that we have packed e into solutions with a total mass of t, i.e. the sum of corresponding λ_i is t.*

Let χ_e be the vector in $\mathbb{R}^E$ with coordinate 1 on e and 0 elsewhere. A packing step will replace some $0 \leq \lambda_i' \leq \lambda_i$ portion of each x^i with $x^i + \chi_e$, where we allow $\lambda_i' > 0$ only when $x^i + \chi_e$ is feasible. Therefore $\sum_i (\lambda_i - \lambda_i') \cdot x^i + \sum_i \lambda_i' \cdot (x^i + \chi_e)$, the result of the packing step, expresses $x + t\chi_e$ as an α-convex combination of integer feasible solutions.

For a packing step to actually be feasible, it is clearly both necessary and sufficient that the set $P = \{i \mid x^i + \chi_e \text{ feasible}\}$ of solutions into which e can be packed must satisfy $\lambda(P) \geq t$.

For the sake of polynomial-time implementation of our final algorithm, note we can ensure at most one i has $\lambda'_i \notin \{0, \lambda_i\}$ in the above argument, so that each packing step increases the number of terms by at most one. Alternatively we could use Carathéodory's theorem which guarantees that any α-convex combination can be rewritten as one with at most $d+1$ terms where d is the number of coordinates.

The basic iterated packing formula starts with a fractional solution x in hand and iteratively constructs an integral solution by starting with an empty hypergraph on V. The edges are processed in some order, and for each edge e, a packing step is performed on e with a target value of x_e. One key fact about iterated packing is that when a target value is larger, packing is easier, hence iterated packing shows how large fractional values facilitate approximation for packing problems much like iterated packing does for covering problems. The basic approach may be refined in several directions. One may start with base integral solution that is non-empty hypergraph. This was explored in [23] to derived an improved approximation for the demand matching problem. Another improvement is to consider a specific ordering of edges.

This key idea driving our algorithm is analyzing an ordering of edges which allows us to obtain a polynomial-time algorithm. Although, as announced in [23], extensions of ideas from [23] may be used to derive an upper bound of $k-1+1/k$ on the integrality gap for the k-hypergraph b-matching problem, the bound is non-constructive and does not give a polynomial-time algorithm. We show that by considering an ordering of edges that was first studied by Chan and Lau [10], we obtain a polynomial-time $k-1+1/k$-approximation. This ordering is based on vertices of small degree in an extreme point solution, which in turns allows one to argue that there is an edge with large fractional value. The lemma below shows that we can find a vertex of sufficiently small degree.

Let $\{A_{v,e}\}_{v,e}$ be the 0-1 incidence matrix for our k-hypergraph: it has rows for vertices and columns for edges, with at most k ones per column. When x^* is an extreme point solution to the matching LP $\{0 \leq x \leq 1 \mid Ax \leq 1\}$, elementary properties of polyhedra show that the incidence matrix of $\{e \mid 0 < x^*_e < 1\}$ has linearly independent columns. This makes the following lemma useful: it was proven by Chan and Lau for the general case, while the bipartite case follows from generalizing their arguments about the k-dimensional case.

Lemma 1. *If the incidence vectors of $\varnothing \neq E' \subseteq E$ are linearly independent, then some vertex in (V, E') has degree between 1 and k. In the bipartite case, the upper bound can be strengthened to $k - 1$.*

Proof. The first part is a counting argument. The incidence matrix retains its rank if we delete the all-zero rows, leaving only those rows corresponding to the set V' of vertices with nonzero degree. The number of such vertices must satisfy $|V'| \geq |E'|$ or else rank $|E'|$ could not be achieved. Since each column has at most k unit entries, there are at most $k|E'|$ unit entries in the whole matrix. So averaging, some row has at most $k|E'|/|V'| \leq k$ nonzeroes, and this gives the desired vertex.

For bipartite hypergraphs, examine the situation in which equality holds. This can only happen if $|E'| = |V'|$ and the matrix has exactly k ones per row and per column. Let U be the subset of vertices so that every hyperedge intersects U exactly once. So, each hyperedge intersects the complement of U exactly $k - 1$ times. Therefore, the vector in $\mathbb{R}^V$ with $(-k - 1)$ entries in U and unit entries elsewhere is orthogonal to all rows, contradicting that the adjacency matrix has full rank.

In order to talk about both the general and bipartite cases in a unified way, define

$$\rho := \begin{cases} k - 1 + \frac{1}{k} & \text{in the general case, and} \\ k - 1 & \text{in the bipartite case.} \end{cases}$$

Additionally, define the degree bound

$$\mu := \begin{cases} k & \text{in the general case, and} \\ k - 1 & \text{in the bipartite case.} \end{cases}$$

3   Non-polynomial Time Algorithm for k-Hypergraph Matching

We now give an alternate proof that k-hypergraph matching has integrality gap of at most $k + 1 - \frac{1}{k}$. The algorithm behind this proof does not run in polynomial time. However, this section also introduces the notation and steps involved in iterated packing, which we will extend in the next section to get our main result.

Lemma 2 [23]. *In k-hypergraph matching, a packing step to bring x to $x + t\chi_e$, where $x + t\chi_e$ is a feasible fractional solution, is possible if $\alpha \geq k - (k - 1)t$.*

Proof. Let Q_v, for each $v \in e$, be the set of solutions i for which $x^i + \chi_e$ is not feasible. We have $\lambda(Q_v) \leq 1 - t$ since $x + t\chi_e$ is feasible[2]. We need room (disjoint in the worst case) for all such Q_v, plus an additional t to pack the new edge in solutions that permit it, giving the bound $k(1 - t) + t = k - (k - 1)t$.

We can indeed get large coordinates using the following strengthening of Lemma 1.

Lemma 3. *Any nonzero extreme point solution x to the k-hypergraph matching polytope has some fractional coordinate at least $1/\mu$.*

---

[2] In detail, the solutions x^i for $i \in Q_v$ have degree 1 at v, so by the definition of a convex combination $(Ax)_v = \lambda(Q_v)$, but $(Ax)_v \leq 1 - t$ since, by feasibility, $1 \geq A(x + t\chi_e)_v = (Ax)_v + t$.

Proof. If $x_e = 1$ for any coordinate then we are done, so suppose otherwise. We know from elementary linear algebra that there is a set V'' of vertices and a set E'' of edges so that x is the unique solution to $x_e = 0, \forall e \notin E''; x(\delta(v)) = 1, \forall v \in V''$. Then the same counting argument as in Lemma 1 (resp. and the same linear independence in the bipartite case) ensures that some $v \in V''$ is incident on at most k (resp. $k - 1$) edges. Since it has $x(\delta(v)) = 1$ the $e \in \delta(v)$ maximizing x_e satisfies the lemma.

Using this, we obtain an iterated packing algorithm for the k-hypergraph matching problem, which is displayed as Algorithm 1. Note that this algorithm is presented as a recursive top-down variant of iterated packing, while the basic version in the previous section was presented as a bottom-up algorithm for ease of exposition. Another more crucial deviation of this algorithm from the basic iterated packing formula is that since our analysis requires an extreme point, we must express each non-extreme solutions as convex combinations of extreme points, and we use that:

A convex combination of α-convex combinations is an α-convex combination.   (1)

In fact this is the reason the algorithm is not guaranteed to run in polynomial time; however, the algorithm does terminate since the number of nonzero coordinates of x decreases in each recursive call.

Algorithm 1. HM$^*(V, E, x)$ // write x as ρ-convex comb. of 0-1 solutions

---
1: If $x = \mathbf{0}$ return the trivial ρ-convex combination $\lambda_1 = \rho, x_1 = \mathbf{0}$.
2: If x is not an extreme point solution to $\{x \in \mathbb{R}_+^E \mid Ax \leq 1\}$,
3:     Write x as a convex combination of extreme point solutions.
4:     Recurse on each extreme point and return their result combined via (1).
5: Pick e so that x_e is maximized and let x' be x except with x'_e set to zero.
6: Recurse: $(x^i, \lambda_i)_i := \text{HM}^*(V, E, x')$.
7: Packing step: pack x_e of e into $(x^i, \lambda_i)_i$ and return the result.

Proposition 2. *Given any LP solution x, Algorithm 1 returns an expression of x as a ρ-convex combination of integral solutions.*

Proof. This follows from Lemmas 2 and 3, since if $\mu = k$ we have $k - (k-1)/\mu = k - 1 + \frac{1}{k} = \rho$, and if $\mu = k - 1$ we have $k - (k-1)/\mu = k - 1 = \rho$.

This completes the non-polynomial time iterated packing proof that the integrality gap for matching is at most ρ. Next, we extend it to b-matching.

4   Iterated Packing and k-Hypergraph b-Matching

In this section, which contains the main new iterated packing technique, we build on the ideas from the previous section. We begin with a non-constructive iterated packing algorithm to show that the integrality gap for k-hypergraph

b-matching is at most ρ. Then, we move to a constructive version via iterated packing that runs in polynomial time.

By Observation 1, we assume unit capacities (simple b-matching). We will use the following statement, whose proof is analogous to Lemma 3.

Lemma 4. *Any nonzero extreme point solution x to the k-hypergraph b-matching polytope has some fractional coordinate at least $1/\mu$.*

The naive adaptation of iterated packing (Algorithm 1) to b-matching would involve writing the input as a convex combination of extreme point solutions to $\{x \in [0,1]^E \mid Ax \le b\}$, working with α-convex combinations of integer 0-1 solutions to $Ax \le b$. However, this approach is unworkable. When we try to mimic Lemma 2, as b gets larger, we cannot bound $\lambda(Q_v)$ by anything less than 1, giving an approximation ratio of k or worse.

To fix this problem, we will enforce two additional conditions. One of these conditions, the main driver of the new proof, is that the *strengthened* degree bound $Ax^i \le \lceil Ax \rceil$ must hold in *every* level of the recursion (rather than the unworkable requirement that solutions merely respect the final target degrees). The second condition is that the λ-mass of solutions meeting this strengthened bound with equality cannot be more than $\langle (Ax)_v \rangle$ (here $\langle \cdot \rangle$ denotes the fractional part), except in the degenerate case that $(Ax)_v$ is integral. Intuitively (i) balances the number of edges packed at a vertex across the solutions x^i, avoiding the trouble that the naive approach would encounter in future iterations, while (ii) helps achieve (i) inductively. A *modified packing step* is a packing step that, given a solution (x, λ) satisfying both of these properties, produces another (x', λ') satisfying both of these properties. Then the definition of the resulting algorithm, Algorithm 2, is as follows.

Algorithm 2. HbM*(V, E, x) // write x as ρ-convex comb. of special 0-1 solutions

1: If $x = \mathbf{0}$ return the trivial ρ-convex combination $\lambda_1 = \rho, x_1 = \mathbf{0}$.   // as before
2: If x is not an extreme point solution to $\{y \in [0,1]^E \mid Ay \le \lceil Ax \rceil\}$,
3:     Write x as a convex combination of extreme point solutions.   // as before
4:     Recurse on each extreme point; return their combination via (1).   // as before
5: Pick e so that x_e is maximized; let x' be x with x'_e set to zero.   // as before
6: Recurse: $(x^i, \lambda_i)_i := $ HbM*(V, E, x').   // as before
7: Modified packing step: pack x_e of e into $(x^i, \lambda_i)_i$ and return the result.

We will prove by induction that the algorithm succeeds in finding packings meeting both conditions.

Lemma 5. *For any $0 \le x \le 1$, HbM* (V, E, x) returns an expression of x as a ρ-convex combination of 0-1 x^i that satisfies (i) $Ax^i \le \lceil Ax \rceil$ for each i, and (ii) for every v such that $(Ax)_v$ is non-integral, $\lambda(\{i \mid (Ax^i)_v = \lceil Ax_v \rceil\}) \le \langle (Ax)_v \rangle$.*

For the proof, it is helpful to realize that we use $(Ax)_v$ interchangeably as $x(\delta(v))$, and that it represents the "degree" of x at v.

Proof. The base case and the non-extreme case are easy; while the extreme points decomposing a non-extreme solution may have smaller values for $\lceil Ax \rceil$, this does not hurt us. So we only need to deal with the case that x is extreme and nonzero, where e is chosen with $x_e \geq 1/\mu$.

To prove that the modified packing step can always be carried out while satisfying (i) and (ii), we again bound a set of unpackable solutions. Specifically, our goal will be to define sets Q_v for each $v \in e$ such that any packing step that avoids adding e to any of the solutions $\bigcup_{v \in e} Q_v$ will satisfy (i) and (ii) for x, and such that the sets Q_v are λ-small enough that e always has room to be added.

For each $v \in e$, there are three cases, the main distinction being whether $\lceil (Ax')_v \rceil = \lceil (Ax)_v \rceil$. Note that these terms are either equal, or differ by one.

- Case (I), $(Ax')_v = 0$. This packing is trivial, set $Q_v = \varnothing$.
- Case (II), $\lceil (Ax)_v \rceil = \lceil (Ax')_v \rceil \neq 0$. Proving (ii) is vacuous when $(Ax)_v$ is integral, and otherwise it follows easily by induction since $\langle (Ax)_v \rangle = \langle (Ax')_v \rangle + x_e$ and at most x_e of λ-mass of solutions will have its degree increased at v. To show (i) is satisfied inductively, just like in Sect. 3, define Q_v to be the set of i with $(Ax^i)_v = \lceil (Ax)_v \rceil$; e can be added to any other x^i without violating the degree constraint. The terms $(Ax)_v$ and $(Ax')_v$ differ by x_e and have the same integer ceiling, so by induction on (ii), we have the bound $\lambda(Q_v) \leq \langle x'(\delta(v)) \rangle \leq 1 - x_e$ showing that Q_v is not too big. This bound will be used later.
- Case (III), $\lceil (Ax)_v \rceil = 1 + \lceil (Ax')_v \rceil$ and $(Ax')_v \neq 0$. Then satisfying (i) at v is easy (since all x^i have degree at most $\lceil (Ax')_v \rceil$ at v) but we must design Q_v so that (ii) is satisfied after the packing step.
If $(Ax)_v$ is integral any packing works (we can take $Q_v = \varnothing$), so assume the opposite. Moreover, when $(Ax')_v$ is integral, by (i) all solutions x^i have degree less than $\lceil (Ax)_v \rceil$ at each $v \in e$, and since we are only packing $x_e = \langle (Ax)_v \rangle$ amount of e, (ii) is also satisfied by any possible packing.
Hence, assume both $(Ax')_v$ and $(Ax)_v$ are non-integral. If we pack e arbitrarily, the total weight of new solutions with degree $\lceil (Ax)_v \rceil$ at v could be too large to satisfy (ii). Therefore, we will define Q_v to exclude some subset of the solutions $Q'_v := \{i \mid (Ax^i)_v = \lceil (Ax')_v \rceil\}$ that could rise to have this degree. We have $\lambda(Q'_v) \leq \langle (Ax')_v \rangle$ from (ii) inductively. We now define Q_v to be some subset of Q'_v with $\lambda(Q_v) = 1 - x_e$. This is not possible if $\lambda(Q'_v) < 1 - x_e$ but in this case we just define $Q_v := Q'_v$. Also, even if no subset of Q'_v has λ-value *exactly* $1 - x_e$ we can split[3] a term of the ρ-convex decomposition to achieve this. The point of this Q_v is that, using (ii) inductively, the post-packing total λ-value of the solutions with degree $\lceil (Ax)_v \rceil$ at v will be at most $\lambda(Q'_v \setminus Q_v) \leq \langle (Ax')_v \rangle - (1 - x_e) = \langle (Ax)_v \rangle$; the latter equality holds since $(Ax)_v = (Ax')_v + x_e$ and by the hypotheses of this case. So these Q_v allow us to inductively satisfy (i) and (ii), on top of which $\lambda(Q_v) \leq 1 - x_e$.

---

[3] Splitting means to replace the term (x^i, λ^i) with two terms $(x^i, p), (x^i, \lambda^i - p)$ with distributed λ-mass on the same integer solution x^i.

In all cases, $\lambda(Q_v) \leq 1 - x_e$. Analogous to Lemma 2 there is enough room to complete the packing step so long as $\rho \geq x_e + \lambda(\bigcup_{v \in e} Q_v)$. By a union bound this would be implied by $\rho \geq x_e + k(1 - x_e)$. This gives the same analysis as before (Proposition 2) in terms of our bounds on x_e and ρ, so the modified packing step succeeds and we are done.

4.1 Polynomial-Time Iterated Packing for k-Hypergraph b-Matching

Finally, we give our main algorithm. It uses modified packing steps and always maintains a ρ-convex combination satisfying the conditions of Lemma 5. As usual, the core algorithm HbM (Algorithm 3) operates on solutions where the incidence matrix A is of full column rank.

Algorithm 3. HbM(V, E, x) // write x as ρ-convex comb. of 0-1 solutions

---

Require: A has its columns linearly independent
1: If $x = \mathbf{0}$ return the trivial ρ-convex combination $\lambda_1 = \rho, x_1 = \mathbf{0}$.
2: Pick a vertex $\widehat{v}$ with minimum nonzero degree.
3: Pick $e \in \delta(\widehat{v})$ such that x_e is maximized.
4: Recurse: $(x^i, \lambda_i)_i := \text{HbM}(V, E \setminus \{e\}, x|_{E \setminus \{e\}})$.
5: Extend each x^i back to $\mathbb{R}^E$ by setting the e-coordinates to 0.
6: Modified packing step: pack x_e of e into $(x^i, \lambda_i)_i$ and return the result.

Lemma 6. *If $0 < x < 1$ and the columns of the incidence matrix A are linearly independent, HbM expresses x as a ρ-convex combination of 0-1 solutions satisfying the same properties as Lemma 5.*

Proof. The proof is very similar to proof of Lemma 5 (except we have linear independence instead of extremeness) and we therefore re-use its notation and some of the observations therein. Our goal is to show that each modified packing step succeeds. Write Q for $\bigcup_{v \in e} Q_v$. For the modified packing step to succeed we need $\lambda(Q) + x_e \leq \rho$ as before. We will use that $\lambda(Q_v) \leq (1 - x_e)$ for each v, which holds as in Lemma 5.

The first case we will handle is $|e| < k$. In this case, $\lambda(Q) + x_e \leq |e|(1 - x_e) + x_e \leq (k - 1)(1 - x_e) + x_e \leq k - 1 \leq \rho$, as needed. So we assume $|e| = k$.

Since Lemma 1 applies to our setting, the degree of $\widehat{v}$ is at most μ. The next case we will handle is $x_e \geq 1/\mu$. In this case, $\lambda(Q) + x_e \leq k(1 - x_e) + x_e = k - (k - 1)x_e \leq k - (k - 1)/\mu = \rho$ (like the proof of Proposition 2). So we may assume $x_e < 1/\mu$.

Likewise, by the definition of μ, we may assume $x(\delta(\widehat{v})) < 1$, since otherwise we fall in to the previous case by our choice of e.

Since $x(\delta(\widehat{v})) < 1$, we can get an exact expression for $Q_{\widehat{v}}$ more specific than that given in the proof of Lemma 5. All solutions x^i in the ρ-convex combination have degree 0 or 1 at v, and the latter are the ones in $Q_{\widehat{v}}$ (blocking e at $\widehat{v}$), and so $\lambda(Q_{\widehat{v}}) = (Ax')_{\widehat{v}} = (Ax)_{\widehat{v}} - x_e = x(\delta(\widehat{v})) - x_e$. This complements the upper bounds $\lambda(Q_v) \leq 1 - x_e$ that hold for all other $v \in e$ with $v \neq \widehat{v}$. This lets us bound the amount of room needed for the modified packing step:

$$x_e + \lambda(Q) \leq x_e + x(\delta(\widehat{v})) - x_e + (k-1)(1 - x_e)$$
$$\leq \mu x_e + (k-1)(1 - x_e) = k - 1 + (\mu - k + 1)x_e$$
$$\leq k - 1 + (\mu - k + 1)/\mu = k - (k-1)/\mu = \rho$$

where the middle inequality used $x(\delta(\widehat{v})) \leq \mu x_e$ and the last used $x_e < \frac{1}{\mu}$.

To complete the proofs of Theorems 1 and 2, we yet again use the approach of starting with an extreme point solution and fixing its integer part (like Observation 1), recursing only on the residual b-matching problem, which has linearly independent rows and $0 < x < 1$.

5   Application: Bounded-Color k-Hypergraph b-Matching

We observe that improved approximations for the bounded-color k-hypergraph b-matching problem, which is defined above Corollary 1, follow directly from our results. The specialization of this problem for the case of matchings in graphs was very recently introduced by Stamoulis [27], who gave a 2-approximation (note that Stamoulis had considered only matchings and not b-matchings). This independent result also leverages a variant of iterated packing. We give a k-approximation for the general case of bounded-color k-hypergraph b-matching, and thus extend the above result to hypergraphs as well as b-matchings.

Stamoulis observed that bounded-color matching is a special case of 3-hypergraph b-matching: for each color class E_i, add a new vertex c_i with capacity w_i. Now replace each edge $\{u, v\}$ with a hyperedge $\{c_i, u, v\}$. This precisely models the bounded-color matching problem. An analogous reduction shows that bounded-color k-hypergraph b-matching is a special case of standard $(k+1)$-hypergraph b-matching. To obtain our approximation, we simply observe that these special instances are bipartite, as the set U consisting of all the c_i vertices intersects every hyperedge exactly once. This gives us a k-approximation since the instance under consideration is a $(k+1)$-hypergraph.

6   Application: Allocations

We will take advantage of the Lavi-Swamy framework [21], which is a fractional version of the well-known Vickrey-Clarke-Groves (VCG) mechanism. We cannot directly use VCG in this setting, because one of the steps in VCG is to compute the allocation which maximizes the total utility of all players, and this problem is NP-complete in our setting for $t \geq 2$, by a reduction from 3-dimensional matching. The main result of Lavi and Swamy is that once we have an *LP-relative ρ-approximation algorithm* with respect to the natural LP, we can get a truthful-in-expectation mechanism, which also maximizes the expected overall utility within a factor of ρ. Minimizing this factor means we are coming closer to a VCG-like mechanism, whereas allocating everyone the empty set is truthful but a bad approximation.

First we define the natural LP relaxation for the allocation problem. Let x_S^i be a fractional indicator variable indicating whether player i will win exactly the set S of items. Then the LP requires that each player wins one set of items, and that each item is allocated at most once, fractionally. Write v_S^i as the valuation of player i for set S. Altogether the fractional allocation LP is:

$$\max \sum_{i,S} x_S^i v_S^i : 0 \le x \le 1; \forall i \in [n] : \sum_S x_S^i = 1; \forall s \in [m] : \sum_i \sum_{S: s \in S} x_S^i \le 1. \qquad (\mathcal{A})$$

We assume the input to the mechanism is an explicit list from each bidder, consisting of their valuation for each set upon which they wish to put a positive bid. The number of variables and constraints in the LP is polynomial in the number of such bids. Although for constant k, any reasonable bid language or oracle can be used, since the number of sets of size $< k$ is polynomial and we can convert everything to an explicit list.

Definition 3. *An ρ-approximate truthful-in-expectation mechanism for the allocation problem is a randomized algorithm of the following form. It takes the values v as inputs; its outputs are a valid allocation of items to players together with prices p_i charged to each player i. It has the following two properties. First, where $S(i)$ denotes the set of items allocated to player i, we have $\sum_i v_{S(i)}^i$ is at least $\sum_i v_{T(i)}^i / \rho$ for every valid allocation T. Second, for every fixed v^{-i}, a player who gives insincere valuations $\widehat{v}^i$ as their input, resulting in random variables $\widehat{p}, \widehat{S}$ compared to the original ones p, S, does not increase their expected net utility:*

$$E[v_{\widehat{S}(i)}^i - \widehat{p}_i] \le E[v_{S(i)}^i - p_i].$$

Moreover, $0 \le E[p_i] \le E[v_{S(i)}^i]$ for all i.

Theorem 4 (Lavi-Swamy [21]). *Given a polynomial-time LP-relative ρ-approximation algorithm for an allocation problem, we can obtain a polynomial-time ρ-approximate truthful-in-expectation mechanism.*

However, the allocation problem here is precisely bipartite k-hypergraph matching: for each bidder and each set of items they could win, create a set out of them all together, and this set has size at most $1 + k - 1 = k$; and each such hyperedge contains exactly one bidder, so the hypergraph is indeed bipartite. So our bipartite extension of the Chan-Lau theorem (Sect. 2) applies and we are done. The LP-relative property is essential; the non-LP relative local search approach from [13] cannot be used with [21].

7   Local Ratio and k-Hypergraph Demand Matching

We recommend [4,6,7] for background on the local ratio method, including its relationship with the primal-dual method. The heart of the local ratio approach is the following lemma:

Lemma 7 (Local Ratio Lemma). *Let x_{OPT} be the (unknown) optimal integral solution. If $w_i \cdot x_{LR} \geq w_i \cdot x_{OPT}$ for all i, and $w = \sum_i w_i$, then $w \cdot x_{LR} \geq w \cdot x_{OPT}$, i.e. x_{LR} is α-approximately optimal.*

Compared with fractional local ratio, we do not start by solving an LP, which is faster. But, we cannot use x^* to guide the algorithm — we have to ensure an oblivious approximation guarantee that holds against the unknown optimal solution.

In this section we briefly outline a reinterpretation of the $2k$-approximation for k-hypergraph demand matching from [23] as a local ratio algorithm. Compared with [23], the new algorithm will be both simpler and faster (as we solve no LPs). The inspiration for this simplified algorithm is a connection between local ratio algorithm and iterated packing elucidated by Bar-Yehuda et al. [5, p. 12].

As before, let A be the incidence matrix, and let $A[d]$ be the same matrix but with the column for each e having its entries multiplied by d_e. Then an ILP formulation for the hypergraph demand matching problem is to find an integral x maximizing wx subject to $A[d]x \leq b$ and $c \geq x \geq 0$. We will assume that $d_e \leq b_v$ whenever $v \in e$. This is without loss of generality for the purposes of approximation, while for bounding the integrality gap this *no-clipping assumption* is needed to even get a constant upper bound (even if $k = 1$, a.k.a. knapsack).

We use the same basic ideas used in [23] but arranged differently. The crux in our case is to show that for every instance, there is a hyperedge e and a weight function satisfying that any feasible solution is either $2k$-approximately optimal or has room for e to be added. With this (Lemma 8) and using the local ratio lemma, we can show that Algorithm 4 is a $2k$-approximation algorithm.

Lemma 8. *Let e be the hyperedge so that d_e is minimal. Define a weight function $\widehat{w}$ on all hyperedges by $\widehat{w}_e = 1$, and for all other f,*

$$\widehat{w}_f := \sum_{v \in e \cap f} \frac{d_f}{\max\{b_v - d_e, d_e\}}. \tag{2}$$

Then (i) every feasible solution (whether or not it contains e) has value at most $2k$ under $\widehat{w}$, (ii) $\widehat{w}_e \geq 1$, and (iii) any feasible subset of $E \setminus \{e\}$ to which e cannot be added has weight at least 1 under $\widehat{w}$.

Algorithm 4. HDM(V, E, d, b, w) // for hypergraph demand matching

---
1: Pick $e \in E$ such that d_e is minimum, or return $\varnothing$ if $E = \varnothing$.
2: Define a new weight function $\widehat{w} \in \mathbb{R}^E$ via $\widehat{w}_e = 1$ and (2) for $f \neq e$.
3: Let $w_e \widehat{w}$ be its scalar multiple by w_e, and $w' := w - w_e \widehat{w}$.   // note $w'_e = 0$
4: Define $E' := \{e \in E \mid w'_e > 0\}$.   // note $e \notin E'$
5: Recurse: $\mathcal{F}' := $ HDM$(V, E', d, b, w'|_{E'})$.
6: If $\mathcal{F}' \cup \{e\}$ is feasible define $\mathcal{F} := \mathcal{F}' \cup \{e\}$, else define $\mathcal{F} := \mathcal{F}'$.
7: Return $\mathcal{F}$.

References

1. Aharoni, R., Kessler, O.: On a possible extension of hall's theorem to bipartite hypergraphs. Discret. Math. **84**(3), 309–313 (1990)
2. Arkin, E.M., Hassin, R.: On local search for weighted k-set packing. Math. Oper. Res. **23**(3), 640–648 (1998)
3. Bansal, N., Korula, N., Nagarajan, V., Srinivasan, A.: On k-column sparse packing programs. In: Eisenbrand, F., Shepherd, F.B. (eds.) IPCO 2010. LNCS, vol. 6080, pp. 369–382. Springer, Heidelberg (2010)
4. Bar-Yehuda, R., Bendel, K., Freund, A., Rawitz, D.: Local ratio: a unified framework for approximation algorithms. ACM Comput. Surv. **36**(4), 422–463 (2004)
5. Bar-Yehuda, R., Halldórsson, M.M., Naor, J., Shachnai, H., Shapira, I.: Scheduling split intervals. SIAM J. Comput. **36**(1), 1–15 (2006)
6. Bar-Yehuda, R., Rawitz, D.: On the equivalence between the primal-dual schema and the local ratio technique. SIAM J. Discret. Math. **19**(3), 762–797 (2005)
7. Bar-Yehuda, R., Rawitz, D.: A tale of two methods. In: Goldreich, O., Rosenberg, A.L., Selman, A.L. (eds.) Theoretical Computer Science. LNCS, vol. 3895, pp. 196–217. Springer, Heidelberg (2006)
8. Berman, P.: A $d/2$ approximation for maximum weight independent set in d-claw free graphs. In: Halldórsson, M.M. (ed.) SWAT 2000. LNCS, vol. 1851, pp. 214–219. Springer, Heidelberg (2000)
9. Carr, R.D., Vempala, S.: Randomized metarounding. Random Struct. Algorithms **20**(3), 343–352 (2002)
10. Chan, Y., Lau, L.: On linear and semidefinite programming relaxations for hypergraph matching. Math. Program. **135**, 123–148 (2011)
11. Chekuri, C., Mydlarz, M., Shepherd, F.B.: Multicommodity demand flow in a tree and packing integer programs. ACM Trans. Algorithms **3**(3) (2007)
12. Feige, U., Singh, M.: Edge coloring and decompositions of weighted graphs. In: Halperin, D., Mehlhorn, K. (eds.) ESA 2008. LNCS, vol. 5193, pp. 405–416. Springer, Heidelberg (2008)
13. Feldman, M., Naor, J.S., Schwartz, R., Ward, J.: Improved approximations for k-exchange systems. In: Demetrescu, C., Halldórsson, M.M. (eds.) ESA 2011. LNCS, vol. 6942, pp. 784–798. Springer, Heidelberg (2011)
14. Füredi, Z.: Maximum degree and fractional matchings in uniform hypergraphs. Combinatorica **1**, 155–162 (1981)
15. Füredi, Z., Kahn, J., Seymour, P.D.: On the fractional matching polytope of a hypergraph. Combinatorica **13**(2), 167–180 (1993)
16. Hazan, E., Safra, S., Schwartz, O.: On the complexity of approximating k-set packing. Comput. Complex. **15**(1), 20–39 (2006)
17. Hurkens, C.A.J., Schrijver, A.: On the size of systems of sets every t of which have an SDR, with an application to the worst-case ratio of heuristics for packing problems. SIAM J. Discret. Math. **2**(1), 68–72 (1989)
18. Kann, V.: Maximum bounded 3-dimensional matching is MAX SNP-complete. Inf. Process. Lett. **37**(1), 27–35 (1991)
19. Koufogiannakis, C., Young, N.E.: Distributed fractional packing and maximum weighted b-matching via tail-recursive duality. In: Keidar, I. (ed.) DISC 2009. LNCS, vol. 5805, pp. 221–238. Springer, Heidelberg (2009)
20. Krysta, P.: Greedy approximation via duality for packing, combinatorial auctions and routing. In: Jedrzejowicz, J., Szepietowski, A. (eds.) MFCS 2005. LNCS, vol. 3618, pp. 615–627. Springer, Heidelberg (2005)

21. Lavi, R., Swamy, C.: Truthful and near-optimal mechanism design via linear programming. In: Proceedings of the 46th FOCS, pp. 595–604 (2005)
22. Nisan, N., Roughgarden, T., Tardos, É., Vazirani, V.V.: Algorithmic Game Theory. Cambridge University Press, New York (2007)
23. Parekh, O.: Iterative packing for demand and hypergraph matching. In: Günlük, O., Woeginger, G.J. (eds.) IPCO 2011. LNCS, vol. 6655, pp. 349–361. Springer, Heidelberg (2011)
24. Person, Y., Schacht, M.: Almost all hypergraphs without fano planes are bipartite. In: Proceedings of the Twentieth Annual ACM-SIAM Symposium on Discrete Algorithms, SODA 2009, pp. 217–226. Society for Industrial and Applied Mathematics, Philadelphia (2009)
25. Schrijver, A.: Combinatorial Optimization. Springer, New York (2003)
26. Shepherd, F.B., Vetta, A.: The demand-matching problem. Math. Oper. Res. **32**(3), 563–578 (2007)
27. Stamoulis, G.: Approximation algorithms for bounded color matchings via convex decompositions. In: Csuhaj-Varjú, E., Dietzfelbinger, M., Ésik, Z. (eds.) MFCS 2014, Part II. LNCS, vol. 8635, pp. 625–636. Springer, Heidelberg (2014)
28. Tutte, W.: A short proof of the factor theorem for finite graphs. Can. J. Math. **6**, 347–352 (1954)

Steiner Trees with Bounded RC-Delay

Rudolf Scheifele$^{(\boxtimes)}$

Research Institute for Discrete Mathematics, University of Bonn,
Lennéstr. 2, 53113 Bonn, Germany
scheifele@or.uni-bonn.de

Abstract. We consider the *Minimum Elmore Delay Steiner Tree Problem*, which arises as a key problem in the routing step in VLSI design: Here, we are given a set of pins located on the chip which have to be connected by metal wires in order to make the propagation of electrical signals possible. Challenging timing constraints require that these electrical signals travel as fast as possible. This is modeled as a problem of constructing a Steiner tree minimizing the *Elmore delay* [9] between a source vertex and a set of sink vertices. The problem is strongly NP-hard even for very restricted special cases, and although it is central in VLSI design (see e.g. [18]), no approximation algorithms were known until today.

In this work, we give the first constant-factor approximation algorithm. The algorithm achieves an approximation ratio of 3.39 in the rectilinear plane and 4.11 in metric graphs. We also demonstrate that our algorithm brings improvements on real world VLSI instances compared to the currently used standard method of computing short Steiner trees.

Keywords: Steiner trees · Approximation algorithm · VLSI design

1 Introduction

Due to its complexity, computing the physical layout of a modern computer chip is a task that is largely performed by automated software tools. In this physical design process many combinatorial optimization problems arise – see Held et al. for an overview [14]. In this work, we consider a problem that occurs in *routing*: Here, circuits located on the chip have to be connected by metal wires in order to allow propagation of computed information. This means that information is computed at one circuit, which sends this information from one of its *output pins* (the source) to a set of other circuits, which receive it at their *input pins* (the sinks). Finding such a connection transmitting the signal can then be formulated as a Steiner tree problem in a weighted graph or the rectlinear plane (wires never run diagonal) with pins as terminals.

Here, a signal can be regarded as a voltage change at the source pin (sender), which triggers a voltage change at the sink pins (receivers). Tight timing constraints on the chip require the difference in time between these two events, called *delay*, to be as small as possible. Since the layout of the Steiner tree

© Springer International Publishing Switzerland 2015
E. Bampis and O. Svensson (Eds.): WAOA 2014, LNCS 8952, pp. 224–235, 2015.
DOI: 10.1007/978-3-319-18263-6_19

connecting the given set of pins has a large influence on signal delay, it is natural to formulate a mathematical optimization problem asking for a Steiner tree that minimizes source-sink delays. There are numerous ways to approximate signal delays ranging from very accurate but computationally expensive simulations to very simple but imprecise estimates (e.g. signal delay is a linear function in the distance between source and sink in the Steiner tree).

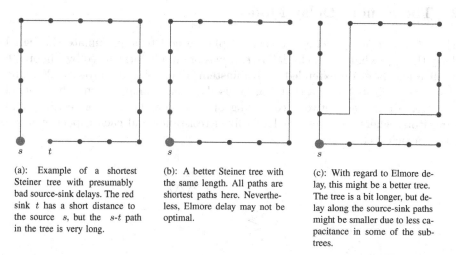

(a): Example of a shortest Steiner tree with presumably bad source-sink delays. The red sink t has a short distance to the source s, but the s-t path in the tree is very long.

(b): A better Steiner tree with the same length. All paths are shortest paths here. Nevertheless, Elmore delay may not be optimal.

(c): With regard to Elmore delay, this might be a better tree. The tree is a bit longer, but delay along the source-sink paths might be smaller due to less capacitance in some of the subtrees.

Fig. 1. Three Steiner trees for the same terminal set with probably very different source-sink delays.

When it comes to getting a fast and reasonably accurate delay approximation, the model that is ubiquitously used in VLSI design is called the Elmore delay model [9]. In a Steiner tree, the Elmore delay between a root vertex s and a sink vertex t depends on the total length of the tree, the square of the length of the path from s to t in the tree, and on the *capacitance* of each subtree rooted at the vertices of this path, which is the sum of edge lengths plus the capacitances of all sink vertices in the subtree. This makes the Elmore delay formula an objective function which is comparatively complicated to state (Fig. 1).

Although the Elmore delay model has been used for decades to evaluate signal delays of given Steiner trees, the problem of constructing a Steiner tree minimizing Elmore delay has only been approached heuristically without achieving any theoretical approximation bounds. Instead, the VLSI design community attacked this problem either by using heuristics without proven performance bounds or by simplifying the objective function to one which is better understood from a theoretical point of view, e.g. to the construction of short Steiner trees with bounded source-sink path lengths. The latter approach results in a significant loss of precision in practice and does not provide any non-trivial performance bounds in theory, as Proposition 3 will prove.

The rest of the paper will be structured as follows: Sect. 2 will contain a short introduction to the Elmore delay model. In Sect. 3 we will formally define the *Minimum Elmore Delay Steiner Tree Problem* and present an overview on previous and related work. Section 4 will then contain the first constant-factor approximation algorithm for constructing Steiner trees minimizing Elmore delay. Finally, Sect. 5 contains experimental results that show that our new algorithm brings significant improvements on real world VLSI instances.

2   The Elmore Delay Model

The *Elmore delay model* is a rather simple method to approximate the signal delay through what is called a *RC tree*. It was originally introduced by Elmore [9] in 1948 and later on extended by Rubinstein, Penfield and Horowitz [26], who also give a simple formula that can be used for fast computation. Their model is a tree structured network consisting of a discrete number of resistors and capacitors, where each resistor has a fixed resistance and each capacitor has a fixed capacitance.

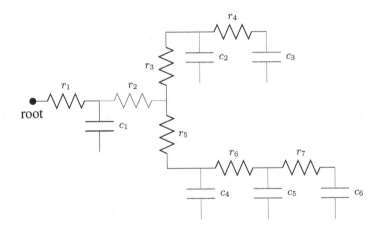

Fig. 2. An RC tree with seven resistors and six capacitors: We have $C_6 = c_5 + c_6$ and $C_2 = c_2 + c_3 + c_4 + c_5 + c_6$. Resistor 2 imposes a delay of $r_2 \cdot C_2$. The capacitors accountable for the downstream capacitance of resistor 2 (green) are shown in red. Resistors closer to the root have a higher downstream capacitance and therefore impose higher delays per resistance unit (Color figure online).

We number the k resistors and n capacitors for some $k, n \in \mathbb{N}$ consecutively with resistances $r_1, ..., r_k$ and capacitances $c_1, ..., c_n$ respectively, and let C_j for $j \in \{1, ..., k\}$ denote the sum of capacitances of all capacitors in the subtree rooted at resistor j. They show that the Elmore delay at capacitor i is then given by $\sum_{j \in I} r_j \cdot C_j$, where $I \subseteq \{1, ..., k\}$ denotes the set of resistors on the path from the root to capacitor i. Figure 2 gives an illustration of this.

We omit their definition of an RC tree at this point but rather give a graph theoretical interpretation with emphasis on our application in VLSI design. In this regard, an RC tree can be modeled as a directed Steiner tree Y with a source

s and a terminal set T, where s is the origin of the signal and the orientation of the edges corresponds to the direction in which the signal propagates. Here, the source s is regarded as a resistor with resistance $r(s) \geq 0$ and the sink vertices $t \in T$ are regarded as capacitors with capacitances $c(t) \geq 0, t \in T$. Each edge in the tree corresponds to a metal wire, which is simultaneously a resistor with resistance $R := r_{\text{wire}} \cdot l$ and capacitor with capacitance $C := c_{\text{wire}} \cdot l$, where l is the length of the wire and $r_{\text{wire}}, c_{\text{wire}} > 0$ are given constants. Steiner points do not have any resistance or capacitance.

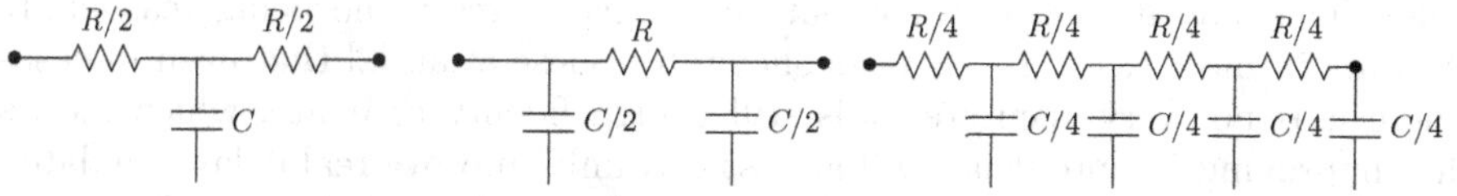

Fig. 3. Left: A wire with resistance R and capacitance C is modeled as two resistors with a capacitor in between. In terms of Elmore delay, it is equivalent to modelling it as two capacitors with a resistor in between (center). It is also the same as modelling it as k alternating resistors and capacitors with resistance R/k and C/k, respectively, and taking the limit for $k \to \infty$ (right shows $k = 4$).

To match the previous model, a wire is divided into two resistors with resistance $R/2$ and one capacitor with capacitance C in between, as shown in Fig. 3. It can be shown that in terms of Elmore delay, this is exactly the limit of dividing the wire into k alternating resistors and capacitors with resistance R/k and C/k, respectively, when k goes to infinity. Using this modelling of RC networks as Steiner trees we arrive at the following mathematical definition of Elmore delay:

Definition 1. *Given a metric space $(M, dist)$, an arborescence $Y = (V, E)$ rooted at $s \in V$ with resistance $r(s) \in \mathbb{R}_{\geq 0}$, a set of sinks $T \subseteq V$ with capacitances $c : T \to \mathbb{R}_{\geq 0}$ and vertex positions $p : V \to M$, we fix the following notation:*

- *For $v, w \in V$ we define $dist(v, w) := dist(p(v), p(w))$.*
- *Let $l(Y) := \sum_{(v,w) \in E} dist(v, w)$ denote the length of Y.*
- *For $v, w \in V$ let $P_Y(v, w)$ denote the v-w path in the underlying undirected graph of Y and $dist_Y(v, w) := \sum_{(x,y) \in P_Y(v,w)} dist(x, y)$ the distance of v and w in Y.*
- *For $v \in V$ let $Y(v)$ denote the subtree rooted at v.*

   Then the Elmore delay to $t \in T$ is defined as

$$d_Y(t) := r(s) \cdot C_Y(s) + \sum_{(v,w) \in E(P_Y(s,t))} dist(v, w) \cdot \left(\frac{dist(v, w)}{2} + C_Y(w) \right),$$

where $C_Y(v) := l(Y(v)) + \sum_{t \in V(Y(v)) \cap T} c(t)$ is said to be the downstream capacitance of $v \in V$.

We partition the Elmore delay into the terms source delay *and* wire delay, *where* $sd(Y) := r(s) \cdot C_Y(s)$ *is the source delay of* Y *and* $wd_Y(t) := \sum_{(v,w)\in E(P_Y(s,t))} dist(v,w) \cdot \left(\frac{dist(v,w)}{2} + C_Y(w)\right)$ *is the wire delay to* t *in* Y.

In our definition of Elmore delay, the constants r_{wire} and c_{wire} do not appear as they can be normalized to be 1 for the sake of mathematical simplicity. A great advantage of the Elmore delay model is that it can be computed in linear time by first computing the downstream capacitances $C_Y(v), v \in V(Y)$, in reverse topological order, and then computing the delay to all vertices in topological order. This way it is fast to compute for a given tree while being reasonably accurate in most cases. It has been shown by Boese et al. [2] that even in cases where it is not very accurate, it is still a high fidelity estimate, which means that improving Elmore delay will almost certainly improve real delay simulated by tools that are too computationally expensive to be called more often than a very few times in the VLSI design flow. For these reasons, the Elmore delay model has been the delay model of choice in VLSI design for the last decades. For more on it, see also Gupta et al. [12] or Peyer [22].

3   The Problem Formulation

Looking at the definition of Elmore delay from Sect. 2, one can see that short Steiner trees produce small source delays, while Steiner trees connecting every sink directly to the source produce small wire delays. The main difficulty is to find a good tradeoff between both extremes. We now give the problem definition:

> **Problem:** Minimum Elmore Delay Steiner Tree Problem (MDST)
>
> **Input:**   A metric space $(M, dist)$, a source s with resistance $r(s) \in \mathbb{R}_{\geq 0}$, a set of sinks T with capacitances $c : T \to \mathbb{R}_{\geq 0}$ and positions $p : \{s\} \cup T \to M$
>
> **Task:**   Find a directed Steiner tree Y rooted at s and positions $p : V(Y)\backslash(\{s\} \cup T) \to M$ minimizing
> (a) $d(Y) := \max_{t \in T} d_Y(t)$ (MAX-MDST),
> (b) $d(Y) := \sum_{t \in T} w(t) \cdot d_Y(t)$ for $w : T \to \mathbb{R}_{\geq 0}$ (SUM-MDST)

We first point out that in general metric spaces an optimum solution of the above problem does not have to exist. However, in the metric spaces that we are mainly interested in, namely metric graphs and $(\mathbb{R}^2, l_1)$, this is trivial for the former and easy to prove for the latter [27]. Secondly, we note that by setting $r(s)$ sufficiently large, the MDST problem degenerates into the Shortest Steiner Tree Problem, which is known to be NP-hard both in metric graphs and $(\mathbb{R}^2, l_1)$ [10,19]. It is actually possible to prove strong NP-hardness even for a very restricted special case of the MDST problem (see [27]):

Theorem 2. *The MDST problem is strongly NP-hard even for $|M| = 2$ or $(M, dist) = (\mathbb{R}^2, l_1)$ and all sinks have the same position.*

Having stated this theorem, we want to remark a small subtlety in the problem formulation. To express a solution, we use a tree structure that we embed into the metric space by the mapping p, which is not required to be injective. There are applications in VLSI design where this model is more useful, e.g. the global routing step, where many vertices of the original routing graph are contracted to a single vertex, resulting in a grid graph that is much smaller than the actual routing graph. In this case it makes sense to allow multiple vertices (including terminals) of the tree to be mapped to the same spot in the metric space. This is not relevant when the goal is to construct a shortest Steiner tree, but it is when trying to minimize Elmore delay. Theorem 2 uses the fact that for $|M| = 2$ the MDST problem can be regarded as a partitioning problem. For more on the VLSI routing problem see e.g. Gester et al. [11].

Previous Work: The special case of this problem where $(M, dist) = (\mathbb{R}^2, l_1)$ has received quite some attention in the past, but almost nothing is known from a theoretical point of view. To give a short summary, Boese et al. show in [4] that for the variant minimizing the weighted sum of source-sink delays there is always an optimum solution using only Steiner points on the Hanan Grid.[1] Therefore, they can solve the problem in exponential time. They also give an example in [3] that even for $|T| = 4$ the existence of optimum solutions on the Hanan Grid is generally not given for the variant minimizing maximum source-sink delay. The works of Kadodi [17] and of Peyer [22] indicate that the same statement can be made for $|T| = 2$. Both show how to solve the problem of minimizing maximum source-sink delay for instances with $|T| \leq 3$ optimally in constant time. Boese et al. [3] also give an overview on some greedy heuristics that have been evaluated in practice, but no performance bounds are proven. Finally, Peyer et al. [23] give heuristics for improving the delay of a given rectilinear Steiner tree without increasing its length. A summary of results is given by the book of Kahng and Robins [18].

Related Work: A related problem with more theoretically founded results is the construction of so called *shallow light* Steiner trees, i.e. short Steiner trees with bounded source-sink path lengths. Here, one has to mention the *Rectilinear Steiner Arborescence Problem*, where the task is to construct a minimum length shortest-path tree in the rectilinear plane for a root vertex and a set of sinks. This problem is NP-hard as was shown by Shi and Su [28], but a 2-factor approximation can be achieved using the algorithm of Rao et al. [24] with the improvements of Córdova and Lee [7]. However, this result is only of minor interest for our purpose since it can produce very long trees. More precisely, Rao et al. [24] give an example that shows that the length of a shortest rectilinear shortest-path tree can be as long as $\Omega(\log(|T|))$ times the length of a shortest rectilinear Steiner tree. A more flexible approach is that of Khuller et al. [20], which also had a highly visible influence on the development of the algorithm we are going to present in this paper. They start with a short Steiner tree Y_0

---

[1] The Hanan Grid is the grid that is induced by the set of x- and y-coordinates of the vertices in $\{s\} \cup T$ – see Hanan [13].

for terminal set $\{s\} \cup T$ and a parameter $\varepsilon > 0$ and compute a tree Y such that $dist_Y(s,t) \leq (1 + \varepsilon) \cdot dist(s,t)$ for all $t \in T$ and $l(Y) \leq (1 + \frac{2}{\varepsilon}) \cdot l(Y_0)$. Their algorithm works for general metric spaces, and in case that the metric space is $(\mathbb{R}^2, l_1)$, Held and Rotter [15] improve this result for small values of ε to produce a tree Y with $dist_Y(s,t) \leq (1+\varepsilon) \cdot dist(s,t)$ for all $t \in T$ and $l(Y) \leq (1 + \frac{2}{\varepsilon}) \cdot l(Y_0)$ for $\varepsilon > 2$ and $l(Y) \leq (2 + \lceil \log(\frac{2}{\varepsilon}) \rceil) \cdot l(Y_0)$ for $0 < \varepsilon \leq 2$. However, we want to give a trivial example showing that none of the above algorithms achieves a non-trivial approximation guarantee when applied to the MDST problem:

Proposition 3. *For any $k \in N$ and $\gamma < 1$ there is an instance of the MDST problem with $|T| = k$ and $(M, dist) = (\mathbb{R}, l_1)$ such that for every shortest Steiner tree Y we have $dist_Y(s,t) = dist(s,t)$ for all $t \in T$ and $d(Y) = \gamma k \cdot OPT$, where $d(Y)$ can be measured in any of the two given objective functions and OPT denotes the optimum objective function value in that respective function.*

4   The Algorithm

We have seen that the MDST problem is strongly NP-hard even for very restricted special cases. Now we present the first constant-factor approximation algorithm. The algorithm will require an initial solution Y_0 and a parameter $\varepsilon > 0$ as additional input, where this initial solution Y_0 should be as short as possible. It is well known that the Shortest Steiner Tree Problem can be approximated efficiently (see e.g. the work of Korte and Vygen [21] for an overview on the topic). Now here comes the description of the algorithm:

Consider an instance I of the MDST problem and let Y_0 and $\varepsilon > 0$ as described above. We may assume that Y_0 is a binary tree with root s such that the leaves of Y_0 are exactly the vertices in T.[2] We also fix the terminology that connecting a subtree $Y(v)$ for a tree Y and $v \in V(Y)$ to s by a shortest path means deleting the incoming edge of v from Y, choosing $x \in V(Y(v))$ with $dist(s,x) = \min_{w \in V(Y(v))} dist(s,w)$, connecting x to s and changing the orientation of the edges in $E(Y(v))$ such that they are directed away from s again.

Algorithm 4. *Let Y be the tree that we are constructing, initially $Y = Y_0$. We traverse the vertices of $V(Y_0) \setminus \{s\}$ in reverse topological order of $V(Y_0)$. Let $w \in V(Y_0) \setminus \{s\}$ be a vertex that we are traversing and v its predecessor in Y. We check whether $C_Y(w) + dist(v,w) \geq \frac{\varepsilon}{2} \min \{ dist(s,x) : x \in V(Y(w)) \cup \{v\} \}$ and connect $Y(w)$ to s by a shortest path if the inequality is true. The algorithm stops when all vertices in $V(Y_0) \setminus \{s\}$ have been traversed.*

We start with an obvious bound for the running time of the algorithm:

Proposition 5. *Algorithm 4 can be implemented in $\mathcal{O}(\tau)$ time, where τ denotes the time it takes to compute $dist(s,v)$ for all $v \in V(Y_0)$.*

In order to analyze the performance guarantee of the algorithm, we first give simple lower bounds that we will use to establish the quality of our solution:

---

[2] Every general Steiner tree can be transformed into such a tree in linear time by adding additional Steiner points and edges of length 0.

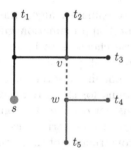 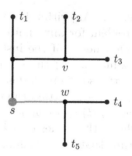

Fig. 4. Reconnection step of Algorithm 4: If $C_Y(w) + dist(v, w)$ is too large, the edge (v, w) is deleted from the tree. Connectivity is then reestablished by connecting s to a vertex of minimum distance in $Y(w)$.

Table 1. Approximation bounds of Algorithm 4 for different metric spaces: "$\mathcal{O}(n \log(n))$" means $\mathcal{O}(|T| \log |T|)$ in $(\mathbb{R}^2, l_1)$ and $\mathcal{O}(|V| \log |V| + |E|)$ if $(M, dist)$ is the metric closure of an edge-weighted graph $G = (V, E)$.

	$(\mathbb{R}^2, l_1)$	Graphs
Polynomial time	3.39 ($\varepsilon = 0.84$)	4.11 ($\varepsilon = 1.025$)
"$\mathcal{O}(n \log(n))$"	4.31 ($\varepsilon = 1.07$)	5.16 ($\varepsilon = 1.27$)

Definition 6. *Given an MDST instance, let $smt(\{s\} \cup T)$ denote the length of a shortest Steiner tree for $\{s\} \cup T$. Then $lb_{sd} := r(s) \cdot (smt(\{s\} \cup T) + \sum_{t \in T} c(t))$ is a lower bound for the source delay and $lb_{wd}(t) := dist(s, t) \cdot \left(\frac{dist(s,t)}{2} + c(t) \right)$ is a lower bound for the wire delay to $t \in T$. The sum $lb(t) := lb_{sd} + lb_{wd}(t)$ is a lower for the total delay to $t \in T$.*

Basically, we will prove that for some functions $f, g : \mathbb{R}_{>0} \to \mathbb{R}_{>0}$ the source delay of the output is bounded by $f(\varepsilon) \cdot r(s) C_{Y_0}(s)$ and the wire delay to each sink $t \in T$ by $g(\varepsilon) \cdot lb_{wd}(t)$, where f will be decreasing while g will be increasing in ε. We get the following main results:

Theorem 7. *Given an instance of the MDST problem, an initial solution Y_0 and $\varepsilon > 0$, Algorithm 4 computes a solution Y such that*

$$- sd_Y \leq \left(1 + \tfrac{2}{\varepsilon}\right) r(s) \cdot C_{Y_0}(s),$$
$$- wd_Y(t) \leq \max \left\{ (1+\varepsilon)^2, \ 1 + \tfrac{1}{16}\varepsilon^3 + \tfrac{3}{4}\varepsilon^2 + 2\varepsilon \right\} \cdot lb_{wd}(t) \ for \ all \ t \in T,$$

in $\mathcal{O}(\tau)$ time, where τ denotes the time it takes to compute $dist(s, v)$ for all $v \in V(Y_0)$.

Corollary 8. *Given an instance of the MDST problem, an initial solution Y_0 with $l(Y_0) \leq \beta \cdot smt(\{s\} \cup T)$ for some $\beta \geq 1$ and $\varepsilon > 0$, Algorithm 4 computes a tree Y such that $d_Y(t) \leq \max \left\{ (1 + \tfrac{2}{\varepsilon})\beta, \ (1+\varepsilon)^2, \ 1 + \tfrac{1}{16}\varepsilon^3 + \tfrac{3}{4}\varepsilon^2 + 2\varepsilon \right\} \cdot lb(t)$ for all $t \in T$.*

By Corollary 8, Algorithm 4 is a constant-factor approximation algorithm for the MDST problem for any choice of ε. To get the best approximation guarantee that is independent of the instance parameters, we choose ε to be (a numerical approximation of) the solution of the equation $(1 + \frac{2}{\varepsilon})\beta = (1 + \varepsilon)^2$, since for $\beta \leq 2$ the solution of this equation is small enough to never let the term $1 + \frac{1}{16}\varepsilon^3 + \frac{3}{4}\varepsilon^2 + 2\varepsilon$ attain the maximum in the bound for the wire delay.[3] This way we get $d_Y(t) \leq \alpha \cdot lb(t)$ for all $t \in T$ for a constant α depending on β, and Table 1 shows the values of α in dependence of our given metric space and the running time that we are willing to spend for the construction of the initial short Steiner tree.

For the row allowing all polynomial time algorithms we make use of the existence of a PTAS for the Shortest Steiner Tree Problem in $(\mathbb{R}^2, l_1)$ (see Arora [1] or Rao and Smith [25]) and use the algorithm of Byrka et al. [5] with an approximation ratio of 1.39 for graphs. As algorithms for the second row we use the fact that a minimum terminal spanning tree yields a 2-approximation for the Shortest Steiner Tree Problem in all metric spaces and, as proven by Hwang [16], a $\frac{3}{2}$-approximation in $(\mathbb{R}^2, l_1)$.[4] Finally, we note that we can achieve an approximation ratio of 3.39 in all metric spaces in case that the input Steiner tree is a shortest Steiner tree (see e.g. Dreyfus and Wagner [8], Vygen [29] and Chu and Wong [6] for algorithms for computing shortest Steiner trees on not too large instances). A lower bound for the best possible maximum ratio between source-sink delay and our lower bound in general metric spaces can be found in [27]. It turns out that our algorithm achieves this bound in case that the input Steiner tree is a shortest Steiner tree.

5   Experimental Results

We ran Algorithm 4 on instances of the rectilinear MDST problem extracted from current chips provided by IBM. In our experiments, we start with a short Steiner tree, apply our algorithm and compare source-sink delays of the initial Steiner tree to the ones of the result of our algorithm. Since computing short Steiner trees is today's method of choice for VLSI routing, we can expose the benefits of our new algorithm this way. Here, the initial short Steiner tree is constructed optimally for $|T| \leq 8$ using the approach of Chu and Wong [6], while it is computed by fast $\frac{3}{2}$-approximation algorithms for larger terminal sets.[5] We then apply Algorithm 4 on this tree for every value of $\varepsilon \in [0.25, 25]$ that is a multiple of 0.25, and take the solution Y with the lowest delay, where we define the delay of a Steiner tree Y as $d(Y) := \max_{t \in T} d_Y(t)/lb(t)$ throughout this section. The running time is not listed in the table because it is very small on every testcase – we can solve the 719690 instances on 45-2 in only 226 s even

---

[3] $\beta \leq 2$ can always be assumed by not using anything worse than a minimum terminal spanning tree as initial solution.

[4] A rectilinear minimum spanning tree can be computed in $\mathcal{O}(|T| \log |T|)$ time using only edges of the Delaunay Triangulation.

[5] The actual algorithm used is depending on the size of the terminal set.

though we call Algorithm 4 100 times on every instance. The machine used for our experiments is an Intel Xeon CPU running at 3.46 GHz.

Table 2. Experimental results of Algorithm 4: For a tree Y the delay of Y is defined as $d(Y) := \max_{t \in T} d_Y(t)/lb(t)$. Y denotes the output of Algorithm 4 while Y_0 denotes the initial short Steiner tree. The avg/min/max/sum values are taken over all instances on the whole chip. Instances with $|T| = 1$ are omitted because they are trivial in our setting. The initial number in a chip name denotes the technology node.

Chip	# Inst	Avg $|T|$	Avg $d(Y_0) \rightarrow d(Y)$	Min $\dfrac{d(Y)}{d(Y_0)}$	Max $d(Y_0) \rightarrow d(Y)$	$\dfrac{\sum l(Y)}{\sum l(Y_0)}$
45-1	56834	4.06	1.06 -> 1.06	0.50	4.69 -> 2.40	1.05
45-2	719690	3.81	1.18 -> 1.12	0.28	7.34 -> 2.62	1.23
32-1	400397	5.25	1.08 -> 1.07	0.40	4.99 -> 2.57	1.04
32-2	474490	4.77	1.05 -> 1.04	0.26	5.51 -> 2.50	1.04
22-1	1042	5.35	1.17 -> 1.12	0.41	3.20 -> 1.88	1.11
22-2	68247	3.88	1.12 -> 1.10	0.37	4.41 -> 2.35	1.10
14-1	29183	3.60	1.12 -> 1.09	0.38	4.38 -> 2.13	1.16
14-2	32159	3.98	1.10 -> 1.09	0.52	2.87 -> 2.09	1.09

As one can see, the average number of terminals per instance is very small on every chip. This together with the fact that the source resistance value $r(s)$ is fairly large on most instances explains why the shortest Steiner tree approach is already very close to the lower bound on average. Nevertheless, our algorithm still produces major improvements, reducing the average ratio between delay and lower bound further.

However, more important than the reduction in average delay is the reduction of the maximum ratio between delay and lower bound. Our algorithm can bound this ratio for every sink by a reasonable number, while we have connections with quite bad delays when using the shortest Steiner tree approach. This is a very desirable behaviour of our algorithm, as such outliers are likely to cause trouble in the design process. On the other hand, one must keep an eye on the increase in wiring length, which may cause routability problems on the chip. Here, one needs a better approach than the one that we used for our experiments (i.e. always taking the tree with the best delay without considering wiring length at all), e.g. bounding the allowed wiring length for a particular tree depending on timing-criticality of the sinks, which is possible in our algorithm by picking the right values of ε. However, in our experiments we just wanted to show the potential benefits of our algorithm when applied in VLSI design, and looking at the numbers in Table 2, Algorithm 4 proves to be a valuable improvement over the existing approach of exclusively using short Steiner trees to route the connections on a chip.

References

1. Arora, S.: Polynomial time approximation schemes for euclidean traveling salesman and other geometric problems. J. ACM **45**, 753–782 (1998)
2. Boese, K.D., Kahng, A.B., McCoy, B.A., Robins, G.: Fidelity and near-optimality of elmore-based routing constructions. In: IEEE International Conference on Computer Design, pp. 81–84 (1993)
3. Boese, K.D., Kahng, A.B., McCoy, B.A., Robins, G.: Rectilinear steiner trees with minimum elmore delay. In: Proceedings of the 31st Annual Design Automation Conference, pp. 381–386. ACM, New York (1994)
4. Boese, K.D., Kahng, A.B., McCoy, B.A., Robins, G.: Near-optimal critical sink routing tree constructions. IEEE Trans. Comput.-Aided Des. Integr. Circuits Syst. **14**, 1417–1436 (1995)
5. Byrka, J., Grandoni, F., Rothvoss, T., Sanità, L.: Steiner tree approximation via iterative randomized rounding. J. ACM **60**, 6:1–6:33 (2013)
6. Chu, C., Wong, Y.C.: FLUTE: fast lookup table based rectilinear steiner minimal tree algorithm for VLSI design. IEEE Trans. Comput.-Aided Des. Integr. Circuits Syst. **27**, 70–83 (2008)
7. Córdova, J., Lee, Y.: A heuristic algorithm for the rectilinear steiner arborescence problem. Technical report, University of Puerto Rico, Computer Science Department (1994)
8. Dreyfus, S., Wagner, R.: The steiner problem in graphs. Networks **1**, 195–207 (1972)
9. Elmore, W.: The transient response of damped linear networks with particular regard to wideband amplifiers. J. Appl. Phys. **19**, 55–63 (1948)
10. Garey, M.R., Johnson, D.S.: The rectilinear steiner tree problem is NP-complete. SIAM J. Appl. Math. **32**, 826–834 (1977)
11. Gester, M., Müller, D., Nieberg, T., Panten, C., Schulte, C., Vygen, J.: BonnRoute: algorithms and data structures for fast and good VLSI routing. ACM Trans. Des. Autom. Electron. Syst. **18**, 32:1–32:24 (2013)
12. Gupta, R., Tutuianu, B., Pileggi, L.: The elmore delay as a bound for RC trees with generalized input signals. IEEE Trans. Comput.-Aided Des. Integr. Circuits Syst. **16**, 95–104 (1997)
13. Hanan, M.: On Steiner's problem with rectilinear distance. SIAM J. Appl. Math. **14**, 255–265 (1966)
14. Held, S., Korte, B., Rautenbach, D., Vygen, J.: Combinatorial optimization in VLSI design. In: Combinatorial Optimization: Methods and Applications, pp. 33–96. IOS Press, Amsterdam (2011)
15. Held, S., Rotter, D.: Shallow-light steiner arborescences with vertex delays. In: Goemans, M., Correa, J. (eds.) IPCO 2013. LNCS, vol. 7801, pp. 229–241. Springer, Heidelberg (2013)
16. Hwang, F.: On steiner minimal trees with rectilinear distance. SIAM J. Appl. Math. **30**, 104–114 (1976)
17. Kadodi, T.: Steiner routing based on elmore delay model for minimizing maximum propagation delay. Master's thesis, Japan Advanced Institute of Science and Technology (1999)
18. Kahng, A., Robins, G.: On Optimal Interconnections for VLSI. Kluwer Academic Publishers, Boston (1995)
19. Karp, R.: Reducibility among combinatorial problems. In: Miller, R., Thatcher, J. (eds.) Complexity of Computer Computations, pp. 85–103. Plenum Press, New York (1972)

20. Khuller, S., Raghavachari, B., Young, N.: Balancing minimum spanning and short-est path trees. In: Proceedings of the Fourth Annual ACM-SIAM Symposium on Discrete Algorithms, pp. 243–250. Society for Industrial and Applied Mathematics, Philadelphia (1993)

21. Korte, B., Vygen, J.: Combinatorial Optimization: Theory and Algorithms, 5th edn. Springer, Heidelberg (2012)

22. Peyer, S.: Elmore-delay-optimale Steinerbäume and VLSI-Design. Diploma's the-sis (in german), Research Institute for Discrete Mathematics, University of Bonn (2000)

23. Peyer, S., Zachariasen, M., Jørgensen, D.G.: Delay-related secondary objectives for rectilinear steiner minimum trees. Discret. Appl. Math. **136**, 271–298 (2004)

24. Rao, S., Sadayappan, P., Hwang, F., Shor, P.: The rectilinear steiner arborescence problem. Algorithmica **7**, 277–288 (1992)

25. Rao, S.B., Smith, W.D.: Approximating geometrical graphs via "spanners" and "banyans". In: Proceedings of the Thirtieth Annual ACM Symposium on Theory of Computing, pp. 540–550. ACM, New York (1998)

26. Rubinstein, J., Penfield, P., Horowitz, M.A.: Signal delay in RC tree networks. IEEE Trans. Comput.-Aided Des. Integr. Circuits Syst. **2**, 202–211 (1983)

27. Scheifele, R.: Steiner trees with bounded elmore delay. Master's thesis, Research Institute for Discrete Mathematics, University of Bonn (2013)

28. Shi, W., Su, C.: The rectilinear steiner arborescence problem is NP-complete. SIAM J. Comput. **35**, 729–740 (2005)

29. Vygen, J.: Faster algorithm for optimum steiner trees. Inf. Process. Lett. **111**, 1075–1079 (2011)

Multiprocessor Jobs, Preemptive Schedules, and One-Competitive Online Algorithms

Jiří Sgall[1]([✉]) and Gerhard J. Woeginger[2]

[1] Computer Science Institute of Charles University, Praha, Czech Republic
sgall@iuuk.mff.cuni.cz
[2] Department of Mathematics and Computer Science,
TU Eindhoven, Eindhoven, The Netherlands
gwoegi@win.tue.nl

Abstract. We study online preemptive makespan minimization on m parallel machines, where the (multiprocessor) jobs arrive over time and have widths from some fixed set $W \subseteq \{1, 2, \ldots, m\}$. For every number m of machines we concisely characterize all the sets W for which there is a 1-competitive *fully* online algorithm and all the sets W for which there is a 1-competitive *nearly* online algorithm.

1 Introduction

In a multiprocessor job system, jobs may occupy several machines in parallel. For instance, concurrent threads of some parallel application may be run simultaneously and thereby block several machines. Multiprocessor scheduling problems have been studied extensively over the last three decades. The papers [4,5] by Drozdowski provide an excellent introduction to the area.

The Considered Scheduling Model. We investigate the following problem of scheduling n multiprocessor jobs $J_1, \ldots, J_n$ on m identical machines. Every job J_j ($j = 1, \ldots, n$) has a *length* $p(J_j)$, a *width* $w(J_j)$, and a *release date* $r(J_j)$. Job J_j enters the system at time $r(J_j)$, and requests simultaneous processing on exactly $w(J_j)$ machines for a total of $p(J_j)$ time units. Preemption is allowed: the processing of a job can be interrupted at any moment in time, and can be resumed at any later moment on the same set of machines or on another set of machines, however this set always has to contain exactly $w(J_j)$ machines. Every machine can process at most one job at a time. The goal is to minimize the largest job completion time, which is called the *makespan* of the schedule. In the three-field notation this problem is denoted $P|\text{pmtn}, \text{size}_j, r_j|C_{\max}$.

The computational complexity of this problem is well-understood. If the number m of machines is a fixed constant, then $Pm|\text{pmtn}, \text{size}_j, r_j|C_{\max}$ can be formulated as a linear program of polynomial size and hence is solvable in polynomial time; see Blazewicz et al. [2]. If the number m of machines is part of the input, then the problem is NP-hard even in the absence of job release dates; see Drozdowski [3]. The survey paper [4] by Drozdowski summarizes the complexity landscape around scheduling multiprocessor jobs. On the approximation

© Springer International Publishing Switzerland 2015
E. Bampis and O. Svensson (Eds.): WAOA 2014, LNCS 8952, pp. 236–247, 2015.
DOI: 10.1007/978-3-319-18263-6_20

side, Johannes [8] and independently Naroska and Schwiegelshohn [11] show that
the List Scheduling algorithm yields an approximation ratio of 2.

Online Algorithms. In the current paper we are mainly interested in the online
version of $P|\text{pmtn}, \text{size}_j, r_j|C_{\max}$ where jobs arrive over time; see for instance
Sgall [14] or Pruhs et al. [12] for surveys of the standard online scheduling scenarios. The scheduler learns about job J_j at time $r(J_j)$, and immediately receives
full knowledge of the length and the width of the job. At any moment in time
the online scheduler decides which of the available jobs should be processed on
which of the machines. An online algorithm is *c-competitive*, if for all instances
the online makespan is at most a multiplicative factor c above the optimal offline
makespan. The *competitive ratio* of an online algorithm is the smallest real c for
which the algorithm is c-competitive.

As jobs may be preempted, there arises a delicate distinction between *fully*
online and *nearly* online algorithms. A *fully* online algorithm makes all its decisions based on the sole knowledge of the jobs that have arrived up to the current
moment. A *nearly* online algorithm additionally knows the arrival time of the
next job arriving in the future. As a nearly online algorithm has more information than a fully online algorithm, nearly online algorithms may possibly reach
better competitive ratios than fully online algorithms. However, the actual difference between these two concepts is very small. If one allows some form of
time-sharing, for example an infinite number of preemptions and infinitesimally
small preempted job pieces, then fully online and nearly online algorithms are
essentially equivalent: The best possible fully online competitive ratio then equals
the best possible nearly online competitive ratio. If on the other hand one only
allows a finite number of preemptions, then fully online algorithms in general
are slightly weaker than nearly online algorithms. Nevertheless, whenever there
exists a c-competitive nearly online algorithm, then for every $\varepsilon > 0$ there also
exists a $(c + \varepsilon)$-competitive fully online algorithm.

Hong and Leung [7] construct a 1-competitive *fully* online algorithm for
$P|\text{pmtn}, r_j|C_{\max}$, that is, for the variant where all widths are 1 and where every
job requests processing on a single machine. (The online algorithm in [7] also
knows the optimal offline makespan from the very beginning, but it only uses
this knowledge to stop in an early stage if it detects an infeasibility.) Labetoulle
et al. [9] give a 1-competitive *nearly* online algorithm for problem $Q|\text{pmtn}, r_j|$
$C_{\max}$, where the machines are uniformly related. Vestjens [15] strengthens this
result: he provides a concise analysis of $Q|\text{pmtn}, r_j|C_{\max}$ and characterizes all
combinations of the machine speeds for which there exists a 1-competitive *fully*
online algorithm. Very recently, Guo and Kang [6] constructed a 1-competitive
fully online algorithm for the two-machine problem $P2|\text{pmtn}, \text{size}_j, r_j|C_{\max}$. On
the negative side, Johannes [8] proves that no (fully or nearly) online algo-
rithm for the general problem $P|\text{pmtn}, \text{size}_j, r_j|C_{\max}$ can have a competitive
ratio below $6/5$.

Contribution of this Paper. We analyze the problem of $P|\text{pmtn}, \text{size}_j, r_j|C_{\max}$
where all job widths belong to some fixed subset $W \subseteq \{1, 2, \ldots, m\}$ a priori
known to the scheduler. For every number m of machines we characterize all the

sets W for which there is a 1-competitive *fully* online algorithm and all the sets W for which there is a 1-competitive *nearly* online algorithm.

This generalizes the two 1-competitive fully online algorithms mentioned above: the algorithm of Hong and Leung [7] which covers the cases with $W = \{1\}$ and the algorithm of Guo and Kang [6] which covers the case $W = \{1, 2\}$ and $m = 2$.

Statement of Main Result. For a number m of machines and a width w, we define the *rank* of w relative to m as $R(w, m) = \lfloor m/w \rfloor$. In other words, $R(w, m)$ denotes the maximum number of jobs of width w that can be processed simultaneously on m machines.

A width w is called *fat* for m machines if $w > m/2$ (so that w has rank 1), and it is called *skinny* if $w \le m/2$ (so that w has rank at least 2). For a set $W \subseteq \{1, 2, \ldots, m\}$, we denote by $W^- \subseteq W$ its skinny elements and by $W^+ \subseteq W$ its fat elements. Jobs are called fat respectively skinny if their width is fat respectively skinny.

Theorem 1.1. *Let $m \ge 1$ be the number of machines and let $W \subseteq \{1, 2, \ldots, m\}$ be the set of possible job widths. There exists a 1-competitive nearly online scheduling algorithm on m identical parallel machines with job widths in W, if and only if the following two conditions are both fulfilled:*

(c1) *All $a, b \in W^-$ satisfy $R(a, m) = R(b, m)$; in other words, all skinny widths in W have the same rank relative to m.*

(c2) *All $a, b \in W^-$ and all $c \in W^+$ satisfy $R(a, m - c) = R(b, m - c)$; in other words, whenever a fat job blocks some of the machines, then all the skinny widths in W have the same rank relative to the number of remaining machines.*

Furthermore there exists a 1-competitive fully online scheduling algorithm on m identical parallel machines with job widths in W, if and only if conditions (c1) and (c2) together with the following condition (c3) are fulfilled:

(c3) *All $c \in W^+$ and all $a \in W^-$ satisfy $R(a, m - c) = 0$ or $R(a, m) = 2$.*

Note that conditions (c1) and (c2) guarantee that all the ranks are independent of a. Then condition (c3) gives only two possibilities: Either every fat job blocks execution of any other job, or no three jobs can be executed together (and then some fat jobs may block all other jobs, while other fat jobs may be scheduled together with any single skinny width job).

The rest of the paper is dedicated to the proof of Theorem 1.1. After providing some technical preliminaries, the four Sects. 2–5 contain the proofs of the if-parts and the only-if-parts for the characterization of the nearly and the fully online case.

Technical Preliminaries. This section collects some tools and observations that will be useful in the rest of the paper. Preemptive makespan minimization of jobs with unit-widths on parallel machines (that is, problem $P|\text{pmtn}|C_{\max}$) can be solved by the wrap-around rule of McNaughton [10]. We will apply McNaughton's classical result in the following equivalent formulation for multi-processor jobs.

Proposition 1.2. *Consider a system with m machines and n multiprocessor jobs $J_1, \ldots, J_n$, where all jobs are available at time 0 and where all job widths have the same rank R relative to m. There is a preemptive schedule that completes all jobs by time t, if and only if*

(i) *The length of every job satisfies $p(J_j) \le t$.*
(ii) *The total length of all jobs satisfies $\sum_{j=1}^{n} p(J_j) \le Rt$.*

Hong and Leung [7] gave a 1-competitive fully online algorithm for the special case of $P|\text{pmtn}, \text{size}_j, r_j|C_{\max}$ where all jobs have width 1. We will use the following equivalent formulation of their result for multiprocessor jobs.

Proposition 1.3. *The online problem of preemptively scheduling multiprocessor jobs on m machines allows a fully online 1-competitive algorithm, if all job widths have the same rank R relative to m.*

Finally, we state some observations on job widths and job ranks. If two widths a and b have the same rank r relative to m, then any combination of r jobs of width a or b can be run in parallel on m machines. We will use the following observation many times implicitly in our arguments.

Observation 1.4. *For $a, b \in \{1, 2, \ldots, m\}$ with $R(a, m) = R(b, m)$ it holds that*

(i) *$b < 2a$ and $a < 2b$;*
(ii) *$R(b, m - ka) = R(b, m) - k$ for $k = 1, \ldots, R(a, m)$.*

2 The Negative Result for Nearly Online Algorithms

In this section we prove that whenever a set W of job widths violates one of the conditions (c1) and (c2) in the statement of Theorem 1.1, then no 1-competitive *nearly* online scheduling algorithm can exist on m machines.

All our arguments are centered around the *utilization* of machines in an adversarially constructed instance: At any moment in time, the total width of the jobs run in an optimal schedule will be at least the total width of the jobs run in the online schedule, and on some non-trivial time interval the optimal total width will be strictly larger. Consequently the online makespan will be strictly worse than the optimal makespan, and the nearly online scheduler cannot be 1-competitive.

Due to the space limit we present only the case when the condition (c1) is violated. The proof when (c2) is violated, i.e., when $R(a, m - c) > R(b, m - c)$ works similarly, as we can simply block c machines by a fat job. An exception is the case when $R(b, m - c) \in \{0, 1\}$, which needs a separate construction.

Throughout this section we assume that condition (c1) is violated, and we consider $a, b \in W^-$ with $R(a, m) > R(b, m) > 1$; note that this implies $b > a$. For simplicity of presentation we introduce $r = R(b, m)$. The proof of the next lemma is omitted.

Lemma 2.1. *Let x_0 and y_0 be integers that maximize the value of $ax + by$ subject to the constraints $ax + by \le m$ and $x, y \ge 0$. Furthermore let x_1 and y_1 be integers that maximize the value of $ax + by$ subject to the constraints $ax + by \le m$ and $x \ge 1$ and $y \ge 0$. Then $x_0 \ge 1$ or $x_1 \ge 2$.*

We present an adversarial argument against an arbitrary nearly online scheduler, which is built around x_0, y_0, x_1, y_1 from Lemma 2.1. The first adversarial phase is as follows: at time 0 we confront the scheduler with $R(a, m - rb) + 1$ jobs of width a and length r, and with $(r - 1)r$ jobs of width b and length 1. The second adversarial phase starts at time $r - 1$, when a group of long jobs arrives that all have the same length $L = r + R(a, m - rb)$. If $x_0 \geq 1$ (see Lemma 2.1), then this group consists of $x_0 - 1$ jobs of width a and of y_0 jobs of width b. If $x_0 = 0$, then this group consists of $x_1 - 1 \geq 1$ jobs of width a and of y_1 jobs of width b.

The optimal schedule S_1 for the first phase continuously processes all the $R(a, m - rb) + 1$ jobs of width a together with some $r - 1$ of the jobs of width b. The makespan of this schedule S_1 equals r. The optimal schedule S_2 for all the jobs from both phases is as follows:

- During the time interval $[0, r - 1]$, schedule S_2 continuously processes $R(a, m - rb)$ jobs of width a (from the first phase) together with some r of the jobs of width b (from the first phase). Then at time $r - 1$ all jobs from the first phase are completed, with the sole exception of $r + R(a, m - rb)$ unprocessed time units of the jobs of width a.
- From time $r - 1$ to time $r - 1 + L$, schedule S_2 continuously processes x_0 jobs of width a and y_0 jobs of width b (in the case where $x_0 \geq 1$) or it continuously processes x_1 jobs of width a and y_1 jobs of width b (in the case where $x_0 = 0$). In either case the makespan of S_2 equals $r - 1 + L$.

During $[0, r - 1]$, schedule S_2 utilizes $a \cdot R(a, m - rb) + br$ machines. During the remaining time, schedule S_2 either utilizes $ax_0 + by_0$ machines (in the case $x_0 \geq 1$) or $ax_1 + by_1$ machines (in the case $x_0 = 0$). According to Lemma 2.1, this is either the globally best possible utilization (if $x_0 \geq 1$), or the best possible utilization subject to the constraint that at least one job of width a is running (if $x_0 = 0$).

How would a 1-competitive nearly online algorithm behave on this instance? As it has no knowledge on the jobs from the second phase, the online algorithm would have to follow the structure of the optimal schedule S_1 during the time interval $[0, r - 1]$. Then it utilizes $a \cdot (R(a, m - rb) + 1) + b \cdot (r - 1)$ machines, which is strictly smaller than the utilization of schedule S_2 during $[0, r - 1]$. If $x_0 \geq 1$, then during the remaining time the online scheduler cannot beat the (globally optimal) utilization of schedule S_2. If $x_0 = 0$, then the online scheduler must process one of the $x_1 - 1 \geq 1$ jobs of width a (from the second phase) from time $r - 1$ to time $r - 1 + L$; by Lemma 2.1 there is no way of beating the utilization of schedule S_2 in this case. All in all, the utilization of the online schedule is sometimes weaker but never better than that of the optimal schedule.

3   The Positive Result for Nearly Online Algorithms

In this section we design and analyze the 1-competitive *nearly* online scheduling algorithm FatMcN for all cases where the job widths in W satisfy conditions (c1) and (c2) in Theorem 1.1. Our approach applies and extends the machinery from the area, as introduced for instance in Schmidt [13], Hong and Leung [7], and Albers and Schmidt [1]. From the technical point of view our arguments are more subtle, and our results seem to reach the very limits of what can be derived for this type of online problem.

Description of the Algorithm. The main idea of our nearly online algorithm FatMcN is (i) to handle the fat jobs (Fat) with highest preference, and (ii) to fit the skinny jobs into the remaining space by using McNaughton's result (Mc) in Proposition 1.2.

Whenever there are fat jobs available, FatMcN selects a fat job of maximum width and runs it. The resulting completion times of the fat jobs together with the arrival times of all (fat or skinny) jobs constitute the so-called *critical* time points $0 = t_0 < t_1 < \cdots < t_s = C_{\max}$. (For technical reasons, we will assume that at the very end of the instance a final trivial job of length 0 is released, so that the last critical time point coincides with the optimal makespan.) For every time slot $[t_k, t_{k+1}]$ we define m_k as the maximum number of skinny jobs that can be processed simultaneously during the slot. Note that by conditions (c1) and (c2) the numbers m_k are well-defined, and that in fact *any* collection of m_k skinny jobs can be processed simultaneously at any time point during the slot. As a nearly online algorithm, FatMcN is always aware of the next critical time point.

The schedule for the skinny jobs during time slot $[t_k, t_{k+1}]$ is determined at time t_k. Let $p_1 \geq p_2 \geq \cdots \geq p_s$ denote the processing times of the skinny job pieces that are available and still need processing at time t_k. Intuitively it is clear that long job pieces should receive more processing than short job pieces. To make this intuition precise, we introduce a threshold τ whose exact value will be fixed later.

– Short job pieces with $p_j \leq \tau$ are not processed during the time slot.
– Long job pieces with $p_j > \tau$ are processed for $\min\{p_j - \tau, t_{k+1} - t_k\}$ time units.

(The value $t_{k+1} - t_k$ in the minimum expression is the length of the time slot and hence imposes a hard upper bound on the processing of any job piece during the slot.) It remains to fix the value of threshold τ. As the length of the processed job pieces $\min\{p_j - \tau, t_{k+1} - t_k\}$ decreases monotonically with τ and as we want to process the jobs as much as possible, we choose τ as the smallest non-negative real number which satisfies

$$\sum_{j:p_j>\tau} \min\{p_j - \tau, t_{k+1} - t_k\} \leq m_k \cdot (t_{k+1} - t_k). \tag{1}$$

The left-hand side of (1) denotes the total job length packed into the slot, and the right-hand side of (1) imposes the upper bound from Proposition 1.2.(ii). By Proposition 1.2 all selected job pieces can indeed be scheduled during the time slot $[t_k, t_{k+1}]$. This completes the description of algorithm FatMcN.

We conclude this section with some observations on the schedule produced by algorithm FatMcN that will be crucial in the analysis. First, we note that FatMcN maximizes the total length of skinny job pieces processed during slot $[t_k, t_{k+1}]$. Secondly, the processing of jobs during slot $[t_k, t_{k+1}]$ maintains their relative ordering with respect to their lengths:

Lemma 3.1. *Let p_i and p_j be the remaining processing times of two jobs at time t_k, and let x_i^{on} and x_j^{on} be the amounts of processing that these jobs receive from algorithm* FatMcN *during slot $[t_k, t_{k+1}]$. If $p_i \leq p_j$, then $x_i^{on} \leq x_j^{on}$ and $p_i - x_i^{on} \leq p_j - x_j^{on}$.*

Correctness of the Algorithm. We will now prove that algorithm FatMcN always minimizes the makespan and hence indeed is 1-competitive. To this end we fix an arbitrary instance and consider an optimal offline schedule S^* and the corresponding online schedule S^{on} for it. The following lemma follows by a switching argument which we omit.

Lemma 3.2. *W.l.o.g. we may assume that the optimal schedule S^* processes the fat jobs during the same time slots as schedule S^{on}.*

By Lemma 3.2 we will assume from now on that the two schedules S^* and S^{on} only differ in their handling of some skinny jobs and hence are governed by the same sequence of critical time points $t_0 < t_1 < \cdots < t_s$. Let $[t_k, t_{k+1}]$ be the earliest time slot during which schedules S^* and S^{on} disagree in processing the skinny jobs, so that at least one skinny job receives different amounts of processing in S^* and S^{on}. If schedule S^{on} processes x^{on} time units of some job during the slot while S^* processes x^* time units of the job, then we say that the two schedules have an overlap of $\min\{x^{on}, x^*\}$ with respect to this job. As a measure of progress we will use the sum of the overlaps taken over all jobs. We will show how to increase this total overlap by restructuring the optimal schedule S^*, without worsening its makespan.

First we observe that the total length of skinny jobs processed during slot $[t_k, t_{k+1}]$ in schedule S^{on} is at least as larger as in S^*: If $\tau > 0$ in the algorithm, then S^{on} has no idle time and S^* cannot fit more. If $\tau = 0$ then S^{on} schedules the maximal possible part of each skinny job and thus S^* cannot complete more, either.

First we claim that we can assume that the total length of skinny jobs processed during slot $[t_k, t_{k+1}]$ is the same in schedules S^{on} and S^*.

Next suppose that schedule S^{on} processes larger total length of skinny jobs than S^*, let the difference be z. It follows that we can move one or more skinny job pieces of a job with $x^* < x^{on}$ from some later time slot into $[t_k, t_{k+1}]$; we choose the total length of the pieces to be $\min\{x^{on} - x^*, z\}$. This increases the overlap and does not violate the conditions of Proposition 1.2, thus S^{on} may be rearranged in $[t_k, t_{k+1}]$ into a valid schedule. After a finitely many steps we have $z = 0$, as we are moving pieces of each job only once.

Now, in the remaining cases, schedules S^* and S^{on} both process exactly the same total length of skinny jobs during the slot. As the schedules differ, there exists a job J_i that during the slot receives more processing in S^* than in S^{on} and there exists another job J_j that receives more processing in S^{on} than in S^*. If we denote the corresponding four job pieces by x_i^{on} and x_i^* (for job J_i) and by x_j^{on} and x_j^* (for job J_j), then this means

$$x_i^{on} < x_i^* \quad \text{and} \quad x_j^{on} > x_j^* \tag{2}$$

We denote the remaining processing time of jobs J_i and J_j at time point t_k by p_i and p_j. As the schedules S^* and S^{on} fully agree up to time t_k, these values are the same in both schedules. The following lemma follows from the properties of FatMcN.

Lemma 3.3. *The jobs J_i and J_j satisfy $p_j - x_j^* > p_i - x_i^*$.*

By Lemma 3.3 schedule S^* must contain a non-trivial time slot $[u, v]$ with $u \geq t_{k+1}$, during which job J_j is processed continuously while job J_i is not processed at all. We choose such an interval where u and v are preemption times of some jobs and let $\varepsilon = \min\{x_i^* - x_i^{on}, v - u\}$. We switch an ε-piece of job J_i from slot $[t_k, t_{k+1}]$ with an ε-piece of job J_j in slot $[u, v]$. While this keeps schedule S^* feasible and optimal, it also improves the total overlap.

We repeatedly perform such switches until eventually the overlap covers all the processing time in the slot, so that schedule S^* agrees with schedule S^{on} on time slot $[t_k, t_{k+1}]$. To see that the process is finite, note that by the choice of u and v there is only a fixed number of intervals $[u, v]$ we can use (the number is given by the number of preemptions in the schedule), thus after a fixed number of switches the schedules S^{on} and S^* must agree on an additional job and eventually on all jobs in the time slot. Then we handle the remaining time slots in the same fashion, and eventually transform the optimal schedule S^* into the online schedule S^{on} without ever worsening the makespan. Consequently the schedule S^{on} produced by algorithm FatMcN has the optimal makespan, which means that FatMcN is 1-competitive.

4   The Negative Result for Fully Online Algorithms

In this section we show that if W violates one of the conditions (c1), (c2), (c3), then there is no 1-competitive *fully* online algorithm on m machines under the width set W. Throughout we may assume that W actually satisfies conditions (c1) and (c2), as otherwise the arguments in Sect. 2 apply and exclude the existence of a nearly and thus also of a fully online algorithm. Hence condition (c3) is violated, so that there exist $a \in W^-$ and $c \in W^+$ with $R(a, m - c) \geq 1$ and $R(a, m) \geq 3$. This condition implies that a job of width c may be replaced by two jobs of width a, or more precisely $R(a, m) \geq R(a, m - c) + 2$. If $R(a, m - c) = 1$ then this follows from $R(a, m) \geq 3$. If $R(a, m - c) \geq 2$ then $c > m/2$ implies $R(a, c) \geq R(a, m - c) \geq 2$, which yields $R(a, m) \geq R(a, c) + R(a, m - c) \geq R(a, m - c) + 2$, and the condition holds as well.

Once again we use an adversarial argument. In all possible cases, the optimal makespan of the resulting job set will be 4. The first adversarial phase confronts the scheduler at time 0 with one fat job of width c and length 2, with two skinny *crucial* jobs of width a and length 1, and with $R(a, m - c) - 1$ skinny *dummy* jobs of width a and length 4. We stress that if $R(a, m - c) = 1$ then there are no dummy jobs. It is easily verified that the optimal offline makespan for this job set is at most 4.

Next the adversary spends some time waiting and observing the actions of the 1-competitive fully online scheduler. Let $t > 0$ be the first moment in time where the online algorithm changes the collection of running jobs (by preempting a job, or by completing a job, or by starting a new job on a previously idle machine).

- During the time interval $[0, t]$, the online scheduler must continuously process all the $R(a, m - c) - 1$ dummy jobs. If $R(a, m - c) = 1$, this statement is trivial. If $R(a, m - c) \geq 2$ and no further jobs arrive, this is the only way to prevent the online makespan from exceeding the optimal makespan of 4.
- During $[0, t]$ the online scheduler must continuously process the fat job of length 2. Otherwise another fat job of width c and length 2 arrives at time 2. The optimal schedule has makespan 4, whereas the online schedule cannot complete both fat jobs by time 4.

As $c + a \cdot (R(a, m - c) - 1) + 2a > m$, the online scheduler does not have sufficient space to process both crucial skinny jobs during the time interval $[0, t]$.

The second adversarial phase starts at time t, when a skinny job of width a and length $4 - t$ arrives together with a fat job of width c and length $1 + t/2$. The optimal offline schedule still has makespan 4. Indeed, the optimal schedule uses $a \cdot (R(a, m - c) - 1)$ machines to continuously process the dummy jobs during $[0, 4]$. It uses a further machines to first process a piece of length $t/2$ of one crucial job, then a piece of length $t/2$ of the other crucial job, and finally the skinny job of length $4 - t$ that arrives at time t. It uses c machines to first process the two fat jobs during $[0, 3 + t/2]$ and then during $[3 + t/2, 4]$ the remaining pieces of length $1 - t/2$ of the two crucial jobs; this is feasible as $R(a, m) \geq R(a, m - c) + 2$.

The online scheduler, however, must block $a \cdot R(a, m - c)$ machines from time t onwards just in order to complete the dummy jobs and the job of length $4 - t$ that arrives at time t. This leaves a fat job of length 2, a fat job of length $1 + t/2$, and a crucial job of length 1 that has not been processed at all before time t. These three jobs cannot be completed on the remaining machines by time 4. Hence the makespan produced by the fully online scheduler will be above 4, and a fully online scheduler cannot be 1-competitive.

5   The Positive Result for Fully Online Algorithms

In this section we construct 1-competitive *fully* online scheduling algorithms for all the cases where the job widths in W satisfy conditions (c1), (c2), and (c3) in Theorem 1.1. We will separately discuss two scenarios. The first scenario has $R(a, m - c) = 0$ for all $a \in W^-$ and $c \in W^+$ in condition (c3). The second scenario has $R(a, m) = 2$ for some $a \in W^-$ in condition (c3).

The First Scenario. If $R(a, m - c) = 0$ holds for all $a \in W^-$ and $c \in W^+$, then no fat job can be processed simultaneously with a skinny job. Furthermore any set of $R = R(a, m)$ skinny jobs can be processed on the machines in parallel.

We sketch an online algorithm for this scenario. Whenever there are fat jobs available, we run an arbitrary fat job. Similarly as in Lemma 3.2 it can be seen that there is an optimal schedule that handles the fat jobs in exactly the same way as our fully online algorithm. However, this time we are in a simpler situation as the processing of fat jobs and the processing of skinny jobs must occur in disjoint time slots, and thus cannot interfere with each other. Hence we may use the 1-competitive fully online algorithm of Hong and Leung [7] as stated in Proposition 1.3 to schedule the skinny jobs. All in all, this yields a fully online algorithm for the first scenario.

The Second Scenario. If $R(a, m) = 2$ holds for some $a \in W^-$, then conditions (c1) and (c2) imply that $R(a, m) = 2$ for all $a \in W^-$ and that the fat jobs can be divided into two types: *very fat jobs* which cannot be processed simultaneously with any skinny job and the remaining *fat jobs* which can be processed with one arbitrary skinny job.

The main idea of our fully online algorithm TwoFatMcN is first to handle the very fat and fat jobs with high preference and then to fit the skinny jobs into the remaining space. This is easier than for FatMcN as we combine at most two jobs and then the only obstacle against balancing them exactly is if one job is longer than the total processing time of the remaining jobs.

Let at any time $p_1 \geq p_2 \geq \cdots \geq p_s$ denote the processing times of the skinny job pieces that are available and still need processing. Let P denote their total remaining time, $P = \sum_{i=1}^{s} p_s$ and let R denote the total processing time of the fat (but not very fat) job pieces that are available and still need processing.

The algorithm TwoFatMcN at each *decision time* determines the schedule for some future interval. However, whenever a new job arrives, the schedule is stopped immediately and a new decision is made. Thus the next decision time is either the next arrival time or the time when the prespecified schedule ends.

(1) If a very fat job is available, run one such job until its completion.
(2) If a fat job is available, run the first such job f (chosen by some canonical ordering). Run also one skinny job (if available) chosen as follows:
   (a) If no skinny job is available, run only f until its completion.
   (b) If $p_1 > p_2$, run the job with remaining time p_1 for time $p_1 - p_2$ or until the completion of f, whatever happens first. Also, if there is a single skinny job, run it for time p_1 or until the completion of f, whatever happens first.
   (c) If $p_1 = p_2$, run the job with remaining time p_2 for time $\min\{p_2, R/2\}$ or until the completion of f, whatever happens first.
(3) If no fat and no very fat job is available:
   (a) If there is a single skinny job run it until its completion.
   (b) Otherwise, if $p_1 > P/2$, run the job with remaining time p_1 together with one other arbitrary job until the completion of the second job.
   (c) Otherwise, create a schedule of length $P/2$ for the skinny jobs using McNaughton's rule and follow it.

It is not immediately clear that the algorithm is finite, as in steps (2b) and (2c) no job may complete or arrive. However, note that while running a single fat job f, each step (2b) is followed by step (2c). Furthermore, a job j can be run in step (2c) only once: The step (2c) takes time $R/2$ and after that, the other job with remaining time p_1 would need to run for $R/2$ before p_2 ties the longest remaining time again. However, this together would take time R which means that f is completed.

Correctness of the Algorithm. We will now prove that algorithm TwoFatMcN always minimizes the makespan and hence indeed is 1-competitive. To this end we fix an arbitrary instance and consider an optimal offline schedule S^* and the corresponding online schedule S^{on} for it. The next lemma is proven by the same exchange argument as Lemma 3.2. We omit the proof.

Lemma 5.1. *W.l.o.g. we may assume that the optimal schedule S^* processes the fat and very fat jobs during the same time slots as schedule S^{on}.*

Let at any time $Z = \max\{R, p_1, (P + R)/2\}$, taking $p_1 = 0$ if no skinny job is available. Note that Z is the length of the optimal schedule for the remaining pieces of skinny and fat jobs if no further jobs arrive. Lemma 5.1 implies that at any time, the remaining pieces of fat jobs in S^* have total length R. Let P^*, p_1^*, and Z^* denote the values P, p_1, and Z with respect to the optimal schedule S^*. The following invariant follows inductively from the definition of the algorithm, details are omitted.

Lemma 5.2. *At any time during the execution of TwoFatMcN we have*

$$P \leq P^* \quad and \quad Z \leq Z^*. \tag{3}$$

Lemmas 5.1 and 5.2 together imply that as long as S^{on} has some unfinished job, also S^* has an unfinished job. Thus the makespan of S^{on} is equal to the makespan of S^{on} and TwoFatMcN is a fully online 1-competitive algorithm.

6   Conclusions

Now that we understand the 1-competitive cases of problem $P|\text{pmtn}, \text{size}_j, r_j|C_{\max}$, the next goal should be to get a better understanding of the competitive ratios in the remaining cases. Determining the best possible competitive ratio for every possible width set W and every possible number m of machines might be a messy and hopeless enterprise. A realistic first step could be to determine the best possible competitive ratio c^* for the general online problem $P|\text{pmtn}, \text{size}_j, r_j|C_{\max}$. Currently, we only know $6/5 \leq c^* \leq 2$ from the work of Johannes [8] and Naroska and Schwiegelshohn [11].

Our main Theorem 1.1 implies that for $m = 2$ and $m = 3$ machines there exist 1-competitive online algorithms. The smallest open problems arise on $m = 4$ machines. What is the optimal competitive ratio for $m = 4$ and $W = \{1, 2\}$? What is the optimal competitive ratio for $m = 4$ and $W = \{1, 2, 3\}$?

And what if we a priori know the optimal makespan? We have observed that the algorithm of Hong and Leung uses this knowledge but not in any significant way. We know that knowing the optimal makespan cannot help us to design a 1-competitive algorithm: in all our constructions, we may as well announce the optimal makespan to be a fixed large value at the beginning and instead of ending the instance, we could release another batch of jobs that fully utilizes the machines till the announced optimal makespan. However, knowing the optimal makespan (intuitively speaking) should improve the competitive ratio.

Acknowledgements. We are grateful to Martin Böhm for the observation on the known optimum. Jiří Sgall acknowledges support by the project 14-10003S of GA ČR. Gerhard Woeginger acknowledges support by the Alexander von Humboldt Foundation, Bonn, Germany.

References

1. Albers, S., Schmidt, G.: Scheduling with unexpected machine breakdowns. Discret. Appl. Math. **110**, 85–99 (2001)
2. Blazewicz, J., Drabowski, M., Weglarz, J.: Scheduling multiprocessor tasks to minimize schedule length. IEEE Trans. Comput. **35**, 389–393 (1986)
3. Drozdowski, M.: Problems and algorithms of multiprocessor tasks scheduling. Ph.D. thesis, Technical University of Poznan, Department of Computer Science (1992)
4. Drozdowski, M.: On complexity of multiprocessor task scheduling. Bull. Pol. Acad. Sci., Tech. Sci. **43**, 381–393 (1995)
5. Drozdowski, M.: Scheduling multiprocessor tasks—an overview. Eur. J. Oper. Res. **94**, 215–230 (1996)
6. Guo, S., Kang, L.: Online scheduling of parallel jobs with preemption on two identical machines. Oper. Res. Lett. **41**, 207–209 (2013)
7. Hong, K.S., Leung, J.Y.-T.: On-line scheduling of real-time tasks. IEEE Trans. Comp. **41**, 1326–1331 (1992)
8. Johannes, B.: Scheduling parallel jobs to minimize the makespan. J. Sched. **9**, 433–452 (2006)
9. Labetoulle, J., Lawler, E.L., Lenstra, J.K., Rinnooy Kan, A.H.G.: Preemptive scheduling of uniform machines subject to release dates. In: Pulleyblank, W.R. (ed.) Progress in Combinatorial Optimization, pp. 245–261. Academic Press, New York (1984)
10. McNaughton, R.: Scheduling with deadlines and loss functions. Manag. Sci. **6**, 1–12 (1959)
11. Naroska, E., Schwiegelshohn, U.: On an on-line scheduling problem for parallel jobs. Inf. Process. Lett. **81**, 297–304 (2002)
12. Pruhs, K., Sgall, J., Torng, E.: Online scheduling. In: Leung, J.Y.T. (ed.) Handbook of Scheduling: Algorithms, Models, and Performance Analysis, Chap. 15, pp. 1–41. Chapman & Hall/CRC, Boca Raton (2004)
13. Schmidt, G.: Scheduling on semi-identical processors. Z. für Oper. Res. **28**, 153–162 (1984)
14. Sgall, J.: On-line scheduling. In: Fiat, A., Woeginger, C.J. (eds.) Online Algorithms 1996. LNCS, vol. 1442, pp. 196–231. Springer, Heidelberg (1998)
15. Vestjens, A.P.A.: Scheduling uniform machines on-line requires nondecreasing speed ratios. Math. Program. **82**, 225–234 (1998)

Routing Under Uncertainty: The *a priori* Traveling Repairman Problem

Martijn van Ee[1]([⊠]) and René Sitters[1,2]

[1] VU University Amsterdam, Amsterdam, The Netherlands
{m.van.ee,r.a.sitters}@vu.nl
[2] Centrum Voor Wiskunde En Informatica (CWI), Amsterdam, The Netherlands
r.a.sitters@cwi.nl

Abstract. The field of *a priori* optimization is an interesting subfield of stochastic combinatorial optimization that is well suited for routing problems. In this setting, there is a probability distribution over active sets, vertices that have to be visited. For a fixed tour, the solution on an active set is obtained by restricting the solution on the active set. In the well-studied *a priori* traveling salesman problem (TSP), the goal is to find a tour that minimizes the expected length. In the *a priori* traveling repairman problem (TRP), the goal is to find a tour that minimizes the expected sum of latencies. In this paper, we give the first constant-factor approximation for *a priori* TRP.

Keywords: *a priori* optimization · Approximation algorithms · Traveling repairman problem

1 Introduction

In the last few decades, a lot of research has been done in stochastic combinatorial optimization. This field is concerned with classical combinatorial optimization problems, like the shortest path problem and the minimum Steiner Tree problem, but with additional uncertainty in the instance. For example, there are situations where the problem instance changes on a daily basis. Instead of reoptimizing every instance, because it might be impossible or undesirable, one can alternatively choose to pick one solution that will be good on average. This is the setting of *a priori* optimization. In this paper, we consider the *a priori* traveling repairman problem (TRP). This is a routing problem, where there is a probability distribution over subsets of the vertices that have to be visited.

In *a priori* routing, we are given a complete weighted graph $G = (V, E)$ and a probability distribution on subsets of V. Depending on the model, this distribution is given either explicitly or by a sampling oracle. It is assumed that the instances are metric. In the first stage, a tour τ on V has to be constructed. In the second stage, an active set $A \subseteq V$ is revealed, which is the set of vertices to be visited. The second-stage tour τ_A is obtained by shortcutting the first-stage tour over the active set. For each active set, the first-stage tour has a second-stage objective value. The goal is to find a first-stage tour that minimizes the

© Springer International Publishing Switzerland 2015
E. Bampis and O. Svensson (Eds.): WAOA 2014, LNCS 8952, pp. 248–259, 2015.
DOI: 10.1007/978-3-319-18263-6_21

expected cost of the second stage tour. When it is clear form the context, we may refer to this expected second stage cost simply as the expected cost of the solution.

In the literature, several models for the probability distribution over the active sets are used. In the *black-box* model, there is no knowledge on the probability distribution. The only instrument available is a sampling oracle, which gives a sample from the distribution on request. In the *scenario* model, the instance contains an explicit list with active sets and their corresponding probabilities. In the independent decision model, each vertex has its own probability of being active, independent of the other vertices. The special case where all probabilities are equal is called the uniform model.

In the *a priori* traveling salesman problem (TSP), the goal is to minimize the expected length of the tour. The problem was introduced in the PhD-theses of Jaillet [1] and Bertsimas [2]. An approximation algorithm was achieved by Schalekamp and Shmoys [3], who showed that there is a $O(\log n)$-approximation algorithm in the black-box model. Later, Gorodezky et al. [4] showed that this bound is tight. Constant-factor approximations were achieved for the first time by Shmoys and Talwar [5], who showed that there exists a randomized 4-approximation and a deterministic 8-approximation in the independent decision model. The deterministic approximation guarantee was later improved to 6.5 by Van Zuylen [6]. It is easy to show that the randomized 4-approximation can be improved to a factor $\alpha + 2$ by replacing the double-tree subroutine in the algorithm of Shmoys and Talwar by an α-approximation algorithm for TSP. Using Christofides' algorithm [7] gives a randomized 3.5-approximation.

This paper is concerned with the *a priori* traveling repairman problem. In the deterministic traveling repairman problem or minimum latency problem, we have a complete graph $G = (V, E)$, a metric cost function c over the edges and a root vertex r. We want to find a tour τ starting at the root which minimizes the sum of latencies. Here, the latency of a vertex v is defined as the length of the path from r to v along τ. The problem is known to be NP-hard in general [8] and it is even NP-hard on weighted trees [9]. The best known approximation guarantees are 3.59 for general metrics [10] and a polynomial time approximation scheme for the Euclidean plane and weighted trees [11].

The *a priori* traveling repairman problem is defined similarly to the *a priori* traveling salesman problem. The goal is to find a first-stage tour which minimizes the expected second-stage sum of latencies. Here, the second-stage sum of latencies for active set A is obtained by shortcutting the first-stage tour over A and summing up the latencies in the second-stage tour. In this paper, we establish a constant-factor approximation for the *a priori* traveling repairman problem in the uniform model. To achieve this result, we consider the *a priori* k-TSP, the prize-collecting tour single-sink rent-or-buy problem and the *a priori* prize-collecting traveling salesman problem. These problems will be defined in their corresponding sections.

In the next section, the basic ideas for our algorithm for the *a priori* traveling repairman will be discussed. After that, it will be shown how the *a priori* k-TSP can be used to obtain a constant-factor approximation for *a priori* TRP on trees. In Sect. 4, we will discuss how to get a constant-factor approximation for the *a priori* TRP on general metrics. In order to get there, we investigate the prize-collecting tour single-sink rent-or-buy problem and the *a priori* prize-collecting traveling salesman problem. Finally, we end with some remarks on open problems.

In this paper, it is assumed that the edge costs are non-negative integers satisfying the triangle inequality. In the following, we denote A for an active set of vertices. When the sets A are drawn from the probability distribution over all sets A, we denote the expectation with respect to this distribution as $\mathbb{E}_A[\cdot]$.

2   The *a Priori* Traveling Repairman Problem

In the *a priori* traveling repairman problem, the goal is to find a first-stage tour, starting at the root, that minimizes the expected sum of latencies. Finding an approximation algorithm for this problem turns out to be much harder than for *a priori* TSP. It is easy to adjust the proof in [4] to show a $\Omega(\log n)$ lower bound on the approximation guarantee in the black-box model. Getting positive results is even non-trivial if all vertices are on a line. In the deterministic setting, this problem can be solved using dynamic programming [12]. This result relies on the fact that vertices will always be visited when the tour comes across them. In the *a priori* setting, this is not true. Consider the example from the scenario model shown in Fig. 1. Here, there is a point at v_1 at distance 1 from the root which is always active. Further, there are 100 points at v_2 at distance 10 from the root which are simultaneously active with probability 0.01, and there are 10 points at v_3 at distance 2 on the other side of the root which are simultaneously active with probability 0.1. Note that this gives four possible scenarios. Here, the optimal *a priori* tour is (v_2, v_3, v_1), meaning that we pass by the point at v_1 twice before visiting it. The intuition behind this is that we do not want to visit v_1 before v_3, but we do want to visit v_2 before v_3. We conjecture that in the independent model skipping is never optimal. If this is true, then dynamic programming may be used to solve this problem.

Fig. 1. Instance of *a priori* TRP in the scenario model. The optimal tour passes the point at v_1 twice before visiting it.

There also are difficulties in the independent decision model. The intuitive approach of using the probabilities as weights, i.e. $w_i = p_i$, and solving the weighted version of TRP turns out to give arbitrary bad solutions. The problem remains easy on star graphs. It can be shown by an interchange argument that the vertices have to be visited in non-increasing order of $\mathbb{E}[N_i]/\mathbb{E}[L_i]$. Here, $\mathbb{E}[N_i]$

is the expected number of clients at vertex i and $\mathbb{E}[L_i]$ is the expected length to vertex i, i.e. the length of the edge times the probability that at least one of the clients at the endpoint has to be visited.

2.1   Preliminaries

Any tour should start in the given root r. For a given tour and active set A, we denote ℓ_v^A as the latency of vertex $v \in A$ in the tour shortcutted over A. If vertex v is not in A, then we define $\ell_v^A = 0$. Each vertex v has probability p_v of being active. If C_v is the expected latency of vertex v given that v is active, our objective becomes minimizing

$$\mathbb{E}_A[\sum_v \ell_v^A] = \sum_v p_v \mathbb{E}_A[\ell_v^A | v \text{ is active}] = \sum_v p_v C_v. \tag{1}$$

Let $d(r,v)$ be the minimum cost of traveling from the root to vertex v. Note that C_v is the expected latency of vertex v, *given* that it is active. Hence, we obtain the following lemma.

Lemma 1. *For any tour and vertex v, we have $C_v \geq d(r,v)$.*

2.2   Algorithm

Our algorithm is based on algorithms for the deterministic TRP [10,13,14]. However, the *a priori* setting makes the problem a lot harder to solve. As explained above, even the problem on the line is non-trivial in the *a priori* setting and not known to be solvable in polynomial time. Our algorithm makes use of an (α, β)-TSP-approximator in the *a priori* setting, which is similar to the one introduced in [13]. Suppose we have an instance of *a priori* TSP and a number L. The goal is to find a tour of expected length at most L which minimizes the number of unvisited vertices. An (α, β)-TSP-approximator in the *a priori* setting will find a tour of expected length at most βL with a number of unvisited vertices at most α times the optimal number of unvisited vertices. The algorithm can be described as follows. Let $L_0 = 1$ and c a parameter to be determined later and define $L_i = L_0 c^i$. Now for each length L_i, we obtain a tour $T(L_i)$ by applying the (α, β)-TSP-approximator in the *a priori* setting. These tours will then be concatenated, i.e. we first traverse tour $T(L_0)$, then we traverse tour $T(L_1)$ and so on until all vertices are visited, where we shortcut already visited vertices. We output the resulting tour.

Theorem 1. *Given an (α, β)-TSP-approximator in the a priori setting, our algorithm with $c = 2$ is a $(8\lceil \alpha \rceil \beta + 1)$-approximation for the a priori traveling repairman problem in the uniform model, i.e. $p_v = p$ for all $v \in V$.*

Proof. Assume that α is an integer, otherwise use its ceiling as upper bound. Partition the vertices of the *algorithm's* tour in blocks of size at most α. If we renumber the vertices in our tour such that the first visited vertex is 1, the second visited vertex is 2, etc., we define the block B_x to be the subset containing the vertices $n - \alpha(x + 1) + 1, n - \alpha(x + 1) + 2, \ldots, n - \alpha x$ for $x = 0, 1, \ldots, \lceil \frac{n}{\alpha} \rceil - 1$. Let C^*_{n-x} denote the expected latency of vertex $n - x$, the $(n - x)$th vertex on the *optimal* tour, given that it is active. Now let S_i be the set of vertices with a conditional expected latency from L_{i-1} until L_i in the optimal tour. Suppose that the $(n - x)$th vertex visited by the optimal tour is in S_i, i.e. $L_{i-1} \leq C^*_{n-x} < L_i$. We know that there exists a tour visiting at least $n - x$ vertices with expected length at most $2C^*_{n-x} \leq 2L_i = L_{i+1}$, so the TSP-approximator finds a tour visiting at least $n - \alpha x$ vertices of expected length at most βL_{i+1}. This implies that each vertex $v \in B_x$ is visited in $T_0 \cup \ldots \cup T(L_{i+1})$. We can bound the conditional expected latency, denoted as C^{Alg}_v, in the following way. To get an upper bound, we assume that v is visited for the first time in $T(L_{i+1})$. Shortcut vertex v on tour $T(L_{i+1})$ and visit it after the vertices of $T(L_{i+1})$. Denote its expected latency in the new tour by C'_v and note that we have $C^{\mathrm{Alg}}_v \leq C'_v$. Finally note that the expected latency in the new tour is bounded by $\beta(L_0 + \ldots + L_{i+1}) + d(r, v)$. If we sum over all vertices in B_x, we get

$$\sum_{v \in B_x} C^{\mathrm{Alg}}_v \leq \alpha(\beta(L_0 + L_1 + \ldots + L_{i+1})) + \sum_{v \in B_x} d(r, v)$$

$$\leq 2\alpha\beta L_{i+1} + \sum_{v \in B_x} d(r, v)$$

$$= 8\alpha\beta L_{i-1} + \sum_{v \in B_x} d(r, v)$$

$$\leq 8\alpha\beta C^*_{n-x} + \sum_{v \in B_x} d(r, v).$$

If we multiply by p and sum over all blocks, we can bound the objective (1) as follows

$$\sum_{x=0}^{\lceil \frac{n}{\alpha} \rceil - 1} \sum_{v \in B_x} pC^{\mathrm{Alg}}_v \leq 8\alpha\beta \sum_{x=0}^{\lceil \frac{n}{\alpha} \rceil - 1} pC^*_{n-x} + \sum_v pd(r, v)$$

$$\leq 8\alpha\beta \sum_v pC^*_v + \sum_v pd(r, v)$$

$$\leq (8\alpha\beta + 1)\mathrm{OPT}. \qquad \square$$

This approximation guarantee might be improved by choosing another value of c, but it turns out that $c = 2$ is optimal for our analysis. We can improve the approximation factor by randomizing the starting length $L_0 = 2c^U$, where U is a random variable uniformly distributed on $[0, 1]$, and optimize over c.

Theorem 2. *Given an (α, β)-TSP-approximator in the a priori setting, our algorithm with $L_0 = 2c^U$ and $c = \mathrm{e}$ is a $(2\mathrm{e}\lceil \alpha \rceil \beta + 1)$-approximation for the a priori traveling repairman problem in the uniform model, where U is a random variable uniformly distributed on $[0, 1]$.*

Proof. Partition the vertices of the resulting tour in blocks of size at most α and renumber vertices as in Theorem 1. Suppose that $C^*_{n-x} = qc^\ell$, where $q < c$. If $q < c^U$, then there exists a path from the root with expected length at most $c^U c^\ell$ visiting at least $n - x$ vertices. This means that $T(L_\ell)$ contains at least $n - \alpha x$ vertices and is of length at most $2\beta c^U c^\ell$. So, for $v \in B_x$, we have $C_v^{\text{Alg}} \leq \beta \sum_{i=0}^{\ell} L_0 c^i + d(O, v) \leq \beta L_0 c^\ell(\frac{c}{c-1}) + d(O, v)$. In the other case, we have $q < c \leq c^U c$, so there exists a path from the root with expected length at most $c^U c^{\ell+1}$. This means that $T(L_{\ell+1})$ contains at least $n - \alpha x$ vertices and is of length at most $2\beta c^U c^{\ell+1}$. So, for $v \in B_x$, we have $C_v^{\text{Alg}} \leq \beta \sum_{i=1}^{\ell+1} L_0 c^i + d(O, v) \leq \beta L_0 c^{\ell+1}(\frac{c}{c-1}) + d(O, v)$. In the first case, we have $\log_c q \leq U \leq 1$ and we have $0 \leq U \leq \log_c q$ in the second case. Taking expectations over U gives

$$C_v^{\text{Alg}} \leq \int_{\log_c q}^{1} \left(\beta L_0 c^\ell \left(\frac{c}{c-1} \right) + d(O, v) \right) dU$$
$$+ \int_0^{\log_c q} \left(\beta L_0 c^{\ell+1} \left(\frac{c}{c-1} \right) + d(O, v) \right) dU$$
$$= \frac{2c\beta}{\ln c} C^*_{n-x} + d(O, v)$$

If we multiply by p and sum over all vertices in B_x and over all B_x, we get a bound of $\frac{2c}{\ln c}\alpha\beta + 1$. Optimizing over c gives $c = e$ and a bound of $2e\alpha\beta + 1$.   □

The algorithm can be derandomized by trying multiple values for U. This will give an approximation guarantee that is arbitrary close to $2e\alpha\beta + 1$ [14]. Note that if $\alpha = 1$, the approximator corresponds to a β-approximation for *a priori* k-TSP, the problem of finding a tour on k vertices of minimum expected length. This yields the following corollary.

Corollary 1. *If there is a γ-approximation for the a priori k-TSP, there is a $(2e\gamma + 1)$-approximation for the a priori traveling repairman problem in the uniform model.*

3   Tree Metrics

To obtain an approximation guarantee for the *a priori* TRP on trees, we use Corollary 1. Note that finding a k-tour in a tree is similar to finding a k-tree in a tree. So, in this case we can solve the *a priori* k-MST problem, in which we have to find a tree spanning k vertices such that the expected cost of the tree is minimized.

Theorem 3. *The a priori k-TSP in the uniform model on tree metrics can be solved to optimality in polynomial time.*

Proof. First, we turn the tree into a binary tree with the original vertices at the leaves, by adding vertices with probability zero and edges with cost zero. Next, we use dynamic programming to solve the *a priori* k-MST problem.

Define the function $t(v, y)$ to be the minimal expected cost of a subtree rooted at v containing y leaves. For all leaves v, we have $t(v, 0) = t(v, 1) = 0$. For a certain state (v, y), the best tree follows from a combination of z vertices of the left subtree and $y - z$ vertices from the right subtree. For a given combination, the expected cost is equal to the sum of the expected cost of the subtrees plus, for each subtree, the cost of the edge connecting v with the subtree times the probability that at least one of the vertices in the subtree is active. If we denote $\ell(v)$ and $q(v)$ for the left and right child of v respectively and $c(v, w)$ as the cost of the edge between v and w, we get the following recursive formula:

$$t(v, y) = \min_{z=0,\dots,y} \{ t(\ell(v), z) + (1 - (1 - p)^z) c(v, \ell(v))$$
$$+ t(q(v), y - z) + (1 - (1 - p)^{y-z}) c(v, q(v)) \}.$$

The optimal tree containing k vertices is the solution corresponding to $t(r, k)$, where r is the root of the tree. Note that the dynamic program needs $O(nk^2)$ time, so *a priori* k-MST (and hence k-TSP) on trees can be solved in polynomial time. □

Corollary 2. *There is a $2e + 1 \approx 6.44$-approximation for the a priori traveling repairman problem in the uniform model on trees.*

It is not clear how to generalize this result to the non-uniform case. The difficulty is that the probability that at least one vertex in the subtree is active can take exponentially many different values. On the other hand, it is easy to extend the DP above to the case where there is a constant number of different probabilities.

4   General Metrics

For general metrics, we show how to obtain an (α, β)-TSP-approximator with some constant α and β. It turns out that finding such an approximator reduces to finding an approximation algorithm for the *a priori* prize-collecting traveling salesman problem. In addition to *a priori* TSP, each vertex i has a penalty π_i. For a given active set and first-stage tour, the objective value is determined by the cost of the shortcutted tour plus the penalties of the unvisited active vertices. Again, the goal is to minimize the expected value over the active sets. Finding an approximation algorithm for this problem reduces to finding an approximation algorithm for the prize-collecting tour single-sink rent-or-buy problem (prize-collecting tour SRoB). In the single-sink rent-or-buy problem (SRoB) [16], we are given a graph $G = (V, E)$ with a metric cost function c_e on the edges. There is a client at every vertex with unit demand. We have to open a facility at some of the vertices and connect the clients to the facilities. We denote c_{ij} as the cost of the shortest path between i and j in G. Connecting facility i with client j costs c_e if $e = (i, j)$ and buying edge e costs Mc_e, where $M \geq 1$. We need to buy edges such that the open facilities are joined by a Steiner tree. The goal is to minimize the sum of connection cost and Steiner cost. In the tour single-sink rent-or-buy problem (tour SRoB), G is a complete graph. Here, edges have to be bought such that the open facilities are joined by a tour. Note that $c_{ij} = c_e$ if $e = (i, j)$. In the prize-collecting tour SRoB, it is not needed to connect every client. If client i is not connected, then we have to pay penalty π_i. The goal is to minimize the sum of connection cost, tour cost and penalty cost. In this section, we give a constant-factor approximation for the prize-collecting tour SRoB which leads to a constant-factor approximation for the *a priori* prize-collecting TSP. Finally, we will show how this results in an (α, β)-TSP-approximator in the *a priori* setting.

4.1   Prize-Collecting Tour SRoB

The prize-collecting tour SRoB has, to the best of our knowledge, not been considered explicitly in the literature. Let us first consider the problem without penalties. We can obtain a randomized 3-approximation for tour SRoB by adjusting the analysis for tour connected facility location (a generalization of tour SRoB) by Eisenbrand et al. [15]. This can be derandomized by adapting the analysis of Van Zuylen [6] to obtain a deterministic 3-approximation. However, it is not clear how to extend these results to prize-collecting SRoB. Therefore, we will use the primal-dual algorithm for SRoB by Swamy and Kumar [16] instead.

Tour SRoB. First, consider SRoB. We assume that a facility is opened at root vertex r. In the ILP-formulation below, we define x_{ij} to be 1 if j is connected to i, which will be on the tree. We define z_e to be 1 if we use edge e in the tree.

$$(P) \quad \min \quad \sum_i \sum_j c_{ij} x_{ij} + M \sum_e c_e z_e$$

$$\text{s.t.} \quad \sum_i x_{ij} \geq 1 \qquad \forall j \in V$$

$$\sum_{i \in S} x_{ij} \leq \sum_{e \in \delta(S)} z_e \qquad \forall S \subseteq V \setminus \{r\}, j \in V$$

$$x_{ij}, z_e \in \{0, 1\} \qquad \forall i, j \in V, e \in E.$$

In any solution for the tour SRoB, each vertex j is connected with some vertex i on the tour (possibly $i = j$). In that case, any cut separating i from r must contain at least two edges. Hence, an LP-relaxation for tour SRoB is obtained by relaxing the integrality constraints and by putting a factor 2 in front of x_{ij} in the second constraint.

   We can now use the primal-dual algorithm for SRoB to obtain an approximation algorithm for tour SRoB. Given an instance of tour SRoB, we divide all edge costs by 2, i.e. $c'_e = c_e/2$ and $c'_{ij} = c_{ij}/2$. To keep the remaining restrictions of the dual and the Steiner costs the same, we also set $M' = 2M$. Secondly, we use the primal-dual algorithm of Swamy and Kumar [16] on the new instance to obtain a solution for SRoB. Finally, we double the tree and shortcut the resulting Eulerian tour. Note that this algorithm and its analysis are similar to the work of

Goemans and Williamson [17], who showed how to obtain a 2-approximation for the prize-collecting TSP using a 2-approximation for the prize-collecting Steiner tree problem. Further note that this ratio is worse than the ratio that can be obtained from [15]. However, that result is based on a sampling approach which we do not know how to extend to the prize-collecting version of the problem.

Theorem 4. *The approach above gives a 5-approximation for the tour SRoB. Moreover, the value is at most 5 times the optimal value of its LP-relaxation.*

Prize-Collecting Version. In this version, it is not needed to connect all vertices. However, a penalty π_i is incurred when vertex i is not connected. For the LP-relaxation of the prize-collecting tour SRoB problem, we add the variable s_j, which is set to 1 if client j is not visited. In an integral solution, the first constraint corresponds to a client being either not visited or connected with an open facility.

$$\text{(P')} \quad \min \quad \sum_i \sum_j c_{ij} x_{ij} + M \sum_e c_e z_e + \sum_j \pi_j s_j$$

$$\text{s.t.} \quad s_j + \sum_i x_{ij} \geq 1 \qquad \forall j \in V$$

$$2 \sum_{i \in S} x_{ij} \leq \sum_{e \in \delta(S)} z_e \quad \forall S \subseteq V \setminus \{r\}, j \in V$$

$$x_{ij}, z_e, s_j \geq 0 \qquad \forall i, j \in V, e \in E.$$

Using the ellipsoid method, the LP-relaxation can be solved in polynomial time. Note that the separation problem can be solved by using a min-cut problem. Solving gives optimal solution (x^*, z^*, s^*). If $s_j^* \geq \delta$, then we set $\hat{s}_j = 1$, else we set $\hat{s}_j = 0$, where $0 \leq \delta \leq 1$ is determined later, and let $T = \{j : \hat{s}_j = 0\}$. The vertices in $V \setminus T$ will not be visited. Next, we obtain a solution of tour SRoB on T by applying the algorithm from Theorem 4. This results in a feasible solution for prize-collecting tour SRoB on V. Partition the optimal LP-value in the connection plus tour cost C_{LP} and penalty cost Π_{LP}.

Lemma 2. *The algorithm above finds a solution for the prize-collecting tour SRoB such that the resulting tour and connection costs are bounded by $5/(1 - \delta)C_{LP}$ and the resulting penalty costs are bounded by $(1/\delta)\Pi_{LP}$.*

Proof. By rounding the solution, we lose at most a factor $1/\delta$ on the penalty cost. By Theorem 4, the connection and tour cost for tour SRoB on T can be bounded by 5 times the optimal solution of its LP-relaxation. We obtain a feasible solution for this LP-relaxation by deleting the s_j's from the LP-relaxation of prize-collecting tour SRoB and multiply all other variables by a factor $1/(1-\delta)$. Combining the two statements, we obtain that the connection and tour cost can be bounded by $5/(1 - \delta)$ times the connection and tour cost of the optimal LP-solution. $\square$

If we choose δ uniformly at random on $[0, \theta]$, with $0 < \theta \le 1$ to be specified later [18], we obtain the following result.

Lemma 3. *Randomization of the algorithm above gives a solution for the prize-collecting tour SRoB such that the resulting tour and connection costs are in expectation bounded by $(5 \ln (1/(1 - \theta)) /\theta) C_{LP}$ and the resulting penalty costs are in expectation bounded by $(1/\theta)\Pi_{LP}$.*

Note that the algorithm can be derandomized by checking all values $s_j^* \in [0, \theta]$ for δ, since the set of unvisited vertices does not change for values in between two consecutive values of s_j^*. So, by checking at most n values, we obtain a deterministic algorithm with the same guarantees. Choosing $\theta = 1 - e^{-1/5}$ gives the following approximation guarantee.

Theorem 5. *There is a 5.52-approximation for the prize-collecting tour SRoB problem.*

4.2 The *a Priori* Prize-Collecting Traveling Salesman Problem

In this subsection, it is shown how to reduce the *a priori* prize-collecting TSP to the prize-collecting tour SRoB and lose a factor 3 in the approximation. First, we omit the penalties.

Lemma 4. *Any approximation algorithm for the tour SRoB problem can be turned into an approximation algorithm for the a priori TSP in the uniform model with loss of at most a factor 3 in the approximation.*

Proof. Given an instance of *a priori* TSP with edge costs c_{ij} and uniform probabilities p, we define an instance of tour SRoB as follows. The edge costs are $c_{ij}' = 2pc_{ij}$ and $M = 1/(2p)$. Given any feasible solution for this instance we get a feasible solution for *a priori* TSP of at most the same cost as follows. Let T be the tour in the SRoB solution. For the *a priori* tour we take T and double all the edges from clients to facilities in the SRoB solution. It is easy to see that, by the scaling factor $2p$, the expected cost of the shortcut TSP solution is at most that of the SRoB solution.

Let OPT_{TSP} and OPT_{SRoB} denote the optimal value of, respectively, the *a priori* TSP and the tour SRoB instance. It remains to show that $\text{OPT}_{\text{SRoB}} \le 3\text{OPT}_{\text{TSP}}$. Select each vertex with probability p and take an optimal tour on the set of selected vertices S. Let this be the tour for the SRoB solution. Connect all other vertices in the cheapest way to S. It follows from the analysis in [5] that the cost of this SRoB solution is at most 3 times the optimal cost of the *a priori* TSP instance, since the construction above is just their algorithm except for the fact that we take an optimal tour on S. Hence, $\text{OPT}_{\text{SRoB}} \le 3\text{OPT}_{\text{TSP}}$. □

The theorem above applies as well in the prize-collecting setting.

Lemma 5. *Any approximation algorithm for the prize-collecting tour SRoB problem can be turned into an approximation algorithm for the a priori prize-collecting TSP in the uniform model with loss of at most a factor 3 in the approximation.*

Combining the lemma above with Theorem 5 we get the following theorem.

Theorem 6. *There is a 16.55-approximation for the a priori prize-collecting TSP in the uniform model.*

4.3 Main Result

The step from *a priori* prize-collecting TSP to the *a priori* TRP is easy and is similar to that of the deterministic setting as in [13]. It even works in the independent decision model. Given any α-approximation for *a priori* prize-collecting TSP we get a $(2\alpha, 2\alpha)$-TSP-approximator in the *a priori* setting. Let L be given and assume that there exists a tour T of expected cost at most L which visits at least $(1 - \epsilon)n$ vertices. Denote the set of unvisited vertices of tour T by Q. We show how to get a tour of expected length at most $2\alpha L$ that visits at least $(1 - 2\alpha\epsilon)n$ vertices. Define an instance of *a priori* prize-collecting TSP by giving vertex i a penalty $\pi_i = L/(p_i \epsilon n)$. The optimal value of this instance is at most that of solution T which is $L + \sum_{i \in Q} p_i \pi_i \leq 2L$. Hence, any α-approximation for the *a priori* prize-collecting TSP instance should return a tour that has expected length at most $2\alpha L$ and also an expected penalty cost of at most $2\alpha L$. The latter implies that it leaves at most $2\alpha L/(p_i \pi_i) = 2\alpha\epsilon n$ vertices unvisited.

Now, a constant-factor approximation algorithm for the *a priori* TRP in the uniform setting follows from Theorems 2 and 6.

Theorem 7. *There is an $O(1)$-approximation for the a priori traveling repairman problem in the uniform model.*

5 Open Problems

There still are many open problems in the field of *a priori* optimization. For the *a priori* traveling repairman problem we were only able to give a constant-factor approximation in the uniform model and the constant is still large. At several points in the proof the uniformity of the probabilities is essential. The problem is wide open in the independent probability and scenario model. Also, it is not known if the uniform problem can be solved efficiently in case all points are on the line. If any optimal solution has the property that no point is passed without visiting it, like in the deterministic problem, then the problem may be solved by dynamic programming. However, a proof is missing and we have shown that this property does not hold in the scenario setting.

In our analysis we used the theory of (α, β)-TSP-approximators and prize-collecting TSP. Better approximations may be obtained by using the *a priori* k-TSP. No constant-factor approximation is known for this problem.

References

1. Jaillet, P.: Probabilistic traveling salesman problems. Ph.D. thesis, Massachusetts Institute of Technology (1985)
2. Bertsimas, D.: Probabilistic combinatorial optimization problems. Ph.D. thesis, Massachusetts Institute of Technology (1988)
3. Schalekamp, F., Shmoys, D.B.: Algorithms for the universal and a priori tsp. Oper. Res. Lett. **36**(1), 1–3 (2008)
4. Gorodezky, I., Kleinberg, R.D., Shmoys, D.B., Spencer, G.: Improved lower bounds for the universal and *a priori* TSP. In: Serna, M., Shaltiel, R., Jansen, K., Rolim, J. (eds.) APPROX and RANDOM 2010. LNCS, vol. 6302, pp. 178–191. Springer, Heidelberg (2010)
5. Shmoys, D.B., Talwar, K.: A constant approximation algorithm for the *a priori* traveling salesman problem. In: Lodi, A., Panconesi, A., Rinaldi, G. (eds.) IPCO 2008. LNCS, vol. 5035, pp. 331–343. Springer, Heidelberg (2008)
6. Zuylen, A.V.: Deterministic sampling algorithms for network design. Algorithmica **60**(1), 110–151 (2011)
7. Christofides, N.: Worst-case analysis of a new heuristic for the travelling salesman problem. Technical report, DTIC Document (1976)
8. Sahni, S., Gonzalez, T.: P-complete approximation problems. J. ACM (JACM) **23**(3), 555–565 (1976)
9. Sitters, R.A.: The minimum latency problem is NP-hard for weighted trees. In: Cook, W.J., Schulz, A.S. (eds.) IPCO 2002. LNCS, vol. 2337, pp. 230–239. Springer, Heidelberg (2002)
10. Chaudhuri, K., Godfrey, B., Rao, S., Talwar, K.: Paths, trees, and minimum latency tours. In: Proceedings of the 44th Annual IEEE Symposium on Foundations of Computer Science, pp. 36–45. IEEE (2003)
11. Sitters, R.: Polynomial time approximation schemes for the traveling repairman and other minimum latency problems. In: Proceedings of the 25th Annual ACM-SIAM Symposium on Discrete Algorithms, SIAM, pp. 604–616 (2014)
12. Afrati, F., Cosmadakis, S., Papadimitriou, C.H., Papageorgiou, G., Papakostantinou, N.: The complexity of the travelling repairman problem. RAIRO Informatique théorique **20**(1), 79–87 (1986)
13. Blum, A., Chalasani, P., Coppersmith, D., Pulleyblank, B., Raghavan, P., Sudan, M.: The minimum latency problem. In: Proceedings of the 26th Annual ACM Symposium on Theory of Computing, pp. 163–171. ACM (1994)
14. Goemans, M.X., Kleinberg, J.: An improved approximation ratio for the minimum latency problem. Math. Program. **82**(1–2), 111–124 (1998)
15. Eisenbrand, F., Grandoni, F., Rothvoß, T., Schäfer, G.: Connected facility location via random facility sampling and core detouring. J. Comput. Syst. Sci. **76**(8), 709–726 (2010)
16. Swamy, C., Kumar, A.: Primal-dual algorithms for connected facility location problems. Algorithmica **40**(4), 245–269 (2004)
17. Goemans, M.X., Williamson, D.P.: A general approximation technique for constrained forest problems. SIAM J. Comput. **24**(2), 296–317 (1995)
18. Williamson, D.P., Shmoys, D.B.: The Design of Approximation Algorithms. Cambridge University Press, New York (2011)

Primal-Dual Algorithms for Precedence Constrained Covering Problems

Andreas Wierz[1]([✉]), Britta Peis[1], and S. Thomas McCormick[2]

[1] RWTH Aachen University, Aachen, Germany
andreas.wierz@oms.rwth-aachen.de
[2] Sauder School of Business, University of British Columbia, Vancouver, Canada

Abstract. A *covering problem* is an integer linear program of type $\min\{c^T x \mid Ax \geq D, \, 0 \leq x \leq d, \, x \text{ integral}\}$ where $A \in \mathbb{Z}_+^{m \times n}$, $D \in \mathbb{Z}_+^m$, and $c, d \in \mathbb{Z}_+^n$. In this paper, we study covering problems with additional precedence constraints $\{x_i \leq x_j \, \forall j \preceq i \in \mathcal{P}\}$, where $\mathcal{P} = ([n], \preceq)$ is some arbitrary, but fixed partial order on the items represented by the column-indices of A. Such *precedence constrained covering problems (PCCP)* are of high theoretical and practical importance even in the special case of the *precedence constrained knapsack problem*, i.e., where $m = 1$ and $d \equiv 1$.

Our main result is a strongly-polynomial primal-dual approximation algorithm for PCCP with $d \equiv 1$. Our approach generalizes the well-known knapsack cover inequalities to obtain an IP formulation which renders any explicit precedence constraints redundant. The approximation ratio of this algorithm is upper bounded by the width of $\mathcal{P}$, i.e., by the size of a maximum antichain in $\mathcal{P}$. Interestingly, this bound is independent of the number of constraints. We are not aware of any other results on approximation algorithms for PCCP on arbitrary posets $\mathcal{P}$. For the general case with $d \not\equiv 1$, we present pseudo-polynomial algorithms.

Keywords: Approximation algorithms · Precedence constraints · Knapsack problem · Capacitated covering

1 Introduction

We consider integer linear programs of type $\min\{c^T x \mid Ax \geq D, \, 0 \leq x \leq d, \, x \text{ integral}\}$, where $A \in \mathbb{Z}_+^{m \times n}$ is a non-negative integral matrix with entries u_i^k, $D \in \mathbb{Z}_+^m$ is a demand vector with entries D^k, and $c, d \in \mathbb{Z}_+^n$. Such integer linear programs are usually called *covering problems*. The name becomes more evident if we interpret a solution vector $x \in \mathbb{Z}_+^n$ as choice of multiplicities x_i of items of type $i \in N = [n]$ that needs to satisfy a set of m covering constraints $\sum_{i \in N} u_i^k x_i \geq D^k$ ($k \in K = [m]$). Each item i is equipped with a per-unit weight u_i^k w.r.t. covering-constraint k, a per-unit cost c_i, and an upper bound d_i on the multiplicity of item i.

For example, the special case of just one covering constraint ($m = 1$), and upper bounds $d_i = 1$ on the multiplicities models the classical KNAPSACK PROBLEM. The knapsack problem is more familiar in the packing variant

© Springer International Publishing Switzerland 2015
E. Bampis and O. Svensson (Eds.): WAOA 2014, LNCS 8952, pp. 260–272, 2015.
DOI: 10.1007/978-3-319-18263-6_22

$\max\{c^T x \mid \sum_{i \in N} u_i x_i \leq B, x \in \{0,1\}^n\}$ which asks for a subset of items in N of maximum c-value that fits into a knapsack of capacity B. This problem can equivalently be formulated in the covering variant $\min\{c^T x \mid \sum_{i \in N} u_i x_i \geq D, x \in \{0,1\}^n\}$, where $D = \sum_{i \in N} u_i - B$, which aims at minimizing the value of items that are *not* packed into the knapsack. The knapsack problem belongs to one of the most studied problems in combinatorial optimization (see, e.g. [10]), and is well-known to be weakly NP-hard [10]. Nevertheless, it is computationally tractable as there exist e.g., an FPTAS [8] and a primal-dual 2-approximation [3,10] for the problem.

However, in various settings of knapsack-type problems, and also in more general covering problems, the items to be chosen need to obey certain precedence relations $\preceq$ in the following sense: the chosen multiplicity of item i may not exceed the chosen multiplicity of item j, whenever $j \prec i$ in some predefined partially ordered set (poset) $\mathcal{P} = (N, \preceq)$. That is, we consider the PRECEDENCE CONSTRAINED COVERING PROBLEM

$$\min\{c^T x \mid Ax \geq D,\ 0 \leq x \leq d,\ x \text{ integral},\ x_i \leq x_j\ \forall j \preceq i \in \mathcal{P}\} \quad \text{(PCCP)}$$

w.r.t. an arbitrary, but fixed poset $\mathcal{P} = (N, \preceq)$. Due to its importance both from a theoretical and practical perspective, we will put special emphasize on the case where $m = 1$ and $d = 1^n$, i.e., the PRECEDENCE CONSTRAINED KNAPSACK PROBLEM (PCKP). For the sake of simplicity, we use D as a scalar whenever $m = 1$.

PCKP and its generalizations are not only very interesting on their own right. Besides, they are also used as subroutine in MAX CLIQUE and SCHEDULING WITH PRECEDENCE CONSTRAINTS (see, e.g., [19]).

Evidently, the complexity of precedence constrained covering problems, already in PCKP, depends strongly on the structure of the underlying poset $\mathcal{P}$. For example, if $\mathcal{P}$ is an *antichain*, i.e., if there is no precedence relation between any two items, then PCKP reduces to the classical and tractable knapsack problem. In the other extreme, where $\mathcal{P}$ is a *chain*, i.e., where all items are comparable, the problem is trivially solvable: the optimal solution consists of the smallest initial subchain whose weight covers D.

Known Results. Due to the two extreme examples of PCKP described above, the reader could be tempted to think that PCKP is rather easy in the presence of arbitrary partial orders as well - but this is not the case.

The packing variant of PCKP restricted to $u_i = c_i$ for all elements $i \in N$ with a bipartite partial order was already proven to be strongly NP-complete by Johnson and Niemi [9]. Hajiaghayi et al. [7] show that the packing version of PCKP is inapproximable within a factor of $2^{\log^\delta(n)}$, for some $\delta > 0$, unless 3SAT $\in$ DTIME($2^{n^{\frac{3}{4}+\epsilon}}$) for all $\epsilon > 0$. Special cases of the packing variant of PCKP can be solved in polynomial time. Johnson and Niemi [9] provide an FPTAS for instances where $\mathcal{P}$ is a tree. Kolliopoulos and Steiner [12] show that there is an FPTAS if $\mathcal{P}$ is bipartite and all items either have zero weight or zero capacity. Pritchard et al. [17] provide an $O(\alpha^2)$-approximation algorithm for the multidimensional precedence constrained packing problem with dilation α, which is defined as the maximal number of constraints any variable appears in. The polyhedral structure of PCKP was investigated by a wide range of papers, including work of Boyd [2], Park and Park [16], Leensel et al. [15] and Boland et al. [1]. The work includes cutting planes as well as preprocessing steps that can be used to reduce the problem size.

Covering problems (CPs) without precedence constraints and with at most p non-zero entries per row are NP-hard to approximate within a factor of $p - 1 - \epsilon$ and $p - \epsilon$ under the unique games conjecture, respectively, due to Dinur et al. [5] and Khot et al. [11]. This bound was also achieved by a primal-dual algorithm due to Fujito et al. [6] and a greedy algorithm due to Koufogiannakis et al. [14].

The special case of CP with dilation α was proven to be inapproximable within any factor below $\Theta(ln(\alpha))$ [18]. This bound was also achieved by an iterative rounding algorithm due to Kolliopoulos and Steiner [13]. For more information on covering problems without precedence constraints, we refer the reader to Pritchard et al. [17]. Since PCCP contains CP as a special case, it inherits its lower bounds. To the best of our knowledge, there are no approximation algorithms known for PCCP under the presence of arbitrary precedence constraints, yet.

Our Contribution and Outline of the Paper. In Sect. 2, we extend the primal-dual 2-approximation algorithm of Carnes and Shmoys [3] for the knapsack problem without precedence constraints (KP) to also work for general bounds d on the multiplicities. As in [3], the algorithm is based on *knapsack cover inequalities* to bound the integrality gap. Though this is probably not too exciting from a theoretical point of view, it helps to understand the more involved algorithmic and analytic techniques in the subsequent sections. In Sect. 3, we investigate the KNAPSACK PROBLEM WITH PRECEDENCE CONSTRAINTS (PCKP). We observe that adding precedence constraints to the knapsack covering inequalities may lead to an unbounded integrality gap. However, we were able to incorporate the poset structure into the knapsack covering inequalities by shifting weights along the poset. This way, a primal-dual approach led us to a $w(\mathcal{P})$-approximate solution, where $w(\mathcal{P})$ denotes the *width* of poset $\mathcal{P}$, i.e., the maximal size of an antichain in $\mathcal{P}$. We give an example showing that this result is tight in the sense that our algorithm may in fact output a solution with performance ratio $w(\mathcal{P})$.

Thereafter, in Sect. 4, we show that the algorithm can be generalized to deal with more than one covering constraint, i.e., to the PRECEDENCE CONSTRAINED COVERING PROBLEM (PCCP), provided that $d \equiv 1$. Surprisingly, the approximation factor $w(\mathcal{P})$ remains the same, i.e., the factor is independent of the number of constraints. As a byproduct we yield a p-approximation for the covering problem without precedence constraints, where p denotes the maximal number of non-zero entries in any row of matrix A, which was already discussed by [6, 14].

For PCCP with general d, we provide a construction that allows to solve the problem within an approximation factor of $w(\mathcal{P})\Delta$ in pseudo-polynomial time, where $\Delta = \max_i d_i$ is the maximum item multiplicity.

Finally, in Sect. 5, we provide an inapproximability result in showing that there is no PTAS for PCKP unless $NP \subseteq \cap_{\epsilon>0} \text{BPTIME}(2^{n^\epsilon})$ even if $c_i = u_i$ for all $i \in N$, and the poset is bipartite.

1.1   Preliminaries

A partially ordered set is a tuple $\mathcal{P} = (N, \preceq)$ consisting of a set of elements N and an order relation $\preceq$ for pairs of elements. In our context, elements are also called *items*. Two elements $i, j \in N$ are *comparable* if $i \preceq j$ or $j \preceq i$ holds, and *incomparable* otherwise. An element $i \in N$ is *smaller or equal*, respectively *larger or equal* to $j \in N$, if $i \preceq j$ or $j \preceq i$ holds, respectively. The elements are equal if i is both, smaller and larger or equal to j. A subset of items $I \subseteq N$ is called a *chain* if $\mathcal{P}$ restricted to I induces a linear order on the set I. Analogously, a subset of items $I \subseteq N$ is called an *antichain* if no pair of items $i, j \in I$ is comparable. The *size* of a maximum chain and antichain, respectively, is denoted by $h(\mathcal{P})$ and $w(\mathcal{P})$. A subset $I \subseteq N$ is called *ideal* if $i \in I, j \preceq i$ implies $j \in I$. An ideal $I \subseteq [n]$ is also said to be *closed* under $\mathcal{P}$. We define the *ideal* $i{\downarrow}$ and *filter* $i{\uparrow}$, respectively, of an element $i \in N$ as the set of all items which are smaller or equal and larger or equal, respectively, to i, that is, $i{\downarrow} := \{j \in N : j \preceq i\}$ and $i{\uparrow} := \{j \in N : i \preceq j\}$.

The set of all ideals of $\mathcal{P}$ is denoted by $\mathcal{L}(\mathcal{P})$. An element $i \in N$ covers $j \in N$, if $j \preceq i$ and there is no element $k \in N$ such that $j \prec k \prec i$. The set $\min \mathcal{P} := \{i \in N : \nexists j \in N \text{ with } j \prec i\}$ denotes the set of minimal items in N. Finally, we define $\mathcal{P}(A)$ for an ideal $A \in \mathcal{L}(\mathcal{P})$ as the poset $\mathcal{P}$ restricted to items which are not in A, i.e. $\mathcal{P}(A) := (N \setminus A, \preceq)$. Notice that the *restriction* of a poset is also a poset. Often it is easier to talk about a poset in terms of a graph, hence, whenever it is more convenient, we use $\mathcal{P}$ as a graph (N, E) with an edge $(i, j) \in E$ if and only if j covers i. This graph is denoted by the *cover graph* of $\mathcal{P}$.

An α-*approximation algorithm* for a minimization problem P is an algorithm with polynomially bounded running time which finds a feasible solution for P with solution cost of at most α times the cost of an optimal solution. A *fully polynomial time approximation scheme* (FPTAS) is a family $\mathcal{F}$ of α-approximation algorithms such that for any $\alpha > 1$ there exists an α-approximation algorithm in $\mathcal{F}$. Analogously, a *pseudo-polynomial α-approximation algorithm* and *polynomial time approximation scheme* (PTAS), respectively, are defined as above however with the relaxation that each algorithm only requires a running time polynomial in the input size and in α for any constant α.

2   Knapsack Cover Inequalities

Carnes and Shmoys [3] introduced the first primal-dual algorithm for the classical knapsack problem in its covering variant (KP) $\min \left\{ \sum_{i \in [n]} c_i x_i : \sum_{i \in [n]} u_i x_i \geq D, x \in \{0, 1\}^n \right\}$, i.e., the special case of PCCP where $m = 1, d \equiv 1$, and $\mathcal{P}$ is an antichain. A simple instance with two items $\{1, 2\}$, weights $u_1 = D - 1, u_2 = D$, costs $c_1 = 0, c_2 = D$ and demand D shows that the LP relaxation of the formulation above has an unbounded integrality gap. The optimal IP solution has to select item two with solution cost D while an optimal LP solution chooses item one and only a fraction $\frac{1}{D}$ of item two, yielding optimal solution costs of one.

The integrality gap can be strengthened with help of the so-called knapsack cover inequalities $\sum_{i \in N \setminus A} u_i(A) x_i \geq D(A)$ for $A \subseteq N$, introduced by Carr et al. [4]. These inequalities read as follows: if set $A \subseteq N$ were selected into the solution, then the remaining items have to cover a residual demand of at least $D(A) := \max\{D - \sum_{i \in A} u_i, 0\}$. Due to integrality constraints for items, the constraint coefficients can be cropped at the right hand side value, hence $u_i(A) := \min\{u_i, D(A)\}$ can be seen as the *effective weight* of an item given that the items in A are part of the solution. The inequalities are also helpful for the design of approximation algorithms for a variety of covering related problems such as facility location [3] or network design [4].

As shown in [3], a primal-dual greedy algorithm now yields a 2-approximation to (KP). The overall goal of this paper is to design and analyze primal-dual algorithms for far reaching generalizations of (KP). Our generalizations advance in three directions: (1.) we consider arbitrary integral bounds $d \in \mathbb{Z}^n$ on the item-multiplicities, (2.) we consider a set of m covering constraints, instead of only one, and (3.) we add precedence constraints.

At this point however, mainly in order to get the reader familiar with the techniques used in this paper, we extend the problem only in direction (1): we present a short analysis of a generalization of the result in [3] towards $d \in \mathbb{Z}_+^N$. Therefore, we redefine the residual demand for a set of items $A \subseteq N$ as $D(A) := \max\{D - \sum_{i \in A} u_i d_i, 0\}$, that is, we assume that whenever an item is part of the solution, its item multiplicity is chosen to be as large as possible. With these definitions, the linear relaxation of the knapsack problem with general $d \in \mathbb{Z}_+^n$ becomes

$$\min\left\{ \sum_{i \in N} c_i x_i \mid \sum_{i \notin A} u_i(A) \cdot x_i \geq D(A) \ \forall A \subseteq N, \ x \geq 0 \right\}. \tag{1}$$

Note that the upper bound $x \leq d$ is given implicitly by the new definition of $D(A)$ and $u_i(A)$ as increasing a variable x_i beyond d_i increases the solution costs without further decreasing the residual demand for constraints for sets

---

Algorithm 1. Primal-Dual Algorithm for PCKP

1: Let $S = \emptyset, y \equiv 0$.
2: **while** $D(S) > 0$ **do**
3:     Increase $y(S)$ until some dual constraint becomes tight for item $i \in N \setminus S$.
4:     $x_i = \min\{\lceil \frac{D(S)}{u_i} \rceil, d_i\}$.
5:     $S = S \cup \{i\}$.
6: **end while**
7: **return** S.

$A + i$ (which is short for $A \cup \{i\}$). For the same reason, the new formulation is infeasible if the knapsack instance was infeasible as $D(N) = D - \sum u_i d_i > 0$ and the constraint for $A = N$ contains no more variables.

The primal-dual algorithm, given as Algorithm 1, works as follows. We start with an integral infeasible primal solution $S = \emptyset$ and a feasible solution $y \equiv 0$ of

$$\max \left\{ \sum_{A \subseteq N} D(A) \cdot y(A) \mid \sum_{\substack{A \subseteq N: \\ i \notin A}} u_i(A) \cdot y(A) \leq c_i \; \forall i \in N, \; y(A) \geq 0 \right\}, \quad (2)$$

which is the dual of (1). In each iteration, we increase the dual variable corresponding to the current partial solution S until a dual constraint becomes tight for some item $i \in N \setminus S$. The item is added to the solution and x_i is either set to its upper bound d_i or to the minimal value such that the residual demand $D(S)$ is exceeded. The algorithm iterates until the solution exceeds the knapsack demand, i.e. until $D(S) \leq 0$.

Theorem 1. *Algorithm 1 is a primal-dual 2-approximation for the knapsack problem with general $d \in \mathbb{Z}_+^n$ (See full version for the proof).*

A straightforward generalization of the results for KP towards PCKP would invoke the following precedence constrained minimum knapsack formulation:

$$\min \left\{ \sum_{i \in N} c_i x_i \mid \sum_{i \notin A} u_i(A) x_i \geq D(A) \; \forall A \subseteq N, \; x_i - x_j \geq 0 \; \forall i \preceq j, \; x \in \{0,1\}^n \right\} \quad (P_{KP2})$$

However, the integrality gap of this LP can be unbounded (see Lemma below). In order to bound the integrality gap, we will incorporate the poset structure into the knapsack cover inequalities, see the next section.

Lemma 1. *Formulation (P_{KP2}) has an unbounded integrality gap (See full version for the proof).*

3   Precedence Constrained Knapsack Cover Inequalities

This section introduces our novel generalization of knapsack cover inequalities towards precedence constraints. For this section we consider only PCKPs, i.e. $m = 1$ and $d \equiv 1$. Instead of independently formulated precedence constraints, we generalize the notion of effective weight towards *precedence constrained effective weight*. Recall that the formulation from Sect. 2 ensures a constraint for each subset $A \subseteq [n]$ of items which lower bounds the weight of selected items in $[n] \setminus A$ given that all items in A are part of the solution. In this section, we reformulate this statement as follows. We introduce a constraint for each ideal in $\mathcal{P}$ and lower bound the effective weight of all selected *minimal* items in $\mathcal{P}(A)$, given that all items in A are part of the solution. Therefore, we define the effective weight $\bar{u}_i(A)$ of an item i and ideal $A \in \mathcal{L}(\mathcal{P})$ of selected items to be a share of the total weight of the filter $i{\uparrow}$, if $i \in \min \mathcal{P}(A)$, or zero otherwise. Recall that $\mathcal{P}(A)$ denotes the partially ordered set $\mathcal{P}$ restricted to elements which are not in A. More precisely, we allocate the weight of each item $j \in \mathcal{P}(A) \setminus \min \mathcal{P}(A)$ uniformly among all minimal items in the ideal of j, that is, $\min j{\downarrow}$. Therefore, we define $X_j(A)$ to be the set of items in $\min j{\downarrow}$ with respect to $\mathcal{P}(A)$, i.e. $X_j(A) := \{i : i \preceq j \text{ and } i \in \min \mathcal{P}(A)\}$. If an item is not a minimal item in $\mathcal{P}(A)$, we set the effective weight to zero, i.e. $\bar{u}_i(A) := 0$. As usual, coefficients exceeding the right hand side of any inequality can be cropped. Hence for items $i \in \min \mathcal{P}(A)$, the effective weight is given by

$$\bar{u}_i(A) := \min \left\{ D(A), u_i + \sum_{j : i \prec j} \frac{u_j(A)}{|X_j(A)|} \right\}.$$

With help of the definitions above, we can formulate PCKP as (P_{PCKP}) with dual of its linear relaxation (D_{PCKP}).

$$\min \left\{ \sum_{i \in N} c_i x_i \;\middle|\; \sum_{i \in \min \mathcal{P}(A)} \bar{u}_i(A) \cdot x_i \geq D(A) \; \forall A \in \mathcal{L}(\mathcal{P}), \; x \in \{0,1\}^n \right\} \quad (P_{PCKP})$$

$$\max \left\{ \sum_{A \in \mathcal{L}(\mathcal{P})} y(A) D(A) \;\middle|\; \sum_{\substack{A \in \mathcal{L}(\mathcal{P}) : \\ i \in \min \mathcal{P}(A)}} \bar{u}_i(A) \cdot y(A) \leq c_i \; \forall i \in N, \; y(A) \geq 0 \; \forall A \in \mathcal{L}(\mathcal{P}) \right\} \quad (D_{PCKP})$$

The primal covering constraints $\sum_{i \in \min \mathcal{P}(A)} \bar{u}_i(A) \cdot x_i \geq D(A)$ for $A \in \mathcal{L}(\mathcal{P})$, which we call *precedence constrained knapsack cover inequalities*, ensure that the selected elements in $\min \mathcal{P}(A)$ satisfy the residual demand $D(A)$. In contrast to the previous section, however, the effective weight $\bar{u}_i(A)$ of an item i has changed to also reflect a share of the item capacities in the filter of i. The new set of constraints has two advantages. (1) The set of partial order constraints is no longer required as it is implicitly enforced by the constraints. (2) The primal-dual greedy algorithm is guided such that it will always generate a solution which is closed under $\mathcal{P}$.

The bad example described in Lemma 1 is no longer valid in (P_{PCKP}). In fact, for this particular instance, the LP optimum now coincides with the IP optimum. Theorem 2 shows that the formulation indeed models the precedence constrained knapsack problem.

Theorem 2. *(P_{PCKP}) is a valid relaxation for PCKP, i.e. any integer feasible solution in (P_{KP2}) is feasible in (P_{PCKP}). Furthermore, any feasible solution for (P_{PCKP}) which is closed under $\mathcal{P}$ is feasible in (P_{KP2}).*

Proof. Let us consider an integer feasible solution x of (P_{KP2}) with support $S := \{i : x_i = 1\}$ and an arbitrary ideal $A \in \mathcal{L}(\mathcal{P})$. Since x is feasible in (P_{KP2}), we know that $\sum_{i \in S \setminus A} u_i(A) \geq D(A)$ holds. We will see that the corresponding covering constraint in (P_{PCKP}) will also be satisfied. If $S \setminus A \subseteq \min \mathcal{P}(A)$, we are done as the elements in both sums coincide and the coefficients in the covering constraint of (P_{PCKP}) dominate the corresponding coefficients in (P_{KP2}).

Now suppose that there is an item $i \in (S \setminus \min \mathcal{P}(A)) \setminus A$. Since the knapsack instance is precedence constrained, all items in $i{\downarrow} \cap \min \mathcal{P}(A) = X_i(A)$ are also contained in S. Since the weight of item i is allocated uniformly among the items in $X_i(A)$ and $X_i(A) \setminus S = \emptyset$ holds, the weight of item i is also present in the covering constraint of (P_{PCKP}). As this observation holds for all items in $(S \setminus \min \mathcal{P}(A)) \setminus A$, we can deduce that $\sum_{i \in S \cap \min \mathcal{P}(A)} \bar{u}_i(A) \geq \sum_{i \in S \setminus A} u_i(A) \geq D(A)$ holds.

For the second implication, we consider an integer feasible solution x of (P_{PCKP}) with support $S = \{i : x_i = 1\}$. Since x is feasible, we know that $\sum_{i \in \min \mathcal{P}(S)} \bar{u}_i(S) x_i = 0 \geq D(S)$ holds. Hence we conclude that $\sum_{i \in S} u_i \geq D$. Since S is assumed to be closed under $\mathcal{P}$ and the knapsack constraint is satisfied, the solution is feasible in (P_{KP2}).

Theorem 3 shows that the primal-dual greedy algorithm described in the previous section (Algorithm 1), finds a feasible primal-dual solution pair (S, y) to (P_{PCKP}) and (D_{PCKP}). The subsequent Theorem 4 derives an upper bound of $w(\mathcal{P})$ on the solution quality. Finally, Theorem 5 shows that this bound is tight.

Theorem 3. *The greedy algorithm finds a feasible primal-dual solution pair (S, y) to (P_{PCKP}) and (D_{PCKP}) in polynomial time. Furthermore, for any item $i \in S$, the corresponding dual constraint is tight.*

Proof. For this proof, we denote the chain of ideals $S_i \in \mathcal{L}(\mathcal{P})$ which are generated by the algorithm by $\emptyset = S_0 \subset S_1 \subset \cdots \subset S_\ell = S$. It is easy to see that the generated primal solution S satisfies the knapsack constraint $D \leq \sum_{i \in S} u_i$ as the algorithm stops only if $D(S) \leq 0$ or $S = [n]$. If the algorithm stopped due to the latter and $D(S) > 0$, it is clear that the instance was infeasible. Otherwise, the solution satisfies the knapsack constraint and it remains to show that the solution is closed under $\mathcal{P}$, which is clear due to an inductive statement. The empty set is closed under $\mathcal{P}$, hence solution S_0 is closed under $\mathcal{P}$. The algorithm adds an item i in iteration k only if it is a minimal item with respect to $\mathcal{P}(S_k)$. Due to the definition of $\mathcal{P}(A)$ it is clear that $S_k \cup \{i\}$ remains an ideal if $i \in \min \mathcal{P}(S_k)$.

Let us consider the chain of ideals for dual feasibility. The algorithm increases dual variables only on the chain $S_0 \subset \cdots \subset S_\ell$. As soon as an item was added to subset S_k, it has no more non-zero constraint coefficients in any later iteration due to the design of our constraints. Recall that a constraint sums the effective precedence constrained weight among all items in $\mathcal{P}(S_k)$, namely no items which are contained in S_k. Since the algorithm makes sure that an item is added to the solution as soon as the first constraint in an iteration became tight, this concludes the proof.

Theorem 4. *A solution found by the greedy algorithm has cost of at most $w(\mathcal{P}) \cdot OPT$, where $w(\mathcal{P})$ denotes the size of a maximum antichain in $\mathcal{P}$.*

Proof. Let S be the set of items in the final solution with integer feasible solution $x = \chi_S$. Then the solution cost can be evaluated as follows:

$$cost(S) = \sum_{i \in S} c_i = \sum_{i \in S} \sum_{\substack{A \in \mathcal{L}(\mathcal{P}): \\ i \in \min \mathcal{P}(A)}} y(A)\bar{u}_i(A) = \sum_{A \in \mathcal{L}(\mathcal{P})} y(A) \sum_{i \in \min \mathcal{P}(A) \cap S} \bar{u}_i(A)$$

$$\leq \sum_{A \in \mathcal{L}(\mathcal{P})} y(A)D(A)|\min \mathcal{P}(A) \cap S|$$

$$\leq \max_{A \in \mathcal{L}(\mathcal{P})} \{|\min \mathcal{P}(A) \cap S|\} \sum_{A \in \mathcal{L}(\mathcal{P})} y(A)D(A) \leq w(\mathcal{P}) \cdot OPT.$$

The second equality is due to the fact that we only add items to S if the corresponding dual constraint is tight. The third equality can be achieved by reordering the terms of the sum. By definition, we know that $\bar{u}_i(A) \leq D(A)$ holds, hence the first inequality holds. The second inequality estimates the sum by taking the maximum among terms. Since minimal elements are incomparable by definition, we know that the cardinality can be bounded by the size of a maximum antichain in $\mathcal{P}$.

Theorem 5. *The bound in Theorem 4 is tight.*

Proof. Consider a poset with $2nk$ items consisting of k parallel chains each of length $2n$ with the following functions representing the i'th item in the ℓ'th chain: $u_i^\ell = 1$ if $i \leq n$ and $u_i^\ell = k - 1$ if $i > n$, $c_i^\ell = 1$ if $i \leq n$ and $c_i^\ell = K$ if $i > n$ for some fixed, large $K \in \mathbb{Z}_+$. If we consider a corresponding PCKP instance with $D = nk$, an optimal solution will consist of the first n items from each of the k chains with a solution cost of nk. Each chain on its own will also correspond to a feasible solution with cost $n + Kn$.

Let us consider the precedence constrained knapsack inequality for $A = \emptyset$ which reads as follows: $\sum_{\ell=1}^{k} u_1^\ell(\emptyset)x_1^\ell = \sum_{\ell=1}^{k} nkx_1^\ell \geq nk$. For the optimum solution, all variables in this constraint are set to one, hence $\sum_{\ell=1}^{k} u_1^\ell(\emptyset)x_1^\ell = knk$ and $k = w(\mathcal{P})$ which yields equality in our argumentation in the proof of Theorem 4. Since our algorithm starts by increasing the corresponding dual variable $y(\emptyset)$, this concludes the proof.

As a remark we want to point out that the modified primal-dual greedy algorithm inherits the primal-dual greedy algorithm of Carnes and Shmoys [3] in the special case where $\mathcal{P}$ is an antichain. Theorem 6 shows that the algorithm yields a constant factor approximation whenever the precedence constrained effective weight is bounded by a constant factor. In the case of an antichain we have $\alpha = 1$, hence a 2-approximation.

Theorem 6. *If there is a constant $\alpha \geq 1$ such that $\bar{u}_i(A) \leq \alpha u_i(A)$ holds for all items $i \in N$ and ideals $A \in \mathcal{L}(\mathcal{P})$, then the primal-dual algorithm finds a solution of cost at most $2\alpha \cdot OPT$ (See full version for the proof).*

Corollary 1. *If $\mathcal{P}$ is an antichain, then the primal-dual greedy algorithm finds a 2-approximation for PCKP.*

4   Precedence Constrained Covering Problems

The precedence constrained knapsack cover inequalities can also be used to reformulate general PCCPs with $d \equiv 1$ by replacing each knapsack constraint $\sum_{i \in N} u_i^k x_i \geq D^k$ individually. Hence, a PCCP instance can be reformulated as (P_{PCCP}) with dual of the linear relaxation (D_{PCCP}).

$$\min \left\{ \sum_{i \in N} c_i x_i \mid \sum_{i \in \min \mathcal{P}(A)} \bar{u}_i^k(A) \cdot x_i \geq D^k(A) \; \forall A \in \mathcal{L}(\mathcal{P}), \; \forall k \in K \; x \in \{0,1\}^n \right\} \quad (P_{PCCP})$$

$$\max \left\{ \sum_{k \in K} \sum_{A \in \mathcal{L}(\mathcal{P})} y^k(A) D^k(A) \mid \sum_{k \in K} \sum_{\substack{A \in \mathcal{L}(\mathcal{P}): \\ i \in \min \mathcal{P}(A)}} \bar{u}_i^k(A) \cdot y^k(A) \leq c_i \; \forall i \in N \; y \geq 0 \right\}. \quad (D_{PCCP})$$

Similar to Theorem 2, it is easy to see that the model is a relaxation for (PCCP). We will now show that a slightly modified version of Algorithm 1 also finds a primal-dual solution pair for (P_{PCCP}) and (D_{PCCP}). Therefore, let us consider an iteration of the algorithm for some partial solution S'. In the one-dimensional case, we increased the corresponding dual variable $y(S')$ until a dual constraint became tight. In this case, we have m variables $y^k(S')$ to choose from. The natural way to increase variables here is to increase each variable proportionally to its residual demand $D^k(S')$ until a constraint becomes tight for item $i \notin S'$. As before, we add item i to S' and iterate until $D^k(S') \leq 0$ for all $k \in K$.

With the same argumentation as in Theorem 3, we can observe that the algorithm finds a primal-dual solution pair (S, y) to (P_{PCCP}) and (D_{PCCP}). Theorem 7 yields the approximation ratio of $w(\mathcal{P})$ which surprisingly coincides with the one-dimensional case.

Theorem 7. *A solution found by the modified primal-dual algorithm has cost of at most $w(\mathcal{P}) \cdot OPT$. (See full version for the proof).*

Corollary 2. *If the poset is an antichain, the modified primal-dual algorithm finds a solution with cost no larger than $p \cdot OPT$, where p denotes the maximal number of non-zero entries in any row of the initial constraint set $Ax \geq D$. (See full version for the proof).*

For the special case of capacitated covering problems without precedence constraints and with $d \equiv 1$, the primal-dual algorithm finds a p-approximation, where p denotes the maximal number of non-zero entries in any row of the initial constraint set $Ax \geq D$. A similar algorithm for this special case was also pointed out by [6,14]. Unfortunately, the results of this section can not be extended towards PCCP with $d \not\equiv 1$ in the way we did this for KP in Sect. 2. Lemma 1 shows that a solution found by the algorithm considered in this section may no longer have bounded cost. Nevertheless, we are able to provide a pseudo-polynomial time approximation algorithm for this case with help of Theorem 2.

Proposition 1. *The approximation ratio of the modified variant of Algorithm 1 for instances of PCCP with $d \not\equiv 1$ is unbounded even if $m = 1$.*

Proof. Consider an instance with two items $N = \{1, 2\}$ with demand D, a partial order $\mathcal{P} = (N, 1 \preceq 2)$ and $c \equiv 1, u_1 = 1, u_2 = D, d_1 = D, d_2 = 1$. Then the optimal IP solution consists of a single copy of the first item and a single copy of the second item, the algorithm however would select D copies of item one.

Proposition 2. *For PCCP with $d \not\equiv 1$ there is a $w(\mathcal{P})$ Δ-approximation algorithm with running time $O(n^2 m \Delta)$, where $\Delta = \max_i \{d_i\}$ (See full version for the proof).*

5   Inapproximability of PCKP

Although we know that the packing variant of PCKP is inapproximable within any constant factor [7], there are no inapproximability results for the covering variant, yet. With Theorem 8, we provide the first such result showing by reduction from bipartite k-clique that PCKP has no PTAS unless $NP \subseteq \cap_{\epsilon > 0} \text{BPTIME}(2^{n^\epsilon})$.

Theorem 8. *There is no PTAS for PCKP unless $NP \subseteq \cap_{\epsilon > 0} BPTIME(2^{n^\epsilon})$ even if $c_i = u_i$, $m = 1$ and the partial order is bipartite (See full version for the proof).*

6   Summary and Outlook

This paper described primal-dual approximation algorithms for several generalizations of the classical minimum knapsack problem. More precisely, we solved variants of PCCP which are of the form

$$\min \left\{ \sum_{i \in N} c_i x_i : Ax \geq D, x \leq d, x \in \mathbb{Z}_+, (\text{POSET}) \right\},$$

where (POSET) describes a set of partial order constraints, $A \in \mathbb{Z}_+^{m \times n}$ and D and d are of appropriate dimension.

As an introduction, we described a 2-approximation algorithm for $d \not\equiv 1$ and $m = 1$ without precedence constraints. The following results were subject to precedence constraints. We introduced a generalized notion of the well-known knapsack cover inequalities towards precedence constraints, denoted by *precedence constrained knapsack cover inequalities*. With help of these inequalities we were able to provide a $w(\mathcal{P})$ approximation algorithm for PCKP, i.e. $d \equiv 1, m = 1$ which we generalized towards $d \equiv 1, m > 1$. As a byproduct, we derived a p-approximation for CP with $d \equiv 1, m > 1$ without precedence constraints which was already discussed in [6,14]. Finally, we provided a construction which allows to solve PCCP with $d \not\equiv 1$ and $m > 1$ within a factor of

Table 1. Summary of the results in this paper and lower bounds on the approximation ratio. Lower bounds are under the assumption of the unique games conjecture, unbounded dilation and $NP \not\subseteq \cap_{\epsilon > 0} \mathrm{BPTIME}(2^{n^\epsilon})$. Bounds in bold are results of this paper.

$m = 1$	$d \equiv 1$	(POSET)	Lower bound	Previous upper bound	Our upper bound
Y	N	N	NP-hard	2, rounding, (FPTAS)	**2**, primal-dual
Y	Y	Y	**No PTAS**	n/a	$\boldsymbol{w(\mathcal{P})}$
N	Y	Y	**No PTAS**	n/a	$\boldsymbol{w(\mathcal{P})}$
Y	N	Y	**No PTAS**	n/a	$\boldsymbol{w(\mathcal{P})\Delta}$ (Pseudo-polynomial)
N	N	Y	**No PTAS**	n/a	$\boldsymbol{w(\mathcal{P})\Delta}$ (Pseudo-polynomial)

$w(\mathcal{P})\Delta$ in pseudo-polynomial time. It remains open to find an algorithm with strongly polynomial bounds. Our results are also summarized in Table 1.

Although we presented the first approximation algorithms for precedence constrained covering problems, there still remains a gap between the lower and upper bounds. It would especially be interesting to provide a lower bound similar to the packing version of PCKP as provided by Hajiaghayi et al. [7]. For the generalization towards $d \not\equiv 1$ we were only able to provide pseudo-polynomial algorithms, hence it remains open to provide strongly polynomial bounds. Since CPs with bounded dilation α are approximable within $\Theta(log(\alpha))$, it might be interesting to consider this case in the presence of precedence constraints as well.

References

1. Boland, N., Bley, A., Fricke, C., Froyland, G., Sotirov, R.: Clique-based facets for the precedence constrained knapsack problem. Math. Program. **133**(1–2), 481–511 (2012)
2. Boyd, A.: Polyhedral results for the precedence-constrained knapsack problem. Discret. Appl. Math. **41**(3), 185–201 (1993)
3. Carnes, T., Shmoys, D.B.: Primal-dual schema for capacitated covering problems. In: Lodi, A., Panconesi, A., Rinaldi, G. (eds.) IPCO 2008. LNCS, vol. 5035, pp. 288–302. Springer, Heidelberg (2008)
4. Carr, R., Fleischer, L., Leung, V., Phillips, C.: Strengthening integrality gaps for capacitated network design and covering problems. In: Proceedings of the Eleventh Annual ACM-SIAM Symposium on Discrete Algorithms, pp. 106–115. Society for Industrial and Applied Mathematics (2000)
5. Dinur, I., Guruswami, V., Khot, S., Regev, O.: A new multilayered PCP and the hardness of hypergraph vertex cover. SIAM J. Comput. **34**(5), 1129–1146 (2005)
6. Fujito, T., Yabuta, T.: Submodular integer cover and its application to production planning. In: Persiano, G., Solis-Oba, R. (eds.) WAOA 2004. LNCS, vol. 3351, pp. 154–166. Springer, Heidelberg (2005)

7. Hajiaghayi, M., Jain, K., Konwar, K., Lau, L., Mandoiu, I., Russell, A., Shvartsman, A., Vazirani, V.: The minimum k-colored subgraph problem in haplotyping and dna primer selection. In: Proceedings of the International Workshop on Bioinformatics Research and Applications (IWBRA) (2006)

8. Ibarra, O.H., Kim, C.E.: Fast approximation algorithms for the knapsack and sum of subset problems. J. ACM (JACM) **22**(4), 463–468 (1975)

9. Johnson, D., Niemi, K.: On knapsacks, partitions, and a new dynamic programming technique for trees. Math. Oper. Res. **8**(1), 1–14 (1983)

10. Kellerer, H., Pferschy, U., Pisinger, D.: Knapsack Problems. Springer, Berlin (2004)

11. Khot, S., Regev, O.: Vertex cover might be hard to approximate to within 2-ε. J. Comput. Syst. Sci. **74**(3), 335–349 (2008)

12. Kolliopoulos, S.G., Steiner, G.: Partially-ordered knapsack and applications to scheduling. In: Möhring, R.H., Raman, R. (eds.) ESA 2002. LNCS, vol. 2461, pp. 612–624. Springer, Heidelberg (2002)

13. Kolliopoulos, S.G., Young, N.E.: Tight approximation results for general covering integer programs. In: Proceedings of the 42nd IEEE Symposium on Foundations of Computer Science, 2001, pp. 522–528. IEEE (2001)

14. Koufogiannakis, C., Young, N.E.: Greedy δ-approximation algorithm for covering with arbitrary constraints and submodular cost. Algorithmica **66**(1), 113–152 (2013)

15. Van de Leensel, R., van Hoesel, C., van de Klundert, J.: Lifting valid inequalities for the precedence constrained knapsack problem. Math. program. **86**(1), 161–185 (1999)

16. Park, K., Park, S.: Lifting cover inequalities for the precedence-constrained knapsack problem. Discret. Appl. Math. **72**(3), 219–241 (1997)

17. Pritchard, D., Chakrabarty, D.: Approximability of sparse integer programs. Algorithmica **61**(1), 75–93 (2011)

18. Trevisan, L.: Non-approximability results for optimization problems on bounded degree instances. In: Proceedings of the Thirty-third Annual ACM Symposium on Theory of Computing, pp. 453–461. ACM (2001)

19. Woeginger, G.: On the approximability of average completion time scheduling under precedence constraints. Discret. Appl. Math. **131**(1), 237–252 (2003)

Author Index

Printed in the United States
By Bookmasters